Sustainable Territorial Management

Sustainable Territorial Management

Special Issue Editors

David Rodríguez-Rodríguez
Javier Martínez-Vega

MDPI • Basel • Beijing • Wuhan • Barcelona • Belgrade

MDPI

Special Issue Editors

David Rodríguez-Rodríguez
Spanish National Research Council (CSIC)
Spain

Javier Martínez-Vega
Spanish National Research Council (CSIC)
Spain

Editorial Office
MDPI
St. Alban-Anlage 66
Basel, Switzerland

This is a reprint of articles from the Special Issue published online in the open access journal *Environments* (ISSN 2076-3298) from 2017 to 2018 (available at: http://www.mdpi.com/journal/environments/special_issues/sus_land)

For citation purposes, cite each article independently as indicated on the article page online and as indicated below:

LastName, A.A.; LastName, B.B.; LastName, C.C. Article Title. *Journal Name* **Year**, *Article Number*, Page Range.

ISBN 978-3-03897-212-9 (Pbk)
ISBN 978-3-03897-213-6 (PDF)

Cover image courtesy of pixabay.com

Contents

About the Special Issue Editors

David Rodríguez-Rodríguez holds an European Mention PhD in Conservation Biology. He is author of over 30 research articles, four book chapters and five monographs on sustainable territorial development and protected areas. He currently develops his research at the Institute of Economy, Geography and Demography of the Spanish National Research Council and is a Visiting Researcher at the European Topic Centre - University of Málaga.

Javier Martínez-Vega holds a PhD in Geography. He is a Senior Scientist at the Institute of Economy, Geography and Demography of the Spanish National Research Council, where he leads the Research Group on Multi-Scale Analysis of Global Change. He has published 53 scientific articles, 15 book chapters and nine monographs on land use changes, forest fires and protected areas.

environments

MDPI

Editorial

Preface: Special Issue on Sustainable Territorial Management

David Rodríguez-Rodríguez [1,2,]* and Javier Martínez-Vega [1]

[1] Institute of Economy, Geography and Demography, Spanish National Research Council (IEGD-CSIC), Associated Unit GEOLAB, Albasanz, 26–28, 28037 Madrid, Spain; javier.martinez@cchs.csic.es
[2] European Topic Centre-University of Malaga, Andalucía Tech, University of Malaga, 29010 Malaga, Spain
* Correspondence: david.rodriguez@csic.es; Tel.: +34-916-022-322 or +34-951-953-102

Received: 30 July 2018; Accepted: 2 August 2018; Published: 5 August 2018

1. Introduction

Human development has made remarkable social and economic progress possible for most of us [1,2], but has also entailed a range of negative consequences on natural resources, local communities, and the economy at multiple scales. Soil sealing [3–5], erosion, land degradation and erosion [6]; air and water pollution [7,8]; forest fires [9], biodiversity homogenisation and loss [10], isolation and fragmentation of habitats [11,12], poverty [13], human migration [14] and health issues [15,16] are among the most common human-made impacts with a clear sustainability and spatial component. They occur almost everywhere in the territory, be it terrestrial [17], aerial [18] or marine [19], where there is human activity.; Thus, achieving sustainable territorial management that combines healthy and prosperous societies with the long-term maintenance of biodiversity and productive ecosystem services [20] remains the biggest challenge of our modern world [21]. Simulation models of future land uses, under different scenarios of change, can help territorial managers in the decision-making process [22].

This Special Issue seeks to collect a coherent set of studies on techniques and experiences (case studies) aimed at increasing the environmental, social, economic &/or institutional sustainability of landscapes and seascapes from a range of geographic and socioeconomic contexts.

2. Highlights

Land use-land cover (LULC) changes towards intensive uses, especially towards artificial uses, are one of the main global threats to biodiversity conservation. Thus, LULC changes are considered a cornerstone in sustainable territorial management (Figure 1), so a number of articles in this Special Issue deal with this topic directly or indirectly. Geographically, ten case studies representing urban areas, rural areas (chiefly protected areas; PAs) and coastal areas from four countries in Europe and Asia are shown.

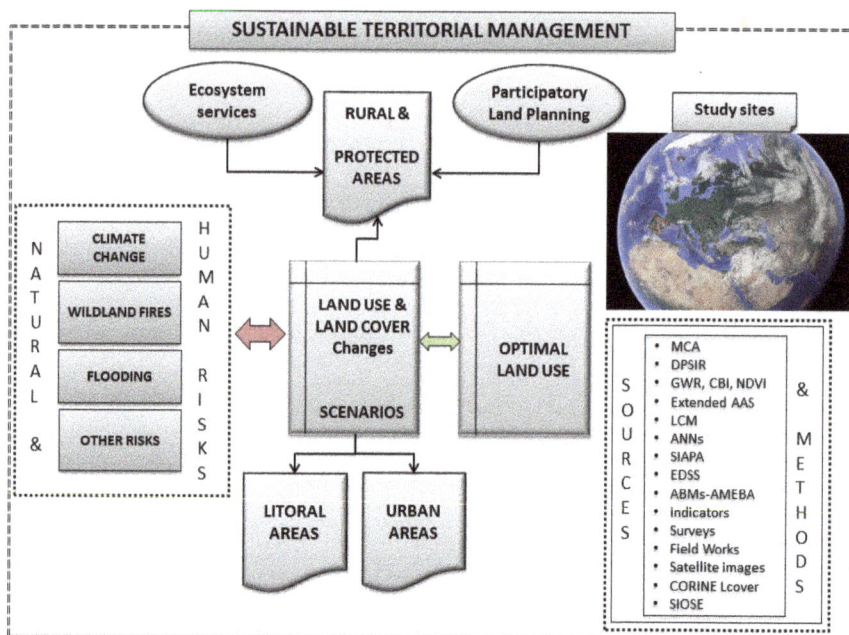

Figure 1. Simplified conceptual framework of ten papers compiled in this Special Issue (picture from Google Earth). MCA: Multi-Criteria Analysis; DPSIR: Driving force-Pressure-State-Impact-Response; GWR: Geographically Weighted Regression; CBI: Composite Burn Index; NDVI: Normalized Difference Vegetation Index; AAS: Agroforestry Accounting System; LCM: Land Change Modeller; ANNs: Artifical Neural Networks; SIAPA: System for the Integrated Assessment of Protected Areas; EDSS: Environmental Decision Support System; ABMs-AMEBA: Agent-Based Models; CORINE Lcover: CORINE Land Cover; SIOSE: Geographic Information System on Land Use-Land Cover of Spain.

Izakovičová et al., 2018 [23] use an integrated approach to attain sustainable LULC management in an agricultural area of western Slovakia. They analyse drivers of LULC change and their impacts on the environment. They propose optimal land uses accounting for interactions among available natural capital, environmental conditions and human needs in the long term to achieve socioeconomic development. They use multi-criteria analysis to guide managers' decision-making.

Forest fires also hamper sustainable territorial management and cause substantial environmental and socioeconomic losses. Viana-Soto et al., 2017 [24] assess regeneration of vegetation in burned areas of the Mediterranean region (Iberian Peninsula). They seek to understand species' recovery dynamics in order to implement suitable restoration actions. Regeneration modelling has been performed through multiple regressions, using ordinary least squares and geographic weight regression. They measure the severity of fire through the composite burn index and a set of environmental variables. They estimate the dynamics of regeneration through the Normalized Difference Vegetation Index obtained from Landsat images.

Furthermore, flooding is a persistent problem in coastal areas. Under scenarios of climate change, it is expected that flooding events will become more frequent and potentially more intense. This risk represents a potential threat to coastal communities that depend on coastal resources to a large extent. Toubes et al., 2017 [25] develop a methodology for coastal flooding risk assessment based on an index that compares 16 hydro-geo-morphological, biophysical, human exposure and resilience indicators, with a specific focus on tourism. They assess the vulnerability to floods of 724 beaches in Galicia (northern Spain). Their results are useful for coastal adaptation and management.

Knowing the value of the services provided by different ecosystems is essential for sustainable territorial management. Nevertheless, the Standard Economic Accounts for Agriculture and Forestry do not measure the ecosystem services and intermediate products embedded in the final products and ignore the private non-commercial intermediate products and self-consumption of private amenities. Campos et al., 2017 [26] apply the Agroforestry Accounting System to simulate sustainable forestry of holm oak and cork oak in Dehesa de la Luz, a Mediterranean tree-grass ecosystem. The net value added is more than 2.3 times greater than the estimated net value using the standard accounts.

Hewitt & Macleod, 2017 [27] use an Environmental Decision Support System (EDSS) to support the management of land and freshwater resources in Scotland, UK, with multiple applications to environmental management. They design a structured participatory process to determine stakeholder requirements, establish principles to meet these requirements and test the prototypes. The resulting specification of this bottom-up process is a free EDSS that is spatially explicit and compatible with portable devices. This application, still under development, does not resemble most existing EDSSs. Its focus on adaptive, stakeholder-centred environmental management strategies based on outcomes offers an opportunity to make better use of these new technologies to aid decision-making processes.

The rapid growth of urban areas close to large metropolises causes negative impacts on natural resources. This Special Issue includes two case studies focused on LULC changes in urban areas. In the first one, Ishtiaque et al., 2017 [28] analyse the increasing urbanization of the Kathmandu Valley (Nepal), in the foothills of the Himalayas. They use four Landsat images of the years 1989, 1999, 2009 and 2016 to compare changes. They relate LULC changes with a set of immediate causes and driving-factors of those changes. They employ a pixel-based hybrid classification approach and analyse the LULC trajectories. The results show that the urban area expanded to 412% in the last three decades.

Cantergiani & Gómez Delgado, 2018 [29] develop AMEBA, a prototype of an exploratory, spatial, agent-based model that considers the main stakeholders involved in the urban development process (urban planners, developers and the population). It consists of three sub-models, one for each agent. The first two are based on a land use allocation technique and the last one, as well as their integration, on an agent-based model approach. The authors describe the conceptualisation and performance of the sub-models that represent urban planners and developers, who are the agents responsible for officially expanding urban land and defining its spatial allocation. The prototype is tested in Corredor del Henares (an urban–industrial area in the Region of Madrid, central Spain), but it is flexible to be adapted to other study areas under different urban growth contexts.

PAs are also affected by LULC changes and other pressures from global change. Some processes such as intensive recreational use, forest fires or the expansion of artificial areas inside and around them jeopardise their environmental sustainability and effectiveness. Martínez-Vega et al., 2017 [30] analyse the LULC changes that took place between 1990 and 2006 in two Spanish national parks (NPs). They also simulate LULC changes between 2006 and 2030 through Artificial Neural Networks, taking into account a business-as-usual scenario and a green scenario. The simulation of LULC changes that are expected in the following decades under different scenarios is a strategic issue to carry out preventive protected area planning and management. Finally, they perform a multi-temporal analysis of natural habitat fragmentation in each NP.

López & Pardo 2018 [31] design an indicator system to monitor and assess the socioeconomic impacts of climate change on Sierra de Guadarrama NP (Spain) that could be used in other PAs in Spain and elsewhere. Indicators assess natural resource use, population change, economic activities and socio-political interactions. They use statistical sources and surveys according to the Driving forces-Pressure-State-Impact-Response framework.

As global biodiversity trends worsen, PA environmental effectiveness evaluation becomes an urgent need to identify strengths and areas to improve. Through a participatory process including PA managers and scientists, Rodríguez-Rodríguez et al., 2017 [32] refine the System for the Integrated Assessment of Protected Areas (SIAPA) in order to increase its legitimacy, credibility and salience to end users in Spain. Then, they test the optimised version of the SIAPA on two emblematic Spanish NPs: Ordesa y Monte Perdido NP, and Sierra de Guadarrama NP. Results show that potential environmental effectiveness is moderate for Ordesa NP and low for Guadarrama NP, according to the indicators that could be evaluated. PA managers and scientists largely coincided in the ratings of SIAPA's indicators and indices.

We hope that the methods developed, the results obtained, and the discussions included in the above-mentioned papers are useful to understand the potential of data modelling techniques, support future research, raise awareness about the complex problems of the territory and provide robust knowledge upon which to base sustainable territorial management.

Conflicts of Interest: The authors declare no conflict of interest.

References

1. Huang, L.; Yan, L.; Wu, J. Assessing urban sustainability of Chinese megacities: 35 years after the economic reform and open-door policy. *Landsc. Urban Plan.* **2016**, *145*, 57–70. [CrossRef]
2. García-Ayllón, S. Rapid development as a factor of imbalance in urban growth of cities in Latin America: A perspective based on territorial indicators. *Habitat Int.* **2016**, *58*, 127–142. [CrossRef]
3. Pons, A.; Rullán, O. Artificialization and Islandness on the Spanish Tourist Coast. *Misc. Geogr.-Reg. Stud. Dev.* **2014**, *18*, 5–16. [CrossRef]
4. De Andrés, M.; Barragán, J.M.; García Sanabria, J. Relationships between coastal urbanization and ecosystems in Spain. *Cities* **2017**, *68*, 8–17. [CrossRef]
5. Rodríguez-Rodríguez, D.; Martínez-Vega, J. Protected area effectiveness against land development in Spain. *J. Environ. Manag.* **2018**, *215*, 345–357. [CrossRef] [PubMed]
6. García-Ruiz, J.M.; Beguería, S.; Nadal-Romero, E.; González-Hidalgo, J.C.; Lana-Renault, N.; Sanjuán, Y. A meta-analysis of soil erosion rates across the world. *Geomorphology* **2015**, *239*, 160–173. [CrossRef]
7. Chen, B.; Song, Y.; Kwan, M.P.; Huang, B.; Xu, B. How do people in different places experience different levels of air pollution? Using worldwide Chinese as a lens. *Environ. Pollut.* **2018**, *238*, 874–883. [CrossRef] [PubMed]
8. Carrascal Incera, A.; Avelino, A.F.T.; Franco Solís, A. Gray water and environmental externalities: International patterns of water pollution through a structural decomposition analysis. *J. Clean. Prod.* **2017**, *165*, 1174–1187. [CrossRef]
9. Chuvieco, E. *Earth Observation of Wildland Fires in Mediterranean Ecosystems*; Springer-Verlag: Berlin/Heidelberg, Germany, 2009.
10. Uchida, K.; Koyanagi, T.F.; Matsumura, T.; Koyama, A. Patterns of plant diversity loss and species turnover resulting from land abandonment and intensification in semi-natural grasslands. *J. Environ. Manag.* **2018**, *218*, 622–629. [CrossRef] [PubMed]
11. Lui, G.V.; Coomes, D.A. Tropical nature reserves are losing their buffer zones, but leakage is not to blame. *Environ. Res.* **2016**, *147*, 580–589. [CrossRef] [PubMed]
12. Piqueray, J.; Bisteau, E.; Cristofoli, S.; Palm, R.; Poschlod, P.; Mahy, G. Plant species extinction debt in a temperate biodiversity hotspot: Community, species and functional traits approaches. *Biol. Conserv.* **2011**, *144*, 1619–1629. [CrossRef]
13. Fardoust, S.; Kanbur, R.; Luo, X.; Sundberg, M. An evaluation of the feedback loops in the poverty focus of world bank operations. *Eval. Program Plan.* **2018**, *67*, 10–18. [CrossRef] [PubMed]
14. Small, C.; Sousa, D.; Yetman, G.; Elvidge, C.; MacManus, K. Decades of urban growth and development on the Asian megadeltas. *Glob. Planet. Chang.* **2018**, *165*, 62–89. [CrossRef]
15. Gao, X.B.; Zhang, F.C.; Wang, C.; Wang, Y.X. Coexistence of High Fluoride Fresh and Saline Groundwaters in the Yuncheng Basin, Northern China. *Procedia Earth Planet. Sci.* **2013**, *7*, 280–283. [CrossRef]
16. Rahman, M.A.; Rahman, A.; Khan, M.Z.K.; Renzaho, A.M.N. Human health risks and socio-economic perspectives of arsenic exposure in Bangladesh: A scoping review. *Ecotoxicol. Environ. Saf.* **2018**, *150*, 335–343. [CrossRef] [PubMed]
17. Almeida, D.; Neto, C.; Esteves, L.S.; Costa, J.C. The impacts of land-use changes on the recovery of saltmarshes in Portugal. *Ocean Coast. Manag.* **2014**, *92*, 40–49. [CrossRef]

18. Pérez-García, J.M.; DeVault, T.L.; Botella, F.; Sánchez-Zapata, J.A. Using risk prediction models and species sensitivity maps for large-scale identification of infrastructure-related wildlife protection areas: The case of bird electrocution. *Biol. Conserv.* **2017**, *210*, 334–342. [CrossRef]
19. Ward-Paige, C.A.; Worm, B. Global evaluation of shark sanctuaries. *Glob. Environ. Chang.* **2017**, *47*, 174–189. [CrossRef]
20. Förster, J.; Barkmann, J.; Fricke, R.; Hotes, S.; Kleyer, M.; Kobbe, S.; Kübler, D.; Rumbaur, C.; Siegmund-Schultze, M.; Seppelt, R.; et al. Assessing ecosystem services for informing land-use decisions: A problem-oriented approach. *Ecol. Soc.* **2015**, *20*, 31. [CrossRef]
21. World Bank. *Sustainable Land Management: Challenges, Opportunities, and Trade-Offs*; Agriculture and Rural Development, World Bank: Washington, DC, USA, 2006. Available online: https://www.openknowledge.worldbank.org/handle/10986/7132 (accessed on 30 July 2018).
22. Camacho Olmedo, M.T.; Paegelow, M.; Mas, J.F.; Escobar, F. *Geomatic Approaches for Modeling Land Change Scenarios*; Lecture Notes in Geoinformation and Cartography; Springer: Berlin, Germany, 2018.
23. Izakovičová, Z.; Špulerová, J.; Petrovič, F. Integrated Approach to Sustainable Land Use Management. *Environments* **2018**, *5*, 37. [CrossRef]
24. Viana-Soto, A.; Aguado, I.; Martínez, S. Assessment of Post-Fire Vegetation Recovery Using Fire Severity and Geographical Data in the Mediterranean Region (Spain). *Environments* **2017**, *4*, 90. [CrossRef]
25. Toubes, D.R.; Gössling, S.; Hall, C.M.; Scott, D. Vulnerability of Coastal Beach Tourism to Flooding: A Case Study of Galicia, Spain. *Environments* **2017**, *4*, 83. [CrossRef]
26. Campos, P.; Mesa, B.; Álvarez, A.; Castaño, F.M.; Pulido, F. Testing Extended Accounts in Scheduled Conservation of Open Woodlands with Permanent Livestock Grazing: Dehesa de la Luz Estate Case Study, Arroyo de la Luz, Spain. *Environments* **2017**, *4*, 82. [CrossRef]
27. Hewitt, R.J.; Macleod, C.J.A. What Do Users Really Need? Participatory Development of Decision Support Tools for Environmental Management Based on Outcomes. *Environments* **2017**, *4*, 88. [CrossRef]
28. Ishtiaque, A.; Shrestha, M.; Chhetri, N. Rapid Urban Growth in the Kathmandu Valley, Nepal: Monitoring Land Use Land Cover Dynamics of a Himalayan City with Landsat Imageries. *Environments* **2017**, *4*, 72. [CrossRef]
29. Cantergiani, C.; Gómez Delgado, M. Urban Land Allocation Model of Territorial Expansion by Urban Planners and Housing Developers. *Environments* **2018**, *5*, 5. [CrossRef]
30. Martínez-Vega, J.; Díaz, A.; Nava, J.M.; Gallardo, M.; Echavarría, P. Assessing Land Use-Cover Changes and Modelling Change Scenarios in Two Mountain Spanish National Parks. *Environments* **2017**, *4*, 79. [CrossRef]
31. López, I.; Pardo, M. Socioeconomic Indicators for the Evaluation and Monitoring of Climate Change in National Parks: An Analysis of the Sierra de Guadarrama National Park (Spain). *Environments* **2018**, *5*, 25. [CrossRef]
32. Rodríguez-Rodríguez, D.; Ibarra, P.; Martínez-Vega, J.; Echeverría, M.; Echavarría, P. Fine-Tuning of a Protected Area Effectiveness Evaluation Tool: Implementation on Two Emblematic Spanish National Parks. *Environments* **2017**, *4*, 68. [CrossRef]

![environments logo] *environments*

MDPI

Article

Integrated Approach to Sustainable Land Use Management

Zita Izakovičová [1], Jana Špulerová [1,*] and František Petrovič [2]

[1]   Institute of Landscape Ecology of the Slovak Academy of Sciences, P.O. Box 254, Štefanikova 3,
      814 99 Bratislava, Slovakia; zita.izakovicova@savba.sk
[2]   Department of Ecology and Environmentalist, Faculty of Natural Sciences, Constantine the Philosopher
      University Nitra, Tr. A. Hlinku 1, 949 01 Nitra, Slovakia; fpetrovic@ukf.sk
*   Correspondence: jana.spulerova@savba.sk; Tel.: +421-2-2092-0341

Received: 21 December 2017; Accepted: 27 February 2018; Published: 1 March 2018

Abstract: This article presents the integrated approach to sustainable land use management based on the assessment of land use and related land cover changes. Land use changes are conditioned by human activities producing changes in landscape cover and initiating processes which cause many environmental problems. It is therefore important to determine the drivers and causality of landscape changes which can then be negated to ensure sustainable land use management. The integrated landscape research approach is based on understanding landscape as a geo-ecosystem with natural, human, cultural, and historical potential. Our aim is to define the aspects of land use management which can regulate social development. The proposal for optimal land use is based on the interaction between natural capital, represented by the supply of natural regional resources and environmental conditions as well as demand represented by community need for development. The conflict between the supply of natural capital and demands lacking respect for landscape resources is an important determining factor in environmental and human problems. The integrated approach is focused on long-term rational utilization of the natural and cultural-historical resources, urban development, and the elimination of current environmental and socioeconomic problems as well as the prevention of new ones. Multi-criteria analysis is required for final environmental decision-making.

Keywords: land use; land cover; sustainable landscape management; geo-ecosystem; environmental problems; landscape processes

---

1. Introduction

Sustainability is an essential precondition for the continued existence of human society. The issue of sustainable land use has increasing importance because of accumulated environmental problems. These include increased demand for natural resources, climate change, regional climate extremes, the threat of environmental pollution, biodiversity loss, disturbed landscape stability, economic globalization, energy security, water supply, and increasing conflicts between sociocultural, political-economic, and environmental goals [1,2].

Approaches and definitions of sustainable land use development on a global scale are numerous, heterogeneous, and based on a variety of aspects. The most frequently quoted definition is the Brundtland Report's "Our Common Future"—"development that meets the needs of the present without compromising the ability of future generations to meet their own needs" [3]. Sustainability is the foundation of today's leading global framework for international cooperation described in the 2030 Agenda for Sustainable Development and its Sustainable Development Goals (SDGs) [4]. Most definitions stress that sustainable development requires socioeconomic development which preserves the principles of sustainable land use and respects the natural and cultural-historical resources and potential of the territory [4–6]. The focus of our research should especially support two of the latest SDG specifics for sustainable land use: (i) goal 15 "Life on land" and (ii) goal 11 "Sustainable cities and communities"; as sustainable land use contributes to halting and reversing land degradation and natural hazards, it also halts biodiversity loss and supports landscape stability. The outputs of the proposal can be applied in spatial and urban planning. Requirements for sustainable land use management issue from:

- the need to ensure and improve spatial stabilization of the territory. The stated criterion here is the demand to achieve biological balance in the country;
- needs for nature protection and rational utilization of natural resources; in particular, the protection of the land, water, forests, and gene pool;
- needs for the protection of cultural and historical resources;
- needs for the regeneration of human resources and the protection of human health;
- demands on the humanization and aesthetic appeal of the landscape.

These requirements incorporate the fundamental principles of sustainable societal development. Sustainable development accentuates caring for the Earth by putting sustainable living principles into practice and integrating conservation and development: the conservation to maintain human actions within the Earth's capacity and the development to enable people everywhere to enjoy long, healthy, and fulfilling lives [7].

Land use and land cover are interconnected, as land use initiates land cover changes [8]. Land cover is continually transformed by anthropogenic land-use influences on the properties, processes, and components of service provision. Changes in land use or management will therefore change service supply, not only for specific services but for the complete array of services provided by that (eco)system [9]. It is therefore important to study not only land use and land cover changes, but also to assess all drivers of land use change; the position and correlation of landscape elements; causality and the consequence and impacts of such changes. The main driving forces of land use are political, economical, cultural, technological, and natural [10–12].

It is not possible to evaluate and propose optimal land use based on one landscape parameter. We must therefore examine the relationships between the different landscape features and emphasize that all decision-making should apply an integrated approach based on understanding landscape as a geo-ecosystem. Landscape is envisaged in integrated scope, combining all layers of the following resources; the geological base, water and soil, climate, and biotic and morphometric parameters [13]. The geo-ecosystem encompasses a complex system of space, position, relief, and all other functionally interconnected physical landscape features of the geo-sphere where man and other organisms live and act. These features of each landscape elements comprise natural, semi-natural, and anthropogenic ecosystems [14,15]. Integrated approaches to sustainable land use management are therefore based on assessing the natural capital and human interaction aspects of landscape structure using appropriate landscape evaluation approaches, as well as multi-scale analysis and modeling [10,13–19]. The effects of individual land use changes over a particular time period determine a study area's rate of sustainability of coexistence between nature and social subsystems. Sustainable land use management must be based on integrated landscape research in the three basic dimensions: environmental, social, and economic. Moreover, analysis of connections and dependencies between these dimensions should aim to define the type of management that will regulate socioeconomic land use development and maintain its natural, human, cultural, and historical potential.

The aim of this study is to develop an integrated approach to sustainable land use management based on the understanding of landscape as geo-ecosystem including different landscape features. The focus is on long-term rational use of the natural and cultural-historical resources, the elimination of current environmental and socioeconomic problems, and the prevention of new ones. These approaches are generally well-recognized, but their application to land use has been inadequate. The specific goals concentrate on developing a method of decision-making for sustainable land use based on limit-setting, establishing the degree of anthropogenic changes, and identifying the type and intensity of environmental problems in a given territory. Determination of these specifics leads to proposals for eliminating the negative factors that influence the area. This methodological approach is applied in the case study of the Trnava region of Slovakia.

2. Methods

2.1. Methodological Approach

The integrated approach to sustainable land use management is a coherent system of interrelated steps, which can be modified based on the type and scale of the study area. The methodology focuses on decision-making processes based on confrontation and subsequent proposal for harmony in (1) the supply of landscape properties as natural capital and complex natural resources and (2) the demands and influences of human activities (Figure 1).

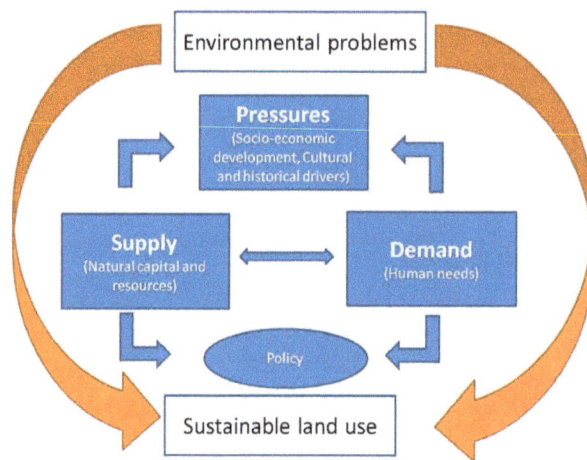

Figure 1. Integrated approach to sustainable land use management.

The approach for applying sustainable land use is therefore based on the methodology of landscape ecological planning [7] (Figure 2). This is one of Agenda 21's recommendations for the integrated protection of natural resources, and has the following steps:

I. Analysis

The principal objective of analysis is to choose, quantify, and describe the main features of landscape elements which define and map the abiotic, biotic, and socioeconomic features of a given territory. The most important analyses are:

- analysis of geomorphological features, geological, hydrological, soil, and climatic conditions. This establishes the properties of abiotic complexes in the territory.
- analysis of fauna and flora and their conditions determines the properties of the biotic complexes.
- analysis of the socioeconomic activities and their negative influences supplies the properties of the socioeconomic complexes.

Data was obtained from several databases, sectoral statistics, and available mapping sources. These features can have different relationships to individual human activities which support their development or restrict or limit it. It is necessary to initially concentrate on the basic selection of landscape features which definitively influence the location of human activities in a given area.

Our choice is inextricably bound to the aims of the task, the degree of processing, and the specifics of the territory. This then provides maps in ArcGIS 10.1., which identify the abiotic, biotic, and socioeconomic conditions of the study area.

Specification of human activities requires detailed analysis of the demands of all forms of human activity on the landscape. These include all activities involved in industry, agriculture, forestry, water economy, urban development, tourism, and nature protection. It is also necessary to specify the results and risks associated with their performance. Miklós and Izakovičová [20] stress the performance of the following individual activities in the landscape:

- areas used for building construction and complexes, industrial and agricultural complexes, and communication lines and facilities.
- extensive use of the landscape for agriculture and forestry.
- definition of functional zones and protected areas. These include recreational zones and areas protected for soil and water resources and nature.
- The pressures connected with the performance of these activities can also initiate atmospheric pollution and soil and water contamination.

II. Synthesis

Synthesis involves the interaction of individual features which create homogeneous areas with different combinations of abiotic, biotic, and socioeconomic features fully integrated in the regional geo-ecosystem. Synthesis herein is achieved by the spatial superimposition of GIS analytic maps.

III. Evaluations

Evaluation establishes the regulations for specific human activities through justification and limit-setting on landscape elements and features involved in human activities. Discrete knowledge of regional landscape vulnerability and specification of regulated environmental limits and restrictions create the basis for decisions to permit specific human activities in a given area, to accept them with provisos, or to exclude them entirely [21]. Comparative research into spatial planning systems typically adopts a structural/legal approach and an integrated perspective embracing system structure and concrete planning practices. Sensitive discourse on planning theory towards culturally-oriented interpretation lies at the heart of appropriate decision-making [22]. The expression of spatial limits confronts landscape ecological complexes with proposed human activity. This enables the mapping of regulations for spatial limits and restrictions on the development of human activities. The limiting values of different landscape features occur in different combinations, where limiting and restricting values from any given combination determine the possibility and advisability of locating a particular activity in a given area. If one landscape feature is above the limit, the particular activity is not possible in the given area. Superimposition of the limiting values of all chosen features provides a comprehensive map of limits which decides:

- activities possible in the given area. This includes multiple ranking of suitability from different perspectives.
- activities not possible in the given area.
- limits and restrictions, including a combination of limits and restrictions, required to exclude particular activities from target localities.

IV. Propositions

The proposal for ecologically optimal land use is as follows. It is necessary to determine functions for each area not limited or restricted by landscape features. This establishes functions harmonious with the natural and socioeconomic conditions of the region and also satisfies societal development needs.

Environmental decision-making involves the identification and comparison of different alternatives based on multiple objectives and criteria. Here, multi-criteria analysis (MCA) provides the framework for integrating factual information on stakeholders' preferences, values, and associated impacts. MCA is increasingly used in combination with GIS spatial multi-criteria analysis (SMCA) [23].

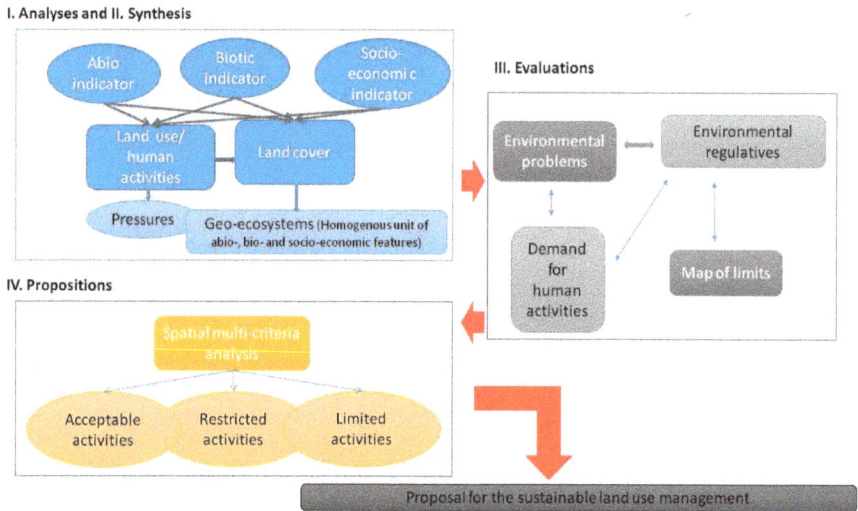

Figure 2. Methodological approach for sustainable land use management.

2.2. Study Area

The Trnava study area (Figure 3) has typical western Slovak agricultural landscape. Administratively, it consists of the Trnava town, surroundings, and 45 village areas. This covers 741 km^2 with 131,167 inhabitants, thus ranking as a medium-sized Slovak district.

Agricultural land covers 53,107 ha and 71.6% of the district area as the dominant landscape element. Up to 93.1% of this is intensively utilized as large-block arable land, with cereals dominating the central and south part of the study area. Forests cover 13,190 ha, 17.7%, and these areas are mainly in sub-mountain villages in the northern part of the area under the Small Carpathian Mountains (Malé Karpaty) Protected Landscape Area. Industrial sites are mainly situated in the Trnava township surroundings, where Peugeot and Samsung complexes have recently been constructed with good transport accessibility and proximity to the capital, Bratislava.

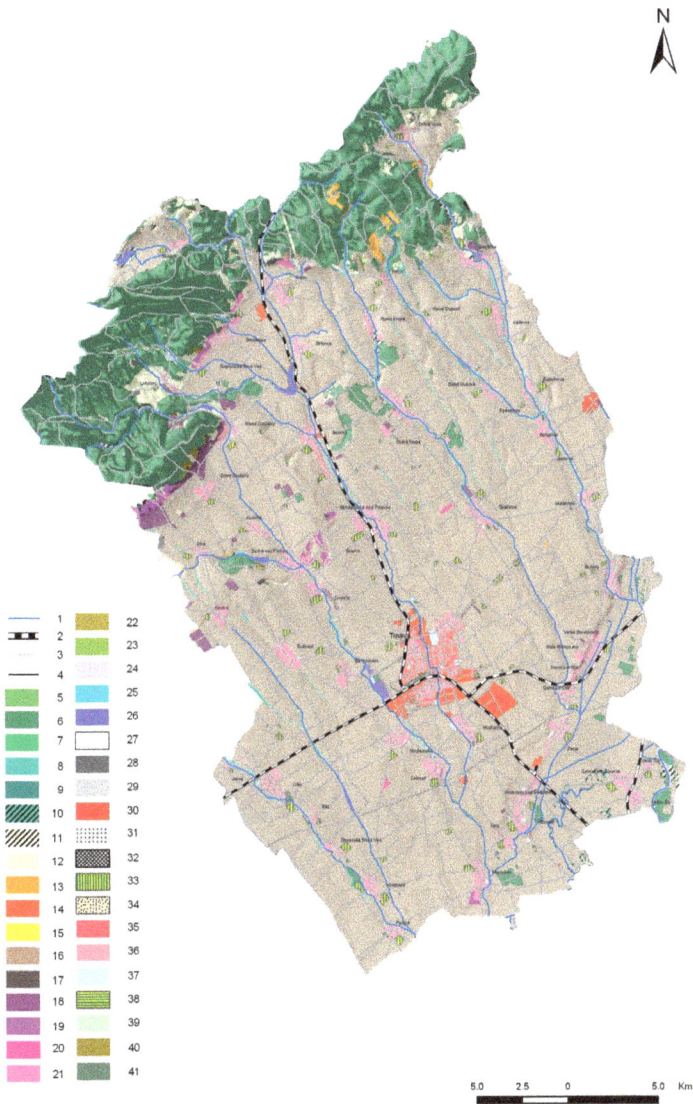

Figure 3. Land use of the study area. Legend: 1. Water flows; 2. Railways; 3. Roads; 4. Highways; 5. Forests; 6. Semi-natural small woodland; 7. Planted small woodland; 8. Riparian vegetation; 9. Lines of trees or shrubs; 10. Wetlands; 11. Wet meadows; 12. Extensively managed grasslands; 13. Intensively managed meadows; 14. Intensively managed pastures; 15. Dry grasslands; 16. Large-block arable land; 17. Small arable fields; 18. Large-block vineyards; 19. Small-block vineyards; 20. Orchards; 21. Gardens; 22. Mosaic of arable lands and grasslands; 23. Wooded grasslands; 24. Natural water body; 25. Channels; 26. Water reservoirs; 27. Rocks; 28. Abandoned fields; 29. Brown fields; 30. Industrial areas; 31. Mining areas; 32. Industrial pond; 33. Agricultural farms; 34. Field airport; 35. Urban areas; 36. Rural settlement; 37. Recreational zone; 38. Garden zones with cottages; 39. Cottages; 40. Abandoned areas; 41. Transport areas.

3. Results

3.1. Landscape Features and Proposed Limits

Analysis provided a set of study area maps identifying the abiotic, biotic, and socioeconomic conditions. Superimposition then established the geo-ecosystem and allowed us to determine environmental limits. This determination enabled objective and scientific decision-making on locating human landscape activities. Limit-setting is very complex and time-consuming, because it demands a multi-disciplinary and synergistic approach. This approach normally begins with the existing natural conditions and societal norms, but when there is no established norm for a given phenomenon or human activity, this process must be assumed by collective experts' and stakeholders' evaluation. The determination of limiting values requires extensive collection of data on the operation of the landscape system as a complex while acknowledging its individual features. This is essential, because the implementation of human activity without conflict depends on the wide variety of environmental conditions where the activity is performed.

We divided the set of regulations enforcing limits and restrictions into the following categories:

A. Abiotic regulations are based on abiotic complexes including geo-mechanical, hydrological, aerodynamic, and soil limitations. The limits have a permanent character including the relative stability of the geological substratum and local climate, and these cannot be easily changed by technology.

B. Biotic regulations are based on the biotic complexes required by living organisms. The gene pool, biodiversity, and landscape ecological stability are threatened by pressures from human activity and land use changes.

C. Anthropogenic regulations result from the competitive requirements and demands of human activities which limit the development of other activities through negative effects or by simply occupying an area. These include technical, hygienic, protective, and other limits, and these indicators are relatively easier to change than in the preceding categories. Although anthropogenic regulations are applied to very serious hygienic and environmental security demands, the limits imposed can be temporary and depend on altered circumstances.

D. Complex landscape regulations are based on the principles of landscape functioning as a complex. They include eco-stabilization, localization, carrying-capacity, behavioral, aesthetic, and cultural-historical limits. These limits are very dynamic as they result from principles of the operation of landscape as an entity and the set limits must strictly respect its historical development.

Landscape limits and restrictions are not isolated but act synergistically, so that the locality of given human activity can be limited or restricted by two or more factors. The determination of limiting and restricting factors for a given activity proceeds from the evaluation of the functional relationships between landscape elements. These center on combining the abiotic, biotic, and socioeconomic complex with requested human activity. Moreover, the process of creating regulations is most frequently performed in conjunction with decision-making tables (Table 1), and this defines three degrees of availability for performance of human activity on an area; acceptable, limited, and restricted activities.

Table 1. Example of creation of environmental regulation based on environmental stress factors.

Stress Factors / Land Use Activities	Air Pollution	Noise Load Area	Soil Contamination	Polluted Water Flows	Damage of Vegetation	Radio Activity	Nature Reserve	Protected Zone of Water Resources
forests (F)	1	1	1	1	1	1	1	1
grassland (G)	1	1	1	1	1	1	1	1
pastures (P)	L	1	L	1	L	L	L	L
vineyards (V)	L	1	L	-	L	L	L	0
forage-crops (C)	L	1	L	-	L	L	L	0
arable land (A)	0	1	0	-	0	0	L	0
orchards (O)	0	1	L	-	L	0	L	0
gardens (GS)	L	1	L	L	L	L	L	0
recreation (R)	L	L	L	L	L	L	L	0
sport areas (S)	L	L	L	-	L	L	L	1
medical areas (M)	L	L	L	-	L	L	L	L
housing areas (H)	L	L	L	-	L	L	L	1
farm animals (FA)	L	1	L	-	0	1	L	L
industrial areas (I)	1	1	1	-	1	1	L	L
transport areas (T)	1	1	1	-	1	1	L	L
Unlimited activities/acceptable activities	I,T,F,G/A,O	F,G,P,V,C,A,O,GS,I,T	I,T,F,G/A	F,G	I,T,F,G/A,FA	F,G,I,T/O,A	F,G/-	F,G,M/V,C,A,O,GS,R,S

Legend: L—environmental limit (limited activities); 0—environmental restriction (restricted activities); 1—acceptable activities.

The determined values and indicators which significantly limit the intensity of a given activity in an area are restrictions and not absolute exclusion. Examples here are; (1) the size of protected areas limits, but does not restrict, the development of recreational space, and (2) although agricultural production is not excluded in areas with water resource protection, its intensity is considerably restricted by recommended crop structure, chemical use, and mechanization.

3.2. The Impact of Land Use Changes

Landscape-ecological evaluation of current land use aims to define landscape-ecologically problem areas where the present land use does not correspond to the criteria. This identifies areas where the present land use is restricted by landscape-ecological limits. Current knowledge [24] enables us to identify the following most significant changes in the study area and interconnected problems.

- Conflicts between the socioeconomic development of nature protection; (1) building stones are extracted in the Small Carpathian Mts. Protected Landscape Area. Extraction is profitable only for entrepreneurs and the employment rate is insignificant. It is therefore deemed necessary to eliminate these mining activities in the protected areas; (2) recreational areas have been developed in the Small Carpathian Mts. and in the Trnavske rybníky fishponds Protected Area. Planned tourist attraction there can negatively affect the natural landscape and especially the avian population. This presents conflict between economical development and nature protection.
- Conflicts in socioeconomical development and natural resource protection; (1) there is competing interest in industrial development and the protection of the most fertile Trnava soils. The recent building boom has appropriated 'green fields', with the best quality soils to be used for industry, industrial parks, and housing, while many existing industrial sites lay abandoned with decreased economic value. It would be advantageous for sustainable development and regional economics if these abandoned sites were refurbished and re-used instead of expanding the industrial occupation of ecologically-valuable green fields; (2) intensive agricultural practices have led to both surface and underground water endangerment; and (3) inappropriate soil management promotes soil degradation, including compaction and erosion.
- Conflicts in nature protection and society; for example, protected areas for hygienic water resource protection and other protective zones limit the land use of some areas. These zones require essential limits in socioeconomic and urban development, including inappropriate property acquisition and utilization rights. However, unsatisfactory compensation and loss of profit create competing interests between nature protection and social justice, and this conflict requires urgent solutions.
- Conflicts in socioeconomic development and environmental quality; while industrial operators are significant employers, fundamental regional economics compete with extreme environmental load. It is currently impossible to close industries because of significant unemployment and regional economical efficiency. It is therefore essential to promote effective technology which limits contaminant production and ensures sustainable development.

3.3. Proposal for Sustainable Land Use Management

The aim of the proposal for optimal land use is to eliminate all problem areas, to anticipate possible new problems, and to create a structure harmonious with the territory's natural and socioeconomic conditions. Here, decision-support methods and tools such as multi-criteria analysis (MCA) and spatial multi-criteria analysis (SMCA) help achieve complex choice settings. These tools collect, organize, and analyze information which supports discussion and value elicitation and enables a better understanding of the implications of different options in sustainable land use (Figure 4).

Figure 4. Example of spatial multi-criteria analysis for optimal allocated recreational activity.

We developed the proposal for sustainable land use management from analysis, synthesis, limit setting, and conflict identification (Figure 5). Moreover, we defined the following principles of limit-setting, which can be generalized for other areas:

- Abiotic conditions are the determining factors in a given area's diversity. They establish appropriate area utilization. The abiotic elements' permanent and unique attributes become determining factors in human development.
- Land use management must reduce the risk factors in sensitive localities which are otherwise predisposed to anthropogenic degradation processes including erosion, subsidence, landslides, and earthquakes.
- It is essential to support development in NATURA 2000 protected territories and ecologically valuable areas of stability. This enables scientific and medical research centers, which encourage appropriate recreational areas and reduce threats to natural landscape units.
- Similarly, the development of human activities with negative impacts must be excluded in areas where natural resources are legally protected, and explicit priority must be given to developing activities which protect individual natural resources.
- All detrimental activities must be excluded from sensitive areas with strong pressure burden. These include areas with air-pollution, soil or water contamination, and noise pollution.
- Areas without pressure loads should be maintained free from activities which can harm current living quality. These areas are suitable for high-quality living development with adequate agricultural, ecological, and recreation services.

The following outcomes result from this decision-making process:

- Selection and exclusion of activities which cannot be located on a given area because of possible landscape-ecological harm.
- Selection and restriction of activities which can be conducted on an area, but which can cause landscape-ecological damage if unrestricted.
- Selection of a hierarchy of activities which maintain the area's optimal landscape-ecological function.
- Selection of complex measurements required to protect the area's nature, natural resources, and environment.

This requires the implementation of effective technology for the following: eliminating excess production of polluting substances, minimizing the allochthonous and other contaminant substance effects on environmental elements, and applying appropriate maintenance technology in agriculture and forestry.

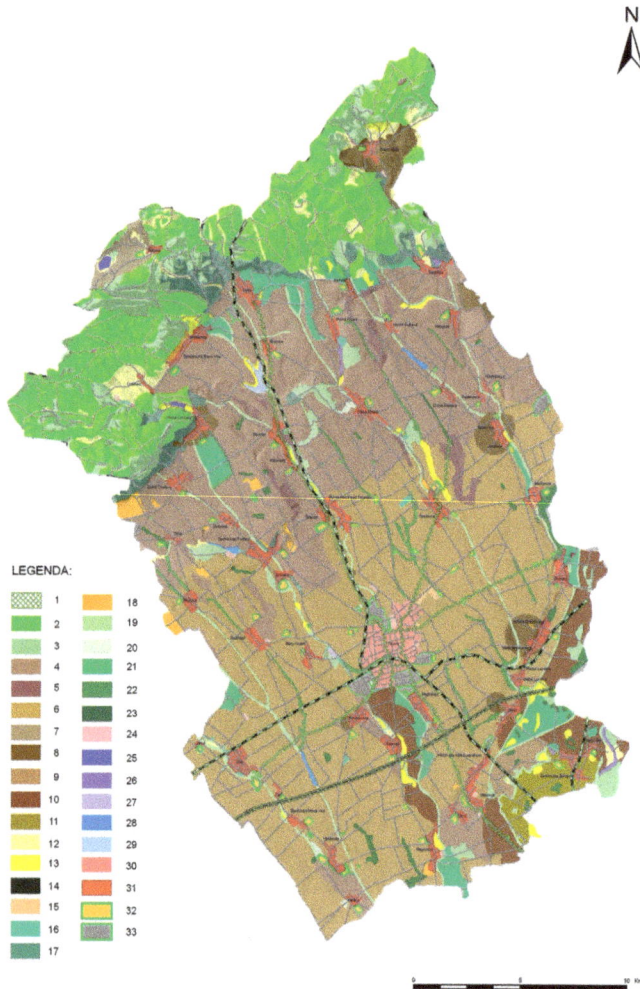

Figure 5. Proposal for optimal land use of the study area. Legend: 1. Green infrastructure of industrial zone; 2. Extensive forest management; 3. Protected forests; 4. Extensive agriculture—small arable fields; 5. Erosion control on small arable fields; 6. Intensive agriculture—large-block arable lands; 7. Extensive agriculture for protected karst areas; 8. Extensive agriculture for water protection; 9. Extensive agriculture for the protection of water resources and mineral resources; 10. Agriculture with special management for contaminated soil; 11. Agriculture with special management for soil contamination and water protection; 12. Intensively managed meadows; 13. Extensively managed wooded grasslands; 14. Recreational park; 15. Extensively managed grasslands; 16. Extensive agriculture in gardens and orchards; 17. Extensive agriculture in mosaics of gardens, arable lands, and grasslands; 18. Extensive agriculture in vineyards; 19. Water protected area—floodplain vegetation; 20. City green infrastructure; 21. Extensive agriculture of mosaics of arable lands and grasslands; 22. Nature protection; 23. Open landscape green infrastructure; 24. Recreation zone with cottages; 25. Recreation zone for water sport; 26. Recreation zone for fishing; 27. Recreation zone for fishing, with nature protection restriction; 28. Water reservoir for irrigation; 29. Living area—block of flats; 30. Living area—individual houses; 31. Built-up area—agricultural buildings; 32. Built-up area—industrial buildings; 33. Industrial zone.

4. Discussion

Sustainable land use management remains a hot topic because it focuses on actual problems and ensures the integration of the natural, cultural-historical, and socioeconomical resources of a given area. Appropriate land use management arises from the necessity to solve both environmental and human existential problems. These include impacts associated with climate change, effects on health and extreme events, such as flooding, which can arise from the prevailing strategies employed in land use and protection [25,26]. An integrated approach to sustainable land use management helps resource users, managers, and stakeholders to manage resources sustainably by considering, reconciling, and synergizing conflicting interests and activities.

Sustainable land use management must be based on recognizing landscape as an integration of natural resources in an individual area. Each point on the Earth's surface presents a specific homogeneous entity of these combined sources. These form the landscape components and its features which satisfy human needs. Understanding the relationship between these natural resources is required to ensure sustainable land use by society. However, it is impossible to satisfy all competing aims, and dangerous to promote the land use and protection of one resource at the expense of another. An example here is the application of intensive soil use in areas with significant groundwater while ignoring the high risk of water contamination. Schulte et al. [27] support this supposition, stressing that the main global policy challenges today are the efficient and prudent use of the world's natural resources and managing the conflicting demands on land use. Labuda [28] and Surova et al. [29] add that sustainable land use must be linked with multi-functionality. This rationale addresses the interdependence of social, economic, and environmental effects of land use, with appropriate consideration of existing commodities and negative and positive external factors. Land and the rural environment provide a variety of functions, with their goods and services covering information, habitat, production, and regulation. Therefore, modifying the landscape to increase multi-functionality and reduce trade-offs with concurrent services will enhance sustainability in human-dominated landscapes [30].

The proposal for optimal land use is based on multi-criteria analysis of the natural capital, represented by the natural resources and environmental condition of the region, as well as demand represented by the community needs for development. Conflict between supply and demand which lacks respect for landscape resources is the determining factor in both environmental and human problems [31]. The proposed approach focuses on overcoming the stated difficulties by eliminating current environmental and socioeconomic problems in addition to preventing new ones. Miklos [17] agrees that this positive action will secure rational long-term utilization of natural and cultural-historical resources, and other authors [32] highlight that the proposal of eco-stabilizing elements must form part of the planned measures for both the agricultural landscape and urban areas.

The application of sustainable development principles in practice contributes to eliminating environmental problems and harmonizing intensified socioeconomical development and natural resources in a given area. This methodological approach to optimal land use has been applied in sectors of Slovakia and in other countries. The most practical result of the agricultural landscape-ecological evaluation is based on the suitability of using abiotic complexes on selected study area crops [33,34]. The conceptual framework for the quantification of supply and demand in agricultural soil-based ecosystem services is taken from Irish agriculture. This involved a case study with proxy-indicators determining the demand for individual soil functions [27]. The localizing precondition of tourism development was then evaluated using complex landscape-ecological geo-database data, land cover, selected morphometric indicators, selected town-planning, and demographical and socioeconomic indicators [35,36]. The pressures were found to be greatest on urban ecosystems, with high population density and multiple activities with different influences on the environment. These can cause unpredictable responses to environmental quality [37]. Investment in conservation, restoration, and sustainable ecosystem use are increasingly considered a "win-win situation" which generates substantial ecological, social, and economic benefits [9].

The optimal landscape-ecological solution for spatial land use is the major outcome of sustainable land use management. This comprises an initial proposal of the most suitable localization of demanded human activities in the given territory, and a subsequent proposal of measurements which ensure the activities' appropriate environmental functioning in that locality. This answers the questions of how and where to provide human activities in the territory that would least conflict with the natural conditions, and how to apply them in the most suitable land use management methods to reduce natural risks and hazards [38–40]. The solution to environmental problems and sustainable land use has (1) aspects of spatial organization, which provide optimal land-use, and (2) aspects ensuring technical expertise in landscape ecology.

The application of limits is most important, because they form the basis for optimal landscape-ecological decision-making processes in land use. The limits are applied to both evaluating the territory's current functional use and establishing proposals for optimal allocation and management of the many different land use options [38,41,42]. Multi-criteria decision analysis determined the multiple well-being dimensions of ecological, economic, cultural, and moral aspects of policy and management problems [43,44]. Complex spatial multi-criteria analysis planning with modeling then created quality outcomes which helped identify impacts on both the environment and residents. These outcomes also have an important function in the decision-making and draft measure phases which mitigate negative impacts on the environment [45].

5. Conclusions

This paper presents the integrated approach to land use management based on limit-setting and other regulative measures, which we developed as a basis for the decision-making process. It can be used to process development documents and strategies from the local scale of cities or municipalities to the regional scale. Integrated land use management is based on landscape research in three basic dimensions: environmental, social, and economic, as well as on examining their interrelationships and contexts. In particular, economic and social benefits are directly dependent on an organization's property including land use, ownership, and other rights, without which any planning activity in the landscape is practically impossible. Our presented method can contribute to the improvement of existing methods of land use assessment, such as land consolidation or the territorial system of ecological stability [46–48]. These methods are aimed at efficient land use and a new land arrangement in accordance with the conditions for improving the environment, soil protection, water management, increasing the ecological stability of the landscape, and improving the quality of rural life. The successful application of sustainable land use management requires multiple social measures at all levels of legislation, economic outcomes, education, and teaching. Successful sustainable development in actual practice demands the following essential measures:

- The regulations for optimal land use must be applied to sector plans—it is unavoidable that the regulated use of particular resources by production and non-production entities favors the development of one area over another and/or fails to avoid conflicts of interest.
- The principles of sustainable development should be implemented with as much population awareness as possible, especially for stakeholders and policymakers—this requirement is based on creating an effective system of education in sustainable development and land use management, because adequate education enhances public acceptance of the principles and criteria for practical sustainable development.
- To ensure the promotion of effective tools for legislative protection and economic outcomes, it is essential that legislative rules and regulations support the rational use of natural resources and protect both the environment and human health. Economic tools such as taxes, duties, and fees support both decision-making and sustainable landscape-ecological policy. Fines imposed for inappropriate land use, environmental pollution, human health endangerment or injury, and breach of regulations help eliminate environmental problems. Finally, subsidies from

rural development programs and other sources help reduce marginality and social disparity in rural communities.

Acknowledgments: This paper is the result of funding from the Slovak Research and Development Agency (Project No. APVV-0866-12, "Evaluation of ecosystem functions and services of the cultural landscape") and Scientific Grant Agency of Ministry of Education of the Slovak Republic (No. 1/0496/16 "Assessment of natural capital, biodiversity and ecosystem services in Slovakia"). We thank Raymond Marshall for English proof-reading.

Author Contributions: The research was conceived, designed and implemented by Zita Izakovičová and Jana Špulerová. František Petrovič contributed with technical knowledge of ArcGIS and with developing methodology scheme. All authors contributed to revision of the article and have given final approval of the version to be published.

Conflicts of Interest: The authors declare no conflicts of interest.

References

1. Axelsson, R.; Angelstam, P.; Elbakidze, M.; Stryamets, N.; Johansson, K.-E. Sustainable Development and Sustainability: Landscape Approach as a Practical Interpretation of Principles and Implementation Concepts. *J. Landsc. Ecol.* **2012**, *4*, 5–30. [CrossRef]
2. Findell, K.L.; Berg, A.; Gentine, P.; Krasting, J.P.; Lintner, B.R.; Malyshev, S.; Santanello, J.A., Jr.; Shevliakova, E. The impact of anthropogenic land use and land cover change on regional climate extremes. *Nat. Commun.* **2017**, *8*. [CrossRef] [PubMed]
3. Brundtland, G.; Khalid, M.; Agnelli, S.; Al-Athel, S.; Casanova, P.G.; Chidzero, B.; Padika, L.; Hauff, V.; Lang, I.; Shijun, M.; et al. *Our Common Future*; Brundtland Report; Oxford University Press: Oslo, Norway, 1987. Available online: www.un-documents.net/our-common-future.pdf (accessed on 2 February 2018).
4. IISD. Sustainable Development. 2016. Available online: http://www.iisd.org/topic/sustainable-development (accessed on 4 December 2017).
5. IUCN. Caring for the Earth. In *A Strategy for Sustainable Living*; IUCN, UNEP, WWF: Glad, Switzerland, 1991.
6. Potschin, M.; Haines-Young, R. Landscapes, sustainability and the place-based analysis of ecosystem services. *Landsc. Ecol.* **2013**, *28*, 1053–1065. [CrossRef]
7. Ružička, M.; Miklós, L. Basic Premises and Methods in Landscape Ecological Planning and Optimization. In *Changing Landscapes: An Ecological Perspective*; Zonneveld, I.S., Forman, R.T.T., Eds.; Springer: New York, NY, USA, 1990; pp. 233–260. Available online: http://link.springer.com/chapter/10.1007/978-1-4612-3304-6_13 (accessed on 8 December 2014).
8. De Sherbinin, A. *A CIESIN Thematic Guide to Land-Use and Land-Cover Change (LUCC)*; Center for International Earth Science Information Network (CIESIN) Columbia University Palisades: Palisades, NY, USA, 2002; Available online: http://sedac.ciesin.columbia.edu/guides (accessed on 28 November 2017).
9. De Groot, R.S.; Alkemade, R.; Braat, L.; Hein, L.; Willemen, L. Challenges in integrating the concept of ecosystem services and values in landscape planning, management and decision making. *Ecol. Complex.* **2010**, *7*, 260–272. [CrossRef]
10. Bürgi, M.; Hersperger, A.M.; Schneeberger, N. Driving forces of landscape change—Current and new directions. *Landsc. Ecol.* **2004**, *19*, 857–868. [CrossRef]
11. Hersperger, A.M.; Buergi, M. Driving Forces of Landscape Change in the Urbanizing Limmat Valley, Switzerland. In *Modelling Land-Use Change: Progress and Applications*; Koomen, E., Stillwell, J., Bakema, A., Scholten, H.J., Eds.; Springer: Dordrecht, The Netherlands, 2007; pp. 45–60.
12. Wohlmeyer, H.F.J. The unconscious driving forces of landscape perception and formation. In *Sustainable Development of Multifunctional Landscapes*; Helming, K., Wiggering, H., Eds.; Springer: New York, NY, USA, 2003; pp. 79–93.
13. Miklós, L.; Izakovičová, Z. *Krajina Ako Geosystém (Landscape as Geo-Ecosystem)*; VEDA Publishing of the Slovak Academy of Sciences: Bratislava, Slovakia, 1997.
14. Braunisch, V.; Patthey, P.; Arlettaz, R.L. Spatially explicit modeling of conflict zones between wildlife and snow sports: Prioritizing areas for winter refuges. *Ecol. Appl.* **2011**, *21*, 955–967. [CrossRef] [PubMed]
15. Frost, P.; Campbell, B.; Medina, G.; Usongo, L. Landscape-Scale Approaches for Integrated Natural Resource Management in Tropical Forest Landscapes. *Ecol. Soc.* **2006**, *11*. Available online: https://www.ecologyandsociety.org/vol11/iss2/art30/ (accessed on 30 November 2017). [CrossRef]

16. Falťan, V.; Krajcirovičová, L.; Petrovič, F.; Khun, M. Detailed Geoecological Research of Terroir with the Focus on Georelief and Soil—A Case Study of Kratke Kesy Vineyards. *Ekologia (Bratislava)* **2017**, *36*, 214–225. [CrossRef]

17. Miklos, L. Landscape-ecological theory and methodology: A goal oriented application of the traditional scientific theory and methodology to a branch of a new quality. *Ekologia (Bratislava)* **1996**, *15*, 377–385.

18. Antrop, M. Why landscapes of the past are important for the future. *Landsc. Urban Plan.* **2005**, *70*, 21–34. [CrossRef]

19. Barančoková, M.; Kenderessy, P. Assessment of Landslide Risk Using Gis and Statistical Methods in Kysuce Region. *Ekologia (Bratislava)* **2014**, *33*, 26–35. [CrossRef]

20. Jakab, I.; Petluš, P. The Use of Viewshed Analysis in Creation of Maps of Potential Visual Exposure. In *GIS OSTRAVA 2013—Geoinformatics for City Transformation*; Ivan, I., Longley, P., Fritsch, D., Horak, J., Cheshire, J., Inspektor, T., Eds.; Publisher VSB-TECH UNIV Ostrava: Ostrava, Czech Republic, 2012; pp. 375–390.

21. Vyskupova, M.; Pavlickova, K.; Baus, P. A landscape vulnerability analysis method proposal and its integration in the EIA. *J. Environ. Plan. Manag.* **2017**, *60*, 1193–1213. [CrossRef]

22. Reimer, M.; Getimis, P.; Blotevogel, H.H. Spatial Planning Systems and Practices in Europe. A comparative perspective. In *Spatial Planning Systems and Practices in Europe: A Comparative Perspetive on Continuity and Changes*; Reimer, M., Getimis, P., Blotevogel, H.H., Eds.; Routledge: Abingdon, UK, 2014; pp. 1–20.

23. Orsi, F.; Geneletti, D.; Borsdorf, A. Mapping wildness for protected area management: A methodological approach and application to the Dolomites UNESCO World Heritage Site (Italy). *Landsc. Urban Plan.* **2013**, *120*, 1–15. [CrossRef]

24. Izakovičová, Z.; Mederly, P.; Petrovič, F. Long-Term Land Use Changes Driven by Urbanisation and Their Environmental Effects (Example of Trnava City, Slovakia). *Sustainability* **2017**, *9*, 1553. [CrossRef]

25. Bastian, O. Ecosystem and Landscape Services: Development and Challenges of Disputed Concepts. In *Landscape and Landscape Ecology*; Halada, L., Baca, A., Boltiziar, M., Eds.; ILE SAS: Bratislava, Slovakia, 2016; pp. 215–226.

26. Izakovičová, Z.; Moyzeová, M.; Oszlányi, J. Problems in Agricultural Landscape Management Arising from Conflicts of Interest—A Study in the Trnava Region, Slovak Republic. In *Innovations in European Rural Landscapes*; Wiggering, H., Ende, H.-P., Knierim, A., Pintar, M., Eds.; Springer: Berlin/Heidelberg, Germany, 2010; pp. 77–95. Available online: http://link.springer.com/chapter/10.1007/978-3-642-04172-3_6 (accessed on 8 December 2014).

27. Schulte, R.P.O.; Creamer, R.E.; Donnellan, T.; Farrelly, N.; Fealy, R.; O'Donoghue, C.; O'hUallachain, D. Functional land management: A framework for managing soil-based ecosystem services for the sustainable intensification of agriculture. *Environ. Sci. Policy* **2014**, *38* (Suppl. C), 45–58. [CrossRef]

28. Labuda, M. Multifunkčné poľnohospodárstvo ako nástroj ekologickej ochrany kultúrnej krajiny (Multifunction Agriculture as the Instrument of Ecological Cultural Landscape Protection). *Životné Prostr.* **2011**, *45*, 38–42.

29. Surova, D.; Surovy, P.; de Ribeiro, N.A.; Pinto-Correia, T. Integrating differentiated landscape preferences in a decision support model for the multifunctional management of the Montado. *Agrofor. Syst.* **2011**, *82*, 225–237. [CrossRef]

30. Sturck, J.; Verburg, P.H. Multifunctionality at what scale? A landscape multifunctionality assessment for the European Union under conditions of land use change. *Landsc. Ecol.* **2017**, *32*, 481–500. [CrossRef]

31. Böhmelt, T.; Bernauer, T.; Buhaug, H.; Gleditsch, N.P.; Tribaldos, T.; Wischnath, G. Demand, supply, and restraint: Determinants of domestic water conflict and cooperation. *Glob. Environ. Chang.* **2014**, *29* (Suppl. C), 337–348. [CrossRef]

32. Reháčková, T.; Pauditšová, E. Evaluation of urban green spaces in Bratislava. *Boreal Environ. Res.* **2004**, *9*, 469–477.

33. Miklós, L.; Miklisova, D.; Reháková, Z. Systematization and Automatization of Decision-Making Process in Landep Method. *Ekologia ČSFR* **1986**, *5*, 203–231.

34. Hrnčiarová, T. Abiotic complexes—An important part of ecological decision making in agricultural landscape. *Ekologia (Bratislava)* **2005**, *24*, 397–410.

35. Drábová-Degro, M.; Krnáčová, Z. Assessment of Natural and Cultural Landscape Capacity to Proposals the Ecological Model of Tourism Development (case Study for the Area of the Zamagurie Region). *Ekologia (Bratislava)* **2017**, *36*, 69–87. [CrossRef]

36. Wade, A.A.; Theobald, D.M.; Laituri, M.J. A multi-scale assessment of local and contextual threats to existing and potential US protected areas. *Landsc. Urban Plan.* **2011**, *101*, 215–227. [CrossRef]

37. Krnáčová, Z.; Hrnčiarova, T. Landscape-ecological planning—A tool of functional optimization of the territory (case study of town Bratislava). *Ekologia (Bratislava)* **2006**, *25*, 53–67.

38. Antrop, M. From holistic landscape synthesis to transdisciplinary landscape management. In *From Landscape Research to Landscape Planning: Aspects of Integration, Education and Application*; Tress, B., Tres, G., Fry, G., Opdam, P., Eds.; Springer: Dordrecht, The Netherlands, 2006; Volume 12, pp. 27–50.

39. Gratton, M.; Morin, S.; Germain, D.; Voiculescu, M.; Ianas, A. Tourism and natural hazards in Balea Glacial area valley, Faragas massif, Romanian Carpathians. *Carpathian J. Earth Environ. Sci.* **2015**, *10*, 19–32.

40. Hrnčiarová, T.; Izakovičová, Z.; Miklós, L.; Lehotský, M.; Tremboš, P.; Durajková, N.; Ot'ahel', J. Environmental-Conditions of Formation and Development of Regions in Slovakia. *Ekologia (Bratislava)* **1994**, *13*, 87–94.

41. Belčáková, I.; Pšenáková, Z. Specifics and Landscape Conditions of Dispersed Settlements in Slovakia—A Case of Natural, Historical and Cultural Heritage. In *Best Practices in Heritage Conservation and Management: From the World*; Piscitelli, M., Ed.; Scuola Pitagora Editrice: Napoli, Italy, 2014; Volume 46, pp. 261–268.

42. Lambin, E.F.; Turner, B.L.; Geist, H.J.; Agbola, S.B.; Angelsen, A.; Bruce, J.W.; Coomes, O.T.; Dirzo, R.; Fischer, G.; Folke, C.; et al. The causes of land-use and land-cover change: Moving beyond the myths. *Glob. Environ. Chang.* **2001**, *11*, 261–269. [CrossRef]

43. Muchova, S.; Svecova, A.; Pavlickova, K.; Zelenakova, M. Evaluation of the development potential in optimisation of the area using. *Ekologia (Bratislava)* **2006**, *25* (Suppl. 1), 179–189.

44. Saarikoski, H.; Mustajoki, J.; Barton, D.N.; Geneletti, D.; Langemeyer, J.; Gomez-Baggethun, E.; Marttunen, M.; Antunes, P.; Keune, H.; Santos, R. Multi-Criteria Decision Analysis and Cost-Benefit Analysis: Comparing alternative frameworks for integrated valuation of ecosystem services. *Ecosyst. Serv.* **2016**, *22*, 238–249. [CrossRef]

45. Pauditšová, E.; Slabeciusova, B. Modelling as a Platform for Landscape Planning. In *Geoconference on Informatics, Geoinformatics and Remote Sensing*; Stef92 Technology Ltd.: Sofia, Bulgaria, 2014; Volume III, pp. 753–760.

46. Sklenicka, P. Applying evaluation criteria for the land consolidation effect to three contrasting study areas in the Czech Republic. *Land Use Policy* **2006**, *23*, 502–510. [CrossRef]

47. Juskova, K.; Muchova, Z. Land Consolidation as an Instrument for Land Ownership Defragmentation. In *MENDELNET 2013*; Skarpa, P., Ryant, P., Cerkal, R., Polak, O., Kovarnik, J., Eds.; Mendel Univ Brno: Brno, Czech Republic, 2013; pp. 444–448.

48. Muchova, Z.; Leitmanova, M.; Petrovic, F. Possibilities of optimal land use as a consequence of lessons learned from land consolidation projects (Slovakia). *Ecol. Eng.* **2016**, *90*, 294–306. [CrossRef]

environments

MDPI

Article

Assessment of Post-Fire Vegetation Recovery Using Fire Severity and Geographical Data in the Mediterranean Region (Spain)

Alba Viana-Soto [1],*, Inmaculada Aguado [2] and Susana Martínez [2]

[1]   Complutum Tecnologías de la Información Geográfica (COMPLUTIG) S.L, University of Alcala,
      Calle Colegios, 2, 28801 Alcalá de Henares, Spain
[2]   Department of Geology, Geography and Environment, University of Alcala, 28801 Alcalá de Henares, Spain;
      inmaculada.aguado@uah.es (I.A.); campoxurado@gmail.com (S.M.)
*   Correspondence: alba.viana@edu.uah.es; Tel.: +34-918-855-264

Received: 11 October 2017; Accepted: 10 December 2017; Published: 12 December 2017

Abstract: Wildfires cause disturbances in ecosystems and generate environmental, economic, and social costs. Studies focused on vegetation regeneration in burned areas acquire interest because of the need to understand the species dynamics and to apply an adequate restoration policy. In this work we intend to study the variables that condition short-term regeneration (5 years) of three species of the genus *Pinus* in the Mediterranean region of the Iberian Peninsula. Regeneration modelling has been performed through multiple regressions, using Ordinary Least Squares (OLS) and Geographic Weight Regression (GWR). The variables used were fire severity, measured through the Composite Burn Index (CBI), and a set of environmental variables (topography, post-fire climate, vegetation type, and state after fire). The regeneration dynamics were measured through the Normalized Difference Vegetation Index (NDVI) obtained from Landsat images. The relationship between fire severity and regeneration dynamics showed consistent results. Short-term regeneration was slowed down when severity was higher. The models generated by GWR showed better results in comparison with OLS (adjusted R^2 = 0.77 for *Pinus nigra* and *Pinus pinaster*; adjusted R^2 = 0.80 for *Pinus halepensis*). Further studies should focus on obtaining more precise variables and considering new factors which help to better explain post-fire vegetation recovery.

Keywords: post-fire regeneration; remote sensing; GIS; burn severity; environmental variables

1. Introduction

Wildfires are one of the most important environmental problems at present. In the European context, Spain is one of the countries which registers the highest fire incidence in number and surface burned [1–3]. These events cause a great amount of damages (i.e., degradation of soils, water and biodiversity), which may further lead to economic and even human costs. From an environmental point of view, wildfires cause alterations of landcover, loss of carbon reserves, and changes in the soil composition and its hydrogeomorphologic behavior [4]. Nevertheless, Mediterranean vegetation is quite adapted to fire disturbance. Mediterranean species have post-fire ecological strategies, like the resprout ability, the seed bank persistency, or the growth or dispersal ability [4]. In this sense, the analysis of Large Forest Fires (LFF) acquires a special interest because the effects they provoke are devastating. For this reason, the application of an adequate restoration policy requires the exhaustive study of the physical environment and dynamic evolution of the affected area [5].

Over the last few years, several studies have underlined the importance of remote sensing in analyzing ecological dynamics following fire and studying post-fire vegetation regeneration [5–7].

Compared to field surveying, satellite images offer a less expensive alternative and provide broader information of burned areas by obtaining biophysical variables of wildfires [7].

Medium-resolution optical sensors such as MSS, TM, ETM+, and OLI of Landsat series enables the monitoring of vegetation in burned areas for more than 40 years [8–10]. There are also researches studying vegetation regeneration using other sensors such as AVIRIS [11], AVHRR [12], MODIS [13], SPOT-VEGETATION [14], or RADAR [15]. According to Hirschmugl et al. [16], time series analysis is the most used approach in the monitoring of forest disturbances and degradation, as well as regeneration processes with the objective to analyze spectral variations in the forest cover [8,11,17].

The factors which define the vegetation regeneration rate after a wildfire are multiple, and their complete identification or modeling is difficult. These factors are related to wildfire characteristics, environmental conditions, and the life history of plant species [18]. Some studies applied to northern regions [10,12,19,20] and Mediterranean regions [8,11,21] have pointed out some of the main drivers of vegetation regeneration following forest fires. Fire severity levels, topography (elevation, slope, and orientation), post-fire climate, or vegetation cover class are the most used in regeneration estimates. Other studies have emphasized the influence of solar radiation on the water availability for vegetation growth [8] or the effects of applying different restoration models [9]. Several studies have revealed that the influence of environmental factors on regeneration can vary across vegetation types [8,20,22].

In order to evaluate fire severity, defined as the magnitude of the ecological change produced by fire [6], different methods have been proposed. Some studies have successfully applied vegetation indices such as the *Normalized Difference Vegetation Index* (NDVI) [23], due to the relationship between the amount of vegetation consumed and fire severity [7]. On the other hand, specific indices have been developed which record with greater spectral contrast the fire effects as the *Normalized Burn Ratio* (NBR) [24] subsequently modified [25,26]. Another method based on parametrized variables to estimate fire effects in vertical strata is the *Composite Burn Index* (CBI) [25], widely and effectively applied with good results (correlation between CBI and dNBR R^2 = 0.83) [27] and modified by De Santis and Chuvieco [28].

To assess vegetation regeneration using time series, several methodologies have also been proposed. The most common is to monitor the vegetation state from spectral indices. Among the most used indices are the *Normalized Difference Vegetation Index* [17,27,29], the *Regeneration Index* [11,14,19], the *Normalized Difference Infrared Index* [30], and the *Soil-Adjusted Vegetation Index* [31]. *Spectral Mixture Analysis* has also effectively been applied in that context [8].

Although several studies have evaluated fire severity and vegetation regeneration using remote sensing data, few have integrated both aspects in a single study. Several studies have identified the significant influence of fire severity [21,32] and environmental factors, such as topography and climate [8,17], in vegetation regeneration. Therefore, the need to integrate both analyses to improve the predictive models of these dynamics has arisen. In recent years, more studies have been undertaken to investigate the factors that determine post-fire regeneration patterns with satisfactory results [19,20].

In this study, we hypothesize that fire severity and environmental variables such as vegetation type, meteorology, and topography determine the post-fire vegetation regeneration. Therefore, regeneration patterns on burned surfaces will vary between areas presenting different severity levels. Furthermore, meteorological post-fire conditions and topography will have different impacts on the regeneration of different forest covers.

The general objective will be to model the short-term vegetation regeneration (five years after fire) in three large forest fires in the Mediterranean region of the Iberian Peninsula, by knowing the fire severity and the interacting environmental variables. In this case, we will model the forest regeneration of the genus *Pinus*, which is widely extended in Mediterranean forests. In addition, *Pinus* communities are one of the most affected by fires due to the high content of resins that promote both fire start and spread [33].

Specific objectives of the study area include: (1) To model the evolution of post-fire vegetation to obtain regrowth patterns through spectral indices; (2) Generate environmental variables involved in regeneration and (3) Identify the variables which are most relevant in explaining the short-term regeneration using multiple regression models and to estimate the short-term regeneration of *Pinus* species.

2. Materials and Methods

2.1. Study Area

The study is based on three large fires that occurred in the summer of 1994 (Figure 1) in the municipalities of Castrocontrigo (Province of León), Uncastillo (Province of Zaragoza), and Moratalla (Province of Murcia). These areas have been selected according to several criteria:

(i)   The burned area exceeds the 500 ha lower limit to be considered a large forest fire in Spain [34].

(ii) Burned areas where post-fire restoration activities were not implemented, and therefore did not alter the natural regeneration of the burned area.

(iii) Burned areas which have not been burned in subsequent fires since 1994.

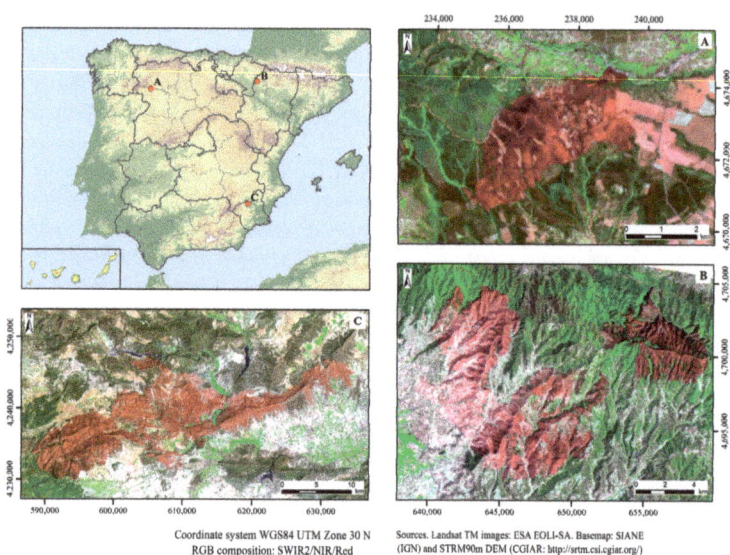

Figure 1. Location of study areas and post-fire images of 1994: (**A**) Castrocontrigo, (**B**) Uncastillo, and (**C**) Moratalla.

These areas are located in the Mediterranean biogeographic region [35]. Topographic and climatic characteristics are quite different; the elevation varies between 1100 m in the highest levels in the Castrocontrigo and Uncastillo fires to 1400 m in the Moratalla fire. On the other hand, mean annual rainfall differs between Moratalla fire (ranging from 300 to 400 mm) and Castrocontrigo or Uncastillo fires (recording from 600 to 700 mm), while mean annual temperatures range from 10 °C in Castrocontrigo, 15 °C in Uncastillo, to 17.5 °C in Moratalla [36]. In the summer season, the scarcity of precipitation and high temperatures, characteristic of the Mediterranean climate, increase the risk of fire.

All study areas are dominated by anthropogenic coniferous forests, mainly with species of the genus *Pinus* along with certain deciduous species of the genus *Quercus*. In the shrub strata, sclerophyll species such as *Rosmarinus*, *Thymus*, or *Juniperus* species have been identified. Nevertheless, there are latitudinal and altitudinal differences between the three burned areas, which generate particularities that are described in the Map of Vegetation Series [35]. The area of Castrocontrigo fire is placed in the Supramesomediterranean series dominated by *Pinus pinaster* and *Pinus sylvestris* among

Quercus pyrenaica. The Uncastillo area is located on the limit between the Supramesomediterranean and Mesomediterranean series. Therefore, most of this area is covered by different species of pines (*Pinus halepensis*, *Pinus nigra*, and *Pinus sylvestris*). The area of Moratalla is in the Mesomediterranean series, an area with a more arid climate than the previous ones, where *Pinus halepensis* is dominant.

As can be seen, the selected study areas represent a broad gradient of environmental conditions representative of the Mediterranean ecoregion. Therefore, the models derived from the analysis of these variables can be applied to a wide variety of Mediterranean environments.

Table 1 shows the forest fires data related to total burned area, type of vegetation affected, cause, and date.

Table 1. Fire data of the studied forest fires [37].

Municipal Term of Origin	Start Date	Total Forest Area (ha)	Wooded Area (ha)	Non-Wooded Area (ha)	Non-Forest Area (ha)	Cause
Castrocontrigo	22/08/1994	1150	500	650	0	Unknown
Uncastillo	16/07/1994	6589	4849	1740	2049	Lightning
Moratalla	04/07/1994	24,817	16,262	9555	2828	Powerline

2.2. Datasets

In this research, several data sources have been used to generate the spatial information used to model the vegetation regeneration. Satellite images have been obtained for the respective fires and spatial data have been collected to generate the environmental variables.

Landsat TM and Landsat ETM+ images were obtained for the pre- and post-fire period between 1994 and 1999 from the EOLI-SA server of the European Space Agency (ESA) (Table 2).

Table 2. List of images used.

Castrocontrigo		Uncastillo		Moratalla		Sensor
Date	Days since Fire	Date	Days since Fire	Date	Days since Fire	
08/07/1994	−45	06/07/1994	−18	29/06/1994	−6	TM
28/08/1994	6	23/08/1994	30	15/08/1994	38	TM
16/08/1995	359	10/08/1995	382	18/07/1995	375	TM
01/08/1996	709	11/07/1996	718	21/08/1996	774	TM
05/08/1997	1078	14/07/1997	1086	07/07/1997	1094	TM
08/09/1998	1477	21/10/1998	1550	07/05/1998	1398	TM
01/07/1999	1773	29/08/1999	1910	21/07/1999	1845	ETM+

Images radiometric resolution is 8 bits with visible spectral information (RGB), near infrared (NIR), and short wave infrared (SWIR1 and SWIR2). Images spatial resolution is 30 m, so all the information used in the study will be adapted to that resolution. The selected scenes belong to the months of July and August, with the exception of images which have been chosen from the end of June, beginning of September, or October. The selection of similar dates was prioritized in order to avoid the influence of seasonal differences in spectral radiation [19]. In addition, the choice of images with zero cloud in the study area was prioritized so as not to interfere in the area of interest [27].

A Digital Elevation Model (DEM) was acquired with a spatial resolution of 5 m from the National Geographic Institute to generate the topographic variables. Meteorological data were obtained from raster models generated from a set of data of average monthly precipitations and average maximum and minimum monthly temperatures for a period of 60 years (1950–2010). Raster models have been generated by interpolating the weather station dataset recorded in the network of AEMET (Agencia Estatal de Meteorología) onto a regular grid (0.1 × 0.1 degree). The grid was developed following an improved version of the interpolation technique described by Brunetti et al. [38]. This improvement combines a radial weight with a Gaussian shape with an angular weight [39,40]. The vegetation cover map was derived from a combination of two vectorial datasets: the Second National Forest

Inventory (INF2) at a scale of 1:50,000 [41] and the Corine Land Cover (CLC) 1990 project at a scale of 1:100,000 [42]. This vegetation cover map only includes forest areas because non-forest areas have been removed due to their lack of interest to our study.

Fire severity was measured using the GeoCBI proposed by De Santis and Chuvieco [28]. It is a method based on the Composite Burn Index [25,43], designed to estimate the damage caused by fire, taking into account five forest strata: soil, herbaceous (<1 m), shrubs (1–5 m), trees (5–20 m), and trees (>20 m). The GeoCBI incorporates two new variables for each plant stratum (percentage of change in the leaf area and vegetation covered area), and considers litter and fuel consumption, changes in soil colour, foliage alteration, canopy mortality, and char height. The severity variable was stratified into five levels to facilitate understanding: 0 = Not burned, >0 to <1 = Low, 1 to <2.5 = Medium, >2.5 to <2.85 = High, and >2.85 to 3 = Very high.

2.3. Methods

The methodology for modelling post-fire regeneration using time series requires a series of processes summarized in Figure 2.

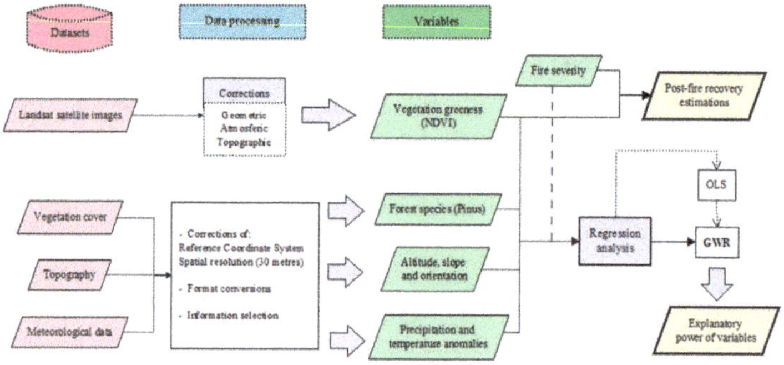

Figure 2. Overall methods implemented in our study. OLS: Ordinary Least Square. GWR: Geographically Weighted Regression.

2.3.1. Image Processing

Satellite image acquisition is subject to errors as a result of alterations in satellite motion, sensing data mechanism, and atmosphere interference. A multitemporal study like this requires radiometric homogenization and geometric correction due to the differences in the conditions in which images were acquired [11].

(A) Geometric correction

This process involves transforming the pixels coordinates in ESA image series so that they match precisely with those of the reference image. In this work, we have taken images from U.S. Geological Survey (USGS) as reference, because they have the greatest quality processing. The WGS84 UTM 30N Coordinate System was established for the whole series. The correction was carried out in ArcGIS 10.2.2 by taking 30 control points distributed evenly throughout the image in order to reduce the geometric error [44]. A second-degree polynomial fit and a cubic resampling were applied, which considers the values of the closest 16 pixels and allows for greater accuracy.

(B) Atmospheric correction

In this step, the digital numbers (DN) are transformed to physical magnitudes, which allow the use of images obtained by different sensors [44]. The DN of each band was transformed to reflectance values (ρ), following the method based on the dark object proposed by Chávez [45].

$$\rho_k = \frac{d^2 \times \pi \times a_{1,k} \left(DN_k - DN_{\min,k} \right)}{E_{0,k} \times \sin \theta_e \times \tau_{k,i}} \tag{1}$$

where ρ is the reflectance value for each pixel, d is the corrector factor of the distance Earth-Sun, which is calculated from the Julian day, $\tau_{k,i}$ is the average transmissivity (the values used are those proposed by Gilbert et al. [46] for each band: 0.7–0.78–0.85–0.91–0.95–0.97), $E_{0,k}$ is the solar irradiance on the top of the atmosphere whose most suitable values are currently available in landsat.usgs.gov/esun; the other parameters, $a_{1,k}$ (multiplicative coefficient of conversion to radiance) and θ_e (sun elevation angle) are obtained from the image metadata.

(C) Topographical shading correction

Topographic correction solves the errors in the reflectance values caused by solar illumination differences due to relief variations. From the resampled DEM to the same images spatial resolution (30 m), the incidence angle (γ) is calculated. The cosine of this angle allows us to measure illumination variations [14]. The topographic shading correction was carried out under the criterion of c-Teillet [47] for each band.

$$\rho_{h,i} = \rho_{h,i} \left(\frac{\cos \theta_i}{\cos \gamma_i} \right) \tag{2}$$

where $\rho_{h,i}$ indicates the reflectance of a pixel i in a horizontal terrain, ρ_i the corresponding slope, θ_i the sun azimut angle of the scene, and γ_i the incidence angle calculated corresponding to that pixel.

2.3.2. Use of Spectral Vegetation Index (NDVI)

Several studies have supported the utility of spectral indices derived from Landsat to study the post-fire regeneration dynamics of vegetation [9,27,48]. The spectral indices help identify the state of the vegetation along time series due to the enhancement in the sensitivity of the different spectral bands. One of the most used indices is the *Normalized Difference Vegetation Index* proposed by Rouse et al. [23], which is calculated by combining the near infrared (NIR) and red (R) bands. The combination of its normalized difference formulation and the use of the higher absorption and reflectance regions of chlorophyll make this index robust in different conditions [9]. In contrast to other vegetation index such as *Enhanced Vegetation Index* (EVI), the NDVI is easier to calculate and has been widely applied to study vegetation regeneration [14,21,49–51].

For this study, it is interesting to model the short-term regeneration of vegetation. For this reason, the NDVI for the complete series (1994–1999) has been calculated to study the regeneration patterns and to generate the variables to be included in the model.

2.3.3. Environmental Variables

The environmental variables which were used to correlate the *Pinus* species recovery, measured by the NDVI in year 5 after fire, have been defined based on accurate data availability.

A vegetation cover map of burned areas was generated by rasterizing the vegetation cover map in vector format. This map was overlaid with severity maps in order to represent the spatial distribution of *Pinus* species on burned areas with the same spatial resolution as satellite images. The *Pinus* species affected by fires were *Pinus pinaster* and *Pinus nigra* in the Castrocontrigo and Uncastillo fires, and *Pinus halepensis* in the Uncastillo and Moratalla fires. Based on the physiological characteristics of each species, as well as the survival and growth capacity of the seeds after fire [33,52], two models have been carried out: one for *Pinus halepensis*, with a greater regenerative ability [53], and other for *Pinus nigra* and *Pinus pinaster*.

Topographic variables (elevation, slope, and aspect) were generated from the DEM mosaic for each fire, resampled to a resolution of 30 m, and projected in the Reference System WGS84 UTM 30N. In order to generate solar orientation variables, the Meng et al. [20] methodology was used. The cosine and sine aspect variables were calculated to provide information on surfaces with northerly and easterly orientations, respectively. Sine and cosine transformations are done so as to convert a nonlinear variable to a linear variable. These variables incorporate information on the time of solar exposure and how it impacts regeneration

With regard to climatic variables, anomalies in precipitation and temperature during the year of the fire (1994) and the successive one (1995), as well as the mean of the anomalies for the 1994–1999 period, were calculated. It is assumed that excess or deficiency in precipitation, as well as abnormally high or low temperatures, may act as limiting factors for vegetation regeneration [20]. Climatic anomalies were calculated for each month at the pixel level. Values were standardized considering the mean and the standard deviation of the 1950–2010 series. Finally, anomalies were obtained by defining thresholds using the first and third quartile of the new Z values.

$$Z = \frac{X - \mu}{\sigma} \tag{3}$$

where X is the pixel value (precipitation values in mm and temperatures in °C), μ is the mean, and σ is the standard deviation. Before calculating the anomalies, extreme values in the datasets were removed (the lower and upper values of 1 and 99 percentiles, respectively).

2.3.4. Post-Fire Regeneration Assessment

Regression analysis is used to model, predict, or explain complex phenomena. In this case, the influence of fire severity and certain environmental variables on the short-term regeneration rate of selected *Pinus* species is explained by multiple regression models (Table 3).

First of all, a regression analysis was carried out with Ordinary Least Squares (OLS) to diagnose if the selected variables for the models are suitable. However, spatial data show properties that fail to carry out the assumptions and requirements of statistical methods such as OLS regression. This global regression technique assumes the existence of spatial stationarity. It means the model has stable predictive capacity throughout the study area. In this case, the explanatory variables show nonstationary relationships. In addition, spatial autocorrelation in the NDVI values for the 1994–1999 period has been detected by the Moran Index [54]. As an alternative, some authors propose to use regression models that incorporate the variation in geographic space, such as the Geographically Weighted Regression (GWR) [55]. For this reason, this method will be used to estimate post-fire regeneration values.

Table 3. List of variables for regression analysis.

Variable		Units	Description
Dependent	NDVI YEAR + 5	Values between −1 and 1	It shows vegetation greenness 5 years after fire.
Explanatory			
Post-fire effects	NDVI YEAR + 1	Values between −1 and 1	Vegetation greenness immediately after fire.
Fire severity	CBI	Values between 0 and 3	A CBI value = 0 indicates unburned surface and CBI = 3, an area of very high severity.
	Elevation	Metres	
	Slope	Grades	
Topography	Northerly	Values between −1 and 1	Obtained by the orientation cosine. Values near 1 indicate North orientation.
	Easterly	Values between −1 and 1	Obtained through the orientation sine. Values near 1 indicate East orientation.
	Precipitations	Z value	Total anomalies in the wet season (months that concentrate 80% of annual precipitation: October–May) in 1994 and 1995 and the mean of the anomalies for 1994–1999 period.
Climatic anomalies	Minimum temperature on January	Z value	Total anomalies in January in 1994 and 1995 and the mean of the anomalies for 1994–1999 period.
	Maximum temperature on July	Z value	Total anomalies in July in 1994 and 1995 and the mean of the anomalies for 1994–1999 period.

NDVI: *Normalized Difference Vegetation Index.*

Pixels used in regression analysis were obtained from a random sample select on the set of burned pixels, using the post-classification functions. Border pixels were previously removed because they were not representative for characterizing the burned surfaces, using a 3 × 3 sum space filter.

$$n = \left(\frac{Z_{\alpha/2} \times \sigma}{E} \right)^2 \tag{4}$$

where α is the confidence level, in this case 5%, $Z_{\alpha/2}$ is the corresponding critical value (1.96), σ is the standard deviation of the dependent variable values, and E is the margin error in estimation. Final samples of 15,814 pixels for *Pinus halepensis* model and a second sample of 7129 pixels for *Pinus nigra* and *Pinus pinaster* model were selected.

3. Results

3.1. Regeneration Evolution According to Fire Severity

First analysis has focused on the evolution of the short-term regeneration of the selected *Pinus* species according to the severity level. According to fire severity level, *Pinus halepensis* evolution on the one hand, and other *Pinus* (*nigra* and *pinaster*) on the other, from the mean NDVI values, has been represented (Figure 3). This separation is based on the unequal germination capacity of the species, being higher in *Pinus halepensis* [53].

Results show a significant decrease in the post-fire values and a later progressive recovery, being faster in the early years in the case of *Pinus nigra* and *pinaster*, while in *Pinus halepensis*, recovery is slower. Fire severity level affects *Pinus* regeneration significantly, being higher NDVI values associated with lower fire severity. Pre-fire values are different between fire severity levels due to the influence of other environmental variables according to the zone. In addition, areas which were drier before fire may have burned more intensely. Some fluctuations in the short-term regeneration may be caused by variations in the images acquisition dates after summer. This could explain the fall in NDVI values in *Pinus nigra* and *Pinus pinaster* between 1997 and 1998.

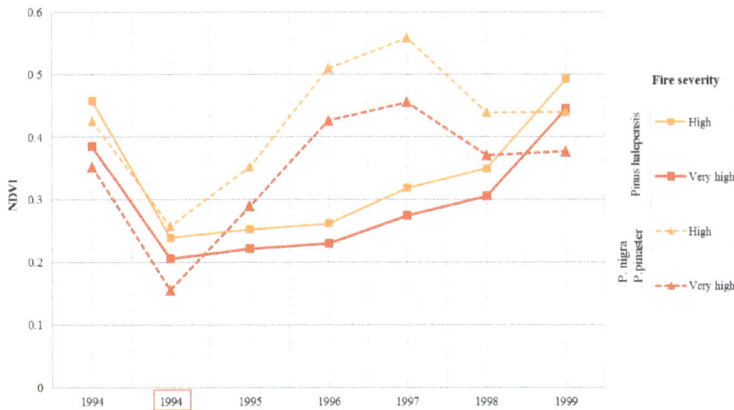

Figure 3. Short-term regeneration of *Pinus* species according to fire severity levels: >2.5 to <2.85 = High; >2.85 to 3 = Very High. NDVI: *Normalized Difference Vegetation Index*.

3.2. Multiple Linear Regression with Ordinary Least Squares

An exploratory regression analysis with OLS allowed for the elimination of some variables initially defined because of multicollinearity problems (spatial autocorrelation) or the low significance for the predictive models.

The average of the climatic anomalies between 1994 and 1999 were dismissed because of their high multicollinearity with post-fire climatic anomalies (1994 and 1995). Moreover, the low spatial resolution of the meteorological data implies a reduced variability of precipitation and temperature data, and there is a high multicollinearity between maximum and minimum temperatures. Therefore, two models were created for each group of species, with one model having maximum temperature, while the other model used minimum temperature. The resulting models were then evaluated and the model with best fit selected. Additionally, the explanatory variables included are the most consistent, that is, those that were significant a greater number of times (>80) when executing the regression with OLS 100 times (Table 4). In this case, the orientation variables have not been significant and have been removed.

Table 4. Frequency of explanatory variables at the 95% level of significance.

Variable	*Pinus halepensis*	Other *Pinus*: *P. nigra* and *P. pinaster*
NDVI YEAR + 1	100	100
Fire severity	100	100
Elevation	100	93
Slope	100	82
Northness	2	0
Eastness	0	3
Prec. anomalies +0 and +1	100	100
Tmax anomalies +1	93	85
Tmin anomalies +1	82	83

Tables 5 and 6 contain the first regression models results. The models that have shown the best results have been those in which the minimum temperatures have not been included because they were redundant (high value of the Variance Inflation Factor—VIF).

Multiple R^2 and Adjusted R^2 values show a good fit for *Pinus halepensis* model (Table 5). In order to assess in detail each explanatory variable, the coefficient, t-statistic, robust probability, and VIF were used.

Table 5. Frequency Ordinary Least Squares (OLS) results for *Pinus halepensis* in the 95% confidence interval.

Variable	Coefficient	Standard Error	t-Statistic	Robust Prob.	VIF
Intercept	0.175275	0.003849	45.536378	0.000000 *	-
Slope	−0.000772	0.000040	−19.111591	0.000000 *	1.326365
Elevation	0.039808	0.002215	17.970745	0.000000 *	2.611230
NDVI+1	0.828052	0.007227	114.571406	0.000000 *	1.489753
Severity	−0.016329	0.001308	−12.486857	0.000000 *	1.173556
Prec. anomalies +0 and +1	0.025462	0.000942	27.018506	0.000000 *	6.863740
Tmax anomalies +0 and +1	−1.104539	0.105490	−10.470555	0.000000 *	6.022914

Multiple R^2: 0.702618; Adjusted R^2: 0.702505; AICc: −58,550.823000; Joint F-Statistic [e]: 6224.47229; Prob. (>F), (6, 15807) degrees of freedom: 0.000000 *; Koenker (BP) Statistic [f]: 2117.39545; Prob. (>chi-squared), (6) degrees of freedom: 0.000000 *; Jarque-Bera Statistic [g]: 23,935.24457; Prob. (>chi-squared), (2) degrees of freedom: 0.000000 *.

Each explanatory variable coefficient shows the relationship between each explanatory and dependent variable. In this case, slope, severity levels, and Tmax anomalies have a negative influence. High slope slows down rooting vegetation while abnormally high temperatures can cause thermal stress and limit regeneration. As anticipated, short-term regeneration rate was lower when severity was higher. Nevertheless, the severity seems to influence less strongly because sample pixels used coincided with high severity values. Consequently, the influence of a low or medium severity on regeneration is not collected, leaving this variable weak against others. On the other hand, the variable with a greater positive impact is the state of the vegetation after fire (NDVI+1). That is, short-term regeneration rate was higher when post-fire vegetation greenness was higher. There is also a positive influence on elevation and precipitation. In the Mediterranean region, above-average rainfall can help the growth of vegetation by increasing soil moisture. The elevation could be related to lower temperatures avoiding thermal stress in summer.

Standard errors allow assessment of the coefficients obtained. Low standard errors across all variables indicate coefficients are consistent. On the other hand, the robust probability or *p*-value showed that all variables are statistically significant and, therefore, important for the regression model. VIF provided information on the possible redundancy in explanatory variables. In this case, the relatively high values for climatic anomalies suggest removing one of the two variables could increase the model fit.

As regards to the regression model for *Pinus nigra* and *pinaster* (Table 6), the model shows lower Multiple R^2 and Adjusted R^2 values due to differences in the vegetative cycle between modelled species. Again, variables with inverse relationships are fire severity and slope, whereas the post-fire vegetation state (greeness) is the variable with more positive influence. Certain above-average rainfall and temperatures higher than usual show a positive coefficient that could be due to compensations in evapotranspiration. Standard errors are reduced and VIF values confirm the absence of variable redundancy in this model.

However, in both models, the significance of Koenker statistic (*p*-value less than 0.05 for a 95% confidence level) indicated biased standard errors due to heteroscedasticity. In addition, the significant *p*-value in the Jarque-Bera statistic showed that the residual values deviated slightly from a normal distribution. Finally, an analysis of the residuals was carried out to study spatial autocorrelation using the Moran Index. In both, autocorrelation has been positive, with values around 0.3. Therefore, it was suitable to run the Geographically Weighted Regression (GWR) analysis.

Table 6. OLS results for other *Pinus* (*P. nigra* and *P. pinaster*) at 95% confidence interval.

Variable	Coefficient	Standard Error	*t*-Statistic	Robust Prob.	VIF
Intercept	0.162278	0.007830	20.725448	0.000000 *	-
Slope	−0.000311	−0.000179	−17.37394	0.008237 *	1.373015
Elevation	0.035865	0.005188	6.912623	0.000000 *	1.293531
NDVI+1	0.528326	0.011330	46.630112	0.000000 *	2.385877
Severity	−0.039403	0.003988	−9.881250	0.000000 *	1.234551
Prec. anomalies +0 and +1	0.118088	0.004990	23.665083	0.000000 *	2.248585
Tmax anomalies +0 and +1	0.077906	0.005814	13.399919	0.000000 *	3.996377

Multiple R^2: 0.457279; Adjusted R^2: 0.456745; AICc: −14,906.559792; Joint F-Statistic [e]: 1000.12498; Prob. (>F), (6, 7122)) degrees of freedom: 0.000000 *; Koenker (BP) Statistic [f]: 285.86905; Prob. (>chi-squared), (6) degrees of freedom: 0.000000 *; Jarque-Bera Statistic [g]: 892.04419; Prob. (>chi-squared), (2) degrees of freedom: 0.000000 *.

3.3. Multiple Linear Regression with Geographically Weighted Regression (GWR)

Regression with GWR allows for the improvement of the adjustments and for the neutralizing of the spatial dependence in residual values. In comparison with a global regression, the coefficients in GWR are functions of spatial location [56]. Regression analyses were carried out using a defined kernel with a two-square function in which the bandwidth was determined by an optimal number of neighbors. The optimal number of neighbors has been 500, where the value of the Akaike Information Criterion (AIC), which is a relative quality indicator of a model, is the lowest. In final models, anomalies in maximum temperatures were removed because of local multicollinearity problems.

Multiple R^2 and adjusted R^2 values obtained in the models using GWR show a significant improvement over the OLS model. In this case, the variables used provide a model explaining 80% of *Pinus halepensis* regeneration and 78% for *Pinus nigra* and *Pinus pinaster* (Table 7). It is noticed that explanatory variable forces and influences have changed over OLS models for *Pinus nigra* and *Pinus pinaster*. Higher slope shows a slightly positive coefficient, as in the study by Meng et al. [20] for spruce forests. In contrast, the elevation has a negative coefficient that could be related to temperatures too low, which can limit regeneration. However, forces and influences are practically maintained in the *Pinus halepensis* regression model. Higher slope and higher fire severity have negative influence, while increasing the elevation and the anomalies in precipitations have a positive influence.

However, the regression models do not have an equal predictive capacity for all areas, so there are regional variations (Figure 4). Models show a better fit for *Pinus halepensis* regeneration in Moratalla and *Pinus pinaster* in Castrocontrigo. These spatial variations could be related to the sampling point distribution by a smaller number of neighbors, because in Uncastillo sampling points are distributed in a more dispersed way, and by the local conditions effects (microclimate, edaphic composition).

Table 7. GWR results at 95% confidence interval.

Variable	Pinus halepensis		Other Pinus (P. nigra and P. pinaster)	
	Coefficient	Standard Error	Coefficient	Standard Error
Intercept	0.177311	0.066181	0.256385	0.122476
Slope	−0.000587	0.000408	0.019479	0.189559
Elevation	0.000038	0.000068	−0.258428	0.052855
NDVI+1	0.738801	0.081561	0.503855	0.050379
Severity	−0.007752	0.011347	−0.003933	0.014675
Prec. anomalies +0 and +1	0.031667	0.035612	0.188268	0.022894

Pinus halepensis: Multiple R^2: 0.802229; Adjusted R^2: 0.801694; AICc: −108.228365; Other *Pinus* (*P. nigra* and *P. pinaster*): Multiple R^2: 0.782399; Adjusted R^2: 0.774602; AICc: −557.066965.

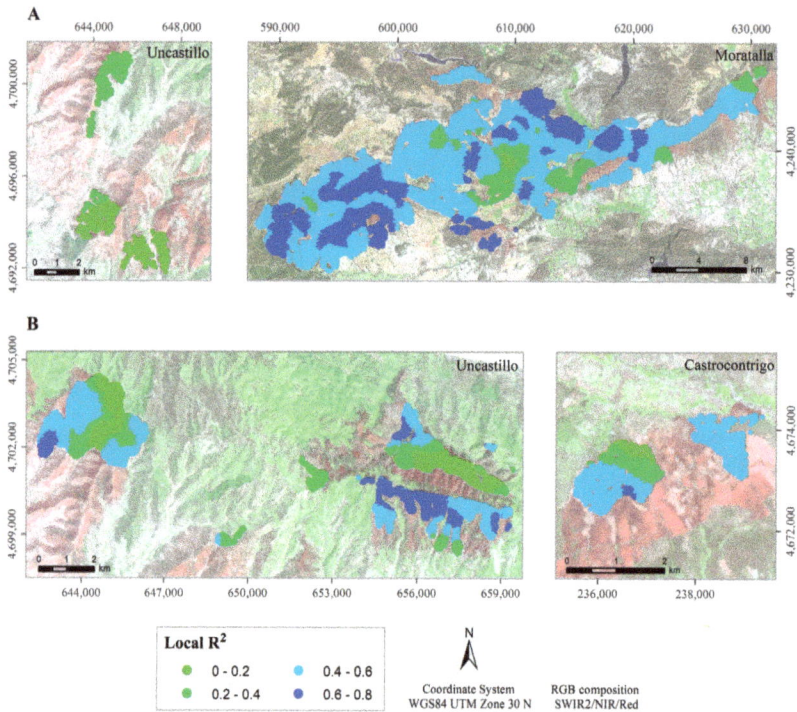

Figure 4. Local R^2 for *Pinus halepensis* (**A**) and *Pinus nigra* and *Pinus pinaster* (**B**) models.

Finally, the relative importance of the GWR explanatory variables were analyzed from the *t*-statistic (Figure 5). As expected, the post-fire vegetation state (NDVI+1) is the most important variable to the model. In contrast, fire severity was found to have a weak influence on the other variables in both models. Elevation is the variable that displayed the most difference among species, and this was attributed to species adaptation to temperatures decreasing with elevations. This decrease can be positive to avoid abnormally high temperatures in summer in Moratalla (excess evapotranspiration), but can act as a restriction in Castrocontrigo in case of too low temperatures (reduced plant activity).

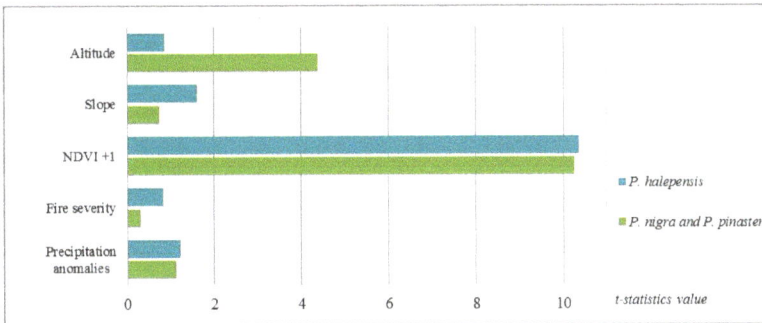

Figure 5. Relative importance of predictor variables according to the absolute value of the *t*-statistic.

Moreover, analysis of the residual values also shows better results in GWR than in OLS. The Moran Index values are much closer to the expected values with values below 0.05, also showing a lower variance and greater probability of random distribution (*p* value and *Z* score).

3.4. Validation

In order to validate the models, we used the root-mean-square error (RMSE) which measures the average of the squared errors (errors among NDVI values on 1999 and the values predicted by OLS and GWR) (Table 8). Results show that GWR models have smaller error, which implies a better adjustment than OLS models. Taking into account that NDVI values fluctuate among −1 and 1, errors are low. Therefore, predictions obtained using GWR could be considered valid.

Table 8. Root-mean-square error (RMSE) from Regression models.

Error	*Pinus halepensis*		Other *Pinus* (*P. nigra* and *P. pinaster*)	
	OLS	**GWR**	**OLS**	**GWR**
RMSE	0.00144253	0.000833933	0.007218565	0.003186855

4. Discussion

Several studies have researched post-fire vegetation regeneration based on growth patterns according to species, or on the relationship among fire severity and regeneration dynamics. Nevertheless, other previous studies have supported the importance of delving into the analysis of factors which determine regeneration [10,19–21]. Therefore, in spite of the difficulties to obtain information to generate certain environmental variables, in this study it has been considered important to model the regeneration considering post-fire climate and topography.

The results show a clear relationship between fire severity levels and regeneration rate values, which are slower when fire severity level is higher [19,27]. Meng et al. [20] and Ireland & Petropoulos [19] agree on the importance that solar radiation levels have on vegetation growth rate. In Mediterranean contexts, north aspect and high slopes present higher regeneration rates due to less evapotranspiration and higher humidity content. However, this phenomenon has not been analyzed because orientation variables have not been significant for the model in the cases studied. With regards to elevation, it has been shown that it could act positively to avoid high temperatures and periods of post-fire drought [53] in more arid environments, such as Moratalla. This is in agreement with Meng et al. [20] for coniferous forests in Sierra Nevada (USA).

With regards to the influence of post-fire weather, a positive relationship between short-term regeneration and above-average precipitation in the months after fire has been found. This relationship was also found for *Pinus halepensis* and *Pinus nigra* in Catalonia [21,22] and *Pinus halepensis* in Ayora in Valencia [8]. Nevertheless, climatic data have a lower spatial resolution compared to Landsat images and topographic data. Consequently, multicollinearity problems have not allowed for a robust analysis of the temperatures or precipitation impact on regeneration. In addition, it would be interesting to include drought indices, which would require greater spatial and temporal resolution data.

Modeling results could also be improved by incorporating new variables that capture local conditions such as landscape configuration, local microclimate, and hydrological processes which can determine vegetation recovery after fire [4,19]. In the study carried out by Chu et al. [10], water soil content showed a positive relationship with the regeneration of the larch forest, being the second conditioning factor in the recovery of this species after fire. Moreover, Röder et al. [8] concluded that water availability, in relation to slope and edaphic composition, is the most important factor that limits vegetation recovery.

In addition, it should be noted that short-term regeneration measured, through the NDVI, represents the relative vegetation cover. A direct measure of *Pinus* forests regeneration could be obtained also considering the vegetation structure using LiDAR techniques [57].

5. Conclusions

Our study provides advances in the analysis of the impact of fire severity and environmental variables on the short-term vegetation regeneration in Mediterranean regions.

Regeneration measurement for different species of *Pinus* from the NDVI has been related to the fire severity levels with good results. Severity degree measured by the CBI indicates that short-term regeneration was slowed down when severity was higher. In addition, the immediate NDVI values after the fire allowed us to verify that when the less damage produced in the vegetation, the short-term regeneration is greater.

From the multiple linear regression models generated, the explanatory capacity of the environmental variables of topography (elevation and slope) and post-fire climate (anomalies in precipitation) in post-fire vegetation recovery after fire has been verified/tested. The impact that each has on regeneration is closely related to the environmental characteristics to which each species is adapted. Thus, the elevation can be a driving factor for the *Pinus nigra* regeneration linked to temperatures too low or, in contrast, can favor *Pinus halepensis* growth, by avoiding high temperatures in summer in the southeast of the Iberian Peninsula.

In contrast to OLS, it has been confirmed that GWR is an important and valid local regression technique to explore spatial heterogeneity in the relations of explanatory variables. With this methodology, it has been possible to model the regeneration obtaining high adjustment values, with adjusted R^2 values of 0.80 for *Pinus halepensis* and 0.77 for *Pinus nigra* and *Pinus pinaster*. The models' improvement should focus on the generation of more precise environmental variables and considering new factors that can increase it explanatory power: lithologic characteristics, alteration of edaphic composition, solar radiation, time elapsed since the last fire, vegetation physical characteristics, etc.

Moreover, the proposed method is an approximation to the modelling of the short-term regeneration for the Iberian Peninsula but exportable to other territories with input variables transformation. The results obtained are useful in improving knowledge about the factors which determine the post-fire regeneration patterns of a forest ecosystem under different environmental and climatic conditions. Therefore, these advances could help decision-makers in determining which areas vegetation will not regenerate naturally after large fires and thus requires the implementation of specific restoration programs.

Environments **2017**, 4, 90

Acknowledgments: The first author would like to thank the teachers from the Geographical Information Technologies (GIT) degree in the University of Alcala (UAH) for the lessons during the period of master training. This study was supported by the project "Severity and Regeneration after large forest fires from satellite remote sensing and geographic information systems" http://www.sergisat.es/), funded by the Ministry of Economy, Industry and Competitiveness of Spain (SERGISAT-Ref. CGL2014-57013-C2-1-R).

Author Contributions: Alba Viana-Soto, Inmaculada Aguado and Susana Martínez designed the study; Alba Viana-Soto and Susana Martínez designed the methods. Susana Martínez downloaded and processed satellite images and meteorological data. Alba Viana-Soto performed the models and validate the results. All the authors analyzed and discussed the results. Alba Viana-Soto wrote the manuscript. All authors performed the final review of the manuscript.

Conflicts of Interest: The authors declare no conflict of interest.

References

1. Martín, P.; Chuvieco, E.; Aguado, I. La incidencia de los incendios forestales en España. *Ser. Geogr.* **1998**, *7*, 23–36.
2. Martínez, J.; Vega-García, C.; Chuvieco, E. Human-caused wildfire risk rating for prevention planning in Spain. *J. Environ. Manag.* **2009**, *90*, 1241–1252. [CrossRef] [PubMed]
3. European Commision. *Forest Fires in Europe, Middle East and North Africa 2015*; Publications Office of the European Union: Luxembourg, 2016; p. 117. ISBN 978-92-79-62959-4.
4. Pérez-Cabello, F.; Echevarría, M.T.; Ibarra, P.; De la Riva, J. Effects of Fire on Vegetation, Soil and Hydrogeomorphological Behavior in Mediterranean Ecosystems. In *Earth Observation of Wildland Fires in Mediterranean Ecosystems*; Chuvieco, E., Ed.; Springer: Berlin/Heidelberg, Germany, 2009; pp. 111–128. ISBN 978-3-642-01753-7.
5. Pérez-Cabello, F.; Echeverría, M.T.; de la Riva, J.; Ibarra, P. Apuntes sobre los efectos de los incendios forestales y restauración ambiental de área quemadas. Estado de la cuestión y principios generales. *Geographicalia* **2011**, *59–60*, 295–308. [CrossRef]
6. Lentile, L.B.; Holden, Z.A.; Smith, A.M.S.; Falkowski, M.J.; Hudak, A.T.; Morgan, P.; Lewis, S.A.; Gessler, P.E.; Benson, N.C. Remote sensing techniques to assess active fire characteristics and post-fire effects. *Int. J. Wildland Fire* **2006**, *15*, 319–345. [CrossRef]
7. Montorio Llovería, R.; Pérez-Cabello, F.; García-Martín, A.; Vlassova, L.; De la Riva Fernández, J. La severidad del fuego: Revisión de conceptos, métodos y efectos ambientales. In *Geoecología, Cambio Ambiental y Paisaje: Homenaje al Profesor José María García Ruiz*; Arnáez, J., González-Sampériz, P., Lasanta, T., Valero Garcés, B.L., Eds.; Instituto Pirenaico de Ecología (CSIC), Universidad de La Rioja: Logroño, Spain, 2014; pp. 427–440. ISBN 978-84-617-3212-8.
8. Röder, A.; Hill, J.; Duguy, B.; Alloza, J.A.; Vallejo, R. Using long time series of Landsat data to monitor fire events and post-fire dynamics and identify driving factors. A case study in the Ayora region (eastern Spain). *Remote Sens. Environ.* **2008**, *112*, 259–273. [CrossRef]
9. Chen, W.; Moriya, K.; Sakai, T.; Koyama, L.; Cao, C. Monitoring of post-fire forest recovery under different restoration modes based on time series Landsat data. *Eur. J. Remote Sens.* **2014**, *47*, 153–168. [CrossRef]
10. Chu, T.; Guo, X.; Takeda, K. Effects of burn severity and environmental conditions on post-fire regeneration in Siberian Larch forest. *Forests* **2017**, *8*, 76. [CrossRef]
11. Riaño, D.; Chuvieco, E.; Ustin, S.; Zomer, R.; Dennison, P.; Roberts, D.; Salas, J. Assessment of vegetation regeneration after fire through multitemporal analysis of AVIRIS images in the Santa Monica Mountains. *Remote Sens. Environ.* **2002**, *79*, 60–71. [CrossRef]
12. Goetz, S.J.; Fiske, G.J.; Bunn, A.G. Using satellite time-series data sets to analyze fire disturbance and forest recovery across Canada. *Remote Sens. Environ.* **2006**, *101*, 352–365. [CrossRef]
13. Van Leeuwen, W.J.; Casady, G.M.; Neary, D.G.; Bautista, S.; Alloza, J.A.; Carmel, Y.; Wittenberg, L.; Malkinson, D.; Orr, B.J. Monitoring post-wildfire vegetation response with remotely sensed time-series data in Spain, USA and Israel. *Int. J. Wildland Fire* **2010**, *19*, 75–93. [CrossRef]
14. Lhermitte, S.; Verbesselt, J.; Verstraeten, W.W.; Veraverbeke, S.; Coppin, P. Assessing intra-annual vegetation regrowth after fire using the pixel based regeneration index. *ISPRS J. Photogramm. Remote Sens.* **2011**, *66*, 17–27. [CrossRef]

15. Tanase, M.; de la Riva, J.; Santoro, M.; Pérez-Cabello, F.; Kasischke, E. Sensitivity of SAR data to post-fire forest regrowth in Mediterranean and boreal forests. *Remote Sens. Environ.* **2011**, *115*, 2075–2085. [CrossRef]

16. Hirschmugl, M.; Gallaun, H.; Dees, M.; Datta, P.; Deutscher, J.; Koutsias, N.; Schardt, M. Methods for Mapping Forest Disturbance and Degradation from Optical Earth Observation Data: A Review. *Curr. For. Rep.* **2017**, *3*, 32–45. [CrossRef]

17. Díaz-Delgado, R.; Pons, X. Spatial patterns of forest fires in Catalonia (NE of Spain) along the period 1975–1995 analysis of vegetation recovery after fire. *For. Ecol. Manag.* **2001**, *147*, 67–74. [CrossRef]

18. Bartels, S.F.; Chen, H.Y.H.; Wulder, M.A.; White, J.C. Trends in post-disturbance recovery rates of Canada's forests following wildfire and harvest. *For. Ecol. Manag.* **2016**, *361*, 194–207. [CrossRef]

19. Ireland, G.; Petropoulos, G.P. Exploring the relationships between post-fire vegetation regeneration dynamics, topography and burn severity: A case study from the Montane Cordillera Ecozones of Western Canada. *Appl. Geogr.* **2015**, *56*, 232–248. [CrossRef]

20. Meng, R.; Dennison, P.E.; Huang, C.; Moritz, M.A.; D'Antonio, C. Effects of fire severity and post-fire climate on short-term vegetation recovery of mixed-conifer and red fir forests in the Sierra Nevada Mountains of California. *Remote Sens. Environ.* **2015**, *171*, 311–325. [CrossRef]

21. Díaz-Delgado, R.; Lloret, F.; Pons, X. Influence of fire severity on plant regeneration by means of remote sensing imagery. *Int. J. Remote Sens.* **2003**, *24*, 1751–1763. [CrossRef]

22. Díaz-Delgado, R.; Lloret, F.; Pons, X.; Terradas, J. Satellite Evidence of Decreasing Resilience in Mediterranean Plant Communities after Recurrent Wildfires. *Ecology* **2002**, *83*, 2293–2303. [CrossRef]

23. Rouse, J.W.; Haas, R.H.; Schell, J.A.; Deering, D.W. *Monitoring the Vernal Advancement and Retrogradation (Green Wave Effect) of Natural Vegetation*; Progress Report RSC 1978-1; Texas A&M University Remote Sensing Center: College Station, TX, USA, 1973; p. 112.

24. López García, M.; Caselles, V. Mapping burns and natural reforestation using Thematic Mapper data. *Geocarto Int.* **1991**, *6*, 31–37. [CrossRef]

25. Key, C.H.; Benson, N.C. Landscape Assessment (LA). Sampling and Analysis Methods. In *FIREMON: Fire Effects Monitoring and Inventory System. Integration of Standardized Field Data Collection Techniques and Sampling Design With Remote Sensing to Assess Fire Effects*; Lutes, D.C., Keane, R.E., Caratti, J.F., Key, C.H., Benson, N.C., Sutherland, S., Gangi, L.J., Eds.; U.S. Department of Agriculture, Forest Service, Rocky Mountain Research Station: Fort Collins, CO, USA, 2006; pp. LA1–LA51. Available online: https://www.fs.fed.us/rm/pubs/rmrs_gtr164.pdf (accessed on 16 September 2017).

26. Miller, J.D.; Thode, A.E. Quantifying burn severity in a heterogeneous landscape with a relative version of the delta Normalized Burn Ratio (dNBR). *Remote Sens. Environ.* **2007**, *109*, 66–80. [CrossRef]

27. Chen, X.; Vogelmann, J.E.; Rollins, M.; Ohlen, D.; Key, C.H.; Yang, L.; Huang, C.; Shi, H. Detecting post-fire burn severity and vegetation recovery using multitemporal remote sensing spectral indices and field-collected composite burn index data in a ponderosa pine forest. *Int. J. Remote Sens.* **2011**, *32*, 7905–7927. [CrossRef]

28. De Santis, A.; Chuvieco, E. GeoCBI: A modified version of the Composite Burn Index for the initial assessment of the short-term burn severity from remotely sensed data. *Remote Sens. Environ.* **2009**, *113*, 554–562. [CrossRef]

29. Viedma, O.; Meliá, J.; Segarra, D.; García-Haro, J. Modeling rates of ecosystem recovery after fires by using landsat TM data. *Remote Sens. Environ.* **1997**, *61*, 383–398. [CrossRef]

30. García Martínez, E.; Pérez-Cabello, F. Análisis de la regeneración vegetal mediante imágenes Landsat-8 y el producto MCD15A2 de MODIS: El caso del incendio de O Pindo. In *Análisis Espacial y Representación Geográfica: Innovación y Aplicación*; de la Riva, J., Ibarra, P., Montorio, R., Rodrigues, M., Eds.; Universidad de Zaragoza: Zaragoza, Spain, 2015; pp. 621–630.

31. Huete, A.R. A soil-adjusted vegetation index (SAVI). *Remote Sens. Environ.* **1988**, *25*, 295–309. [CrossRef]

32. White, J.D.; Ryan, K.C.; Key, C.C.; Running, S.W. Remote Sensing of Forest Fire Severity and Vegetation Recovery. *Int. J. Wildland Fire* **1996**, *6*, 125–136. [CrossRef]

33. Álvarez, R.; Valbuena, L.; Calvo, L. Effect of high temperatures on seed germination and seedling survival in three pine species (*Pinus pinaster*, *P. sylvestris* and *P. nigra*). *Int. J. Wildland Fire* **2007**, *16*, 63–70. [CrossRef]

34. Ministerio de Agricultura, Alimentación y Medio Ambiente. *Los Incendios Forestales en España Decenio 2001–2010*; ICONA, Ed.; Publicaciones del Ministerio de Agricultura Pesca y Alimentación: Madrid, Spain, 2012.

35. Rivas Martínez, S.; Gandullo, J.M.; Serrada, R.; Allué, J.L.; Montero, J.L.; González, J.L. *Mapa de Series de Vegetación de España y Memoria*; ICONA, Ed.; Publicaciones del Ministerio de Agricultura Pesca y Alimentación: Madrid, Spain, 1987; ISBN 84-85496-25-6. Available online: https://floramontiberica.files. wordpress.com/2012/09/mapa_series_vegetacion_1987.pdf (accessed on 16 September 2017).

36. Agencia Estatal de Meteorología; Instituto De Meteorologia. I.P. *Atlás Climático Ibérico/Iberian Climate Atlas*; Agencia Estatal de Meteorología, Ministerio de Medio Ambiente y Rural y Marino; Instituto de Meteorologia de Portugal: Madrid, Spain, 2011; ISBN 978-84-7837-079-5.

37. Ministerio de Agricultura y Pesca, Alimentación y Medio Ambiente (MAPAMA). *Los Incendios Forestales en España Durante 1994*; ICONA, Ed.; Publicaciones del Ministerio de Agricultura Pesca y Alimentación: Madrid, Spain, 1994. Available online: http://www.mapama.gob.es/es/desarrollo-rural/estadisticas/ incendios_forestales_espania_1994_tcm7--349105.pdf (accessed on 16 September 2017).

38. Brunetti, M.; Maugeri, M.; Monti, F.; Nanni, T. Temperature and precipitation variability in Italy during the last two centuries from homogenized instrumental time series. *Int. J. Climatol.* **2006**, *26*, 345–381. [CrossRef]

39. González-Hidalgo, J.C.; Brunetti, M.; de Luis, M. A new tool for monthly precipitation analysis in Spain: MOPREDAS database (monthly precipitation trends December 1945–November 2005). *Int. J. Climatol.* **2011**, *31*, 715–731. [CrossRef]

40. Gonzalez-Hidalgo, J.C.; Peña-Angulo, D.; Brunetti, M.; Cortesi, N. MOTEDAS: A new monthly temperature database for mainland Spain and the trend in temperature (1951–2010). *Int. J. Climatol.* **2015**, *35*, 4444–4463. [CrossRef]

41. Ministerio de Agricultura y Pesca, Alimentación y Medio Ambiente (MAPAMA). Segundo Inventario Forestal Nacional. Available online: http://www.mapama.gob.es/es/biodiversidad/servicios/banco-datos- naturaleza/informacion-disponible/ifn2_cartografia_26_50.aspx (accessed on 11 October 2017).

42. European Environment Agency. Corine Land Cover 1990 (CLC1990) and Corine Land Cover Changes 1975–1990 in a 10 km Zone around the Coast of Europe. Available online: https://www.eea.europa.eu/data-and-maps/ data/corine-land-cover-1990-clc1990-and-corine-land-cover-changes-1975-1990-in-a-10-km-zone-around- the-coast-of-europe (accessed on 11 October 2017).

43. De Santis, A.; Chuvieco, E. Burn severity estimation from remotely sensed data: Performance of simulation versus empirical models. *Remote Sens. Environ.* **2007**, *108*, 422–435. [CrossRef]

44. Chuvieco, E. *Teledetección Ambiental. La Observación de la Tierra Desde el Espacio*; Ariel: Barcelona, Spain, 2010; ISBN 13:978-8434434981.

45. Chavez, P. Image-based atmospheric corrections—Revisited and improved. *Photogramm. Eng. Remote Sens.* **1996**, *62*, 1025–1035.

46. Gilbert, M.A.; Conese, C.; Maselli, F. An atmospheric correction method for the automatic retrieval of surface reflectances from TM images. *Int. J. Remote Sens.* **1994**, *15*, 2065–2086. [CrossRef]

47. Teillet, P.; Guindon, B.; Goodenough, D. On the slope-aspect correction of 696 multispectral scanner data. *Can. J. Remote Sens.* **1982**, *8*, 84–106. [CrossRef]

48. Pascual, M.; Moreno, V. Estudio de la regeneración post-incendio del ecosistema forestal mediterráneo mediante imágenes landsat. In *Teledetección, Medio Ambiente y Cambio Global*; Asociación Española de Teledetección: Paterna, Spain, 2001; pp. 137–140. ISBN 978-8497430012.

49. Escuin, S.; Navarro, R.; Fernandez, P. Fire severity assessment by using NBR (Normalized Burn Ratio) and NDVI (Normalized Difference Vegetation Index) derived from LANDSAT TM/ETM images. *Int. J. Remote Sens.* **2008**, *29*, 1053–1073. [CrossRef]

50. Sever, L.; Leach, J.; Bren, L. Remote sensing of post-fire vegetation recovery; a study using Landsat 5 TM imagery and NDVI in North-East Victoria. *J. Spat. Sci.* **2012**, *57*, 175–191. [CrossRef]

51. Vicente-Serrano, S.M.; Pérez-Cabello, F.; Lasanta, T. *Pinus halepensis* regeneration after a wildfire in a semiarid environment: Assessment using multitemporal Landsat images. *Int. J. Wildland Fire* **2011**, *20*, 195–208. [CrossRef]

52. Pausas, J.G.; Ouadah, N.; Ferran, A.; Gimeno, T.; Vallejo, R. Fire severity and seedling establishment in *Pinus halepensis* woodlands, Eastern Iberian Peninsula. *Plant Ecol.* **2003**, *169*, 205–213. [CrossRef]

53. Daskalakou, E.N.; Thanos, C.A. Aleppo Pine (*Pinus halepensis*) Postfire Regeneration: The Role of Canopy and Soil Seed Canopy and Soil Seed Banks. *Int. J. Wildland Fire* **1996**, *6*, 59–66. [CrossRef]

54. Moran, P.A.P. Notes on continuous stochastic phenomena. *Biometrika* **1950**, *37*, 17–23. [CrossRef] [PubMed]

55. Fotheringham, A.S.; Brunsdon, C.; Charlton, M. *Geographically Weighted Regression: The Analysis of Spatially Varying Relationships*; Wiley: Chichester, UK, 2002; ISBN 978-0-471-49616-8.
56. Charlton, M.; Fotheringham, A.S. *Geographically Weighted Regression. White Paper*; National Centre for Geocomputation, National University of Ireland Maynooth: Maynooth, Ireland, 2009; Available online: https://www.geos.ed.ac.uk/~gisteac/fspat/gwr/gwr_arcgis/GWR_WhitePaper.pdf (accessed on 16 September 2017).
57. Chu, T.; Guo, X. Remote sensing techniques in monitoring post-fire effects and patterns of forest recovery in boreal forest regions: A review. *Remote Sens.* **2014**, *6*, 470–520. [CrossRef]

![environments]

Article

Vulnerability of Coastal Beach Tourism to Flooding: A Case Study of Galicia, Spain

Diego R. Toubes [1,*], Stefan Gössling [2,3,4], C. Michael Hall [5] and Daniel Scott [6]

[1] Department of Business Organization, Business Administration and Tourism School, University of Vigo, 32004 Ourense, Spain
[2] School of Business and Economics, Linnaeus University, 39182 Kalmar, Sweden; sgo@vestforsk.no
[3] Service Management and Service Studies, Lund University, 25108 Helsingborg, Sweden
[4] Western Norway Research Institute, 6851 Sogndal, Norway
[5] Department of Management, Marketing & Entrepreneurship, University of Canterbury, Christchurch 8140, New Zealand; michael.hall@canterbury.ac.nz
[6] Department of Geography and Environmental Management, University of Waterloo, Waterloo, ON N2L 3G1, Canada; daniel.scott@uwaterloo.ca
* Correspondence: drtoubes@uvigo.es; Tel.: +34-988-368-747

Received: 27 September 2017; Accepted: 11 November 2017; Published: 16 November 2017

Abstract: Flooding, as a result of heavy rains and/or storm surges, is a persistent problem in coastal areas. Under scenarios of climate change, there are expectations that flooding events will become more frequent in some areas and potentially more intense. This poses a potential threat to coastal communities relying heavily on coastal resources, such as beaches for tourism. This paper develops a methodology for the assessment of coastal flooding risks, based on an index that compares 16 hydrogeomorphological, biophysical, human exposure and resilience indicators, with a specific focus on tourism. The paper then uses an existing flood vulnerability assessment of 724 beaches in Galicia (Spain) to test the index for tourism. Results indicate that approximately 10% of tourism beaches are at high risk to flooding, including 10 urban and 36 rural beaches. Implications for adaptation and coastal management are discussed.

Keywords: beaches; climate change; flooding; tourism; vulnerability

---

1. Introduction

Coastal and marine environments attract hundreds of millions of tourists every year and are a mainstay of the economy for many coastal communities [1]. Mediterranean countries attract almost a third of international tourist arrivals and, including domestic tourism, coastal zones in the Mediterranean receive an estimated 250 million visitors each year [2]. Tourism also has great economic significance in the majority of small island developing states (SIDS) [3]. Coastal zones are central features of these islands and are used for a wide range of tourism and leisure activities including fishing, swimming, snorkeling, windsurfing, water skiing, jet skiing, boating and yachting [1].

Coastal resources are increasingly threatened. External and tourism-related pressures on coastal zones include urbanization and industrial developments, water pollution, loss of mangroves, as well as overuse of fresh water and marine resources [4]. Climate change exacerbates existing problems in coastal zones, as it affects resources of central value to tourism [5] and can lead to more extreme weather events, increased run-off and sedimentation, sea level rise, salinity and acidification [4]. Fresh water stress is projected to affect many coastal regions in the world, with summer water flows being expected to decline by up to 80% in southern Europe and sea level rise leading to the loss of up to 20% of coastal wetlands in many parts of the Mediterranean [6,7]. The temporal patterns and interactions between these impacts, as well as tourist responses to these compounding impacts remain insufficiently understood [8].

Flood events can negatively affect beach resources through erosion, but have so far received only limited attention in the literature. Globally, the frequency of coastal flooding is expected to double in lower latitudes within decades due to sea-level rise [9], causing significant economic damage [10]. Floods can be caused by rivers, as a result of changing rainfall patterns, as well as by storm surges linked to wave patterns, tides and coastal features [11]. Where storm surges and river floods coincide, this can have negative consequences for beaches and recreational areas, in particular those located in estuaries and river mouths [12].

To provide new insights into these processes and their consequences for tourism, this paper develops a method for the assessment of coastal beach vulnerability to flooding, based on a case study of Galicia, Spain. Droughts and floods have been identified as a specific threat for the region that is located in the north of the country and heavily dependent on tourism [13].

2. Flooding and Tourism in Spain

Floods are the most prevalent and economically significant natural disaster in Europe [14]. For example, in the summer of 2002, floods in central Europe affected 4.2 million people and caused economic losses in excess of €18 billion [15], an estimate that excludes the social cost of disrupted health care or schooling [16]. In Spain, torrential rains and drought are the two major natural hazards [17]. Floods kill on average 20 people per year and lead to economic losses in the order of €800 million [18,19].

Climate change is anticipated to increase the frequency and intensity of rainfall events that will, correspondingly, cause more intense and frequent river flooding [20–22]. The IPCC [11] emphasizes that Europe faces significant risks of river and coastal flooding related to changes in land use, sea level rise, coastal erosion and extreme rainfall events (see also [23,24]). In Galicia exposure risks are also exacerbated by [25]:

- Infrastructure that limits channeling capacity;
- Drainage systems with insufficient capacity;
- Lack of maintenance of drainage systems;
- Ecosystem modification;
- Inappropriate land management and land use, including development in flood zones;
- Forest fires that reduce forest water storage capacity as a result of vegetation loss.

Tourism is a sector of central economic importance in Spain, a country that is ranked third in the world for international tourist arrivals (65 million) and second in tourism revenues ($65.2 billion) [26]. The sector accounts for 11.1% of Spain's GDP and 13.0% of its total employment [27]. In Galicia, the sector contributes 11.1% to GDP and employs 12.0% of the workforce [28].

Impacts of climate change on tourism in Spain have been addressed in various studies (e.g., [29–31]), with most focusing on demand responses under scenarios of warming [32–37]. There is only one study of tourism in Galicia that has focused on the main market for the region: domestic visitors from Madrid [35]. The results of this survey (n = 430) suggest that travel motives of a significant share of tourists (35%) are not influenced by climate variables. For visitors with weather-related motives, mild temperatures are the most attractive and risk of rain is the most aversive. Vazquez and Prada [35] conclude that climate change will actually lead to an increase in arrivals, as a result of a concentration of rainfall events in autumn and increasing temperatures in spring and summer. The study does not consider other vulnerabilities, however. Against this background, this paper seeks to present additional insights regarding the assessment of in particular flood events for tourism in the region.

3. Methodology for the Assessment of Beach Vulnerability

Indicator-based flood vulnerability indices are applicable over different spatial scales, including river basins, sub-catchments and urban areas [38–40] and have been used in a growing number of publications focusing on coastal flood vulnerability analysis [41–45]. Coastal tourism is at risk of flooding, which can be compounded by seasonal river discharges in combination with waves and storm surges [44,46–48]. The first part of this paper is consequently focused on the development of a methodology for beach vulnerability assessments in the context of tourism.

The methodology is set up as follows: The first part of the paper presents data to be included in a compound index for flooding risks. This index includes hydrological, geomorphological-historical, hydraulic as well as exposure indicators and thus goes beyond the use of indicators as outlined in the European Directive 2007/60/EC for the assessment and management of flood risks. The paper then summarizes available data for the different indicators, as these apply to Galicia, Spain. Given the scarcity of data for several areas, the paper then relies on an existing flood risk assessment for the region as provided by Aguas de Galicia [25]. Based on this identification of at-risk beaches, a specific assessment of tourism vulnerabilities is based on three indicators: Level of visitation, tourism facilities and beach width. These are assessed in terms of being at low, medium, or high risk and assigned a corresponding score on a scale from 1 to 3. Although most floods currently occur in the tourist low season, the analysis is relevant for tourism because infrastructure damage may be difficult to repair in time for the high season. Floods can also change coast morphology or erode beaches and lead to the loss of quality certifications (Blue Flag). Flood risks also need to be seen in light of plans to promote tourism in spring and autumn, when flooding risks are greater. Analysis shows the areas in which such flood risks are particularly high.

3.1. Vulnerability Analysis

Vulnerability is the degree to which geophysical, biological and socio-economic systems are susceptible to and unable to cope with, adverse impacts originating out of environmental, social or territorial elements [6]. Floods are caused by extreme weather phenomena and exacerbated by human activities such as urbanization, land clearance and alterations of coastlines that lead to flood susceptibility. Nicholls et al. [46] identified three key drivers of coastal floods:

1. Climate change, which affects sea levels, as well as rainfall and storm frequencies and intensities;
2. Sediment supply, which influences flood pathways, coastal geomorphology and ecosystems;
3. Socio-economic change, which can alter the type and extent of human activities and behaviors within the floodplain.

To assess beach flood vulnerability at the local scale, relevant variables need to be included in a theoretical framework covering exposure, susceptibility and resilience [42] with regard to hydro-geomorphological, socio-economic & administrative and institutional subsystems [44]. *Hydro-geomorphological subsystems* comprise exposure indicators that represent coast or catchment basin characteristics, including storm surges, rainfall, sea level rise, river discharge, soil subsidence and elevation above sea. Specific beach structures can make coastlines more resilient. *Socio-economic subsystems* consider elements that increase the instability of beaches or increase adverse impacts of flooding. *Political-administrative subsystems* refer to legal and regulatory context, preparedness, coping and recovering strategies, as well as contingency plans and nourishment actions.

To assess flood vulnerability, appropriate indicators need to be identified for each subsystem [39,49]. These have been derived from the literature [38,40,43,44,50] and integrated into a comprehensive model for assessment (Figure 1).

3.2. Exposure Indicators for the Hydro-Geomorphological Subsystem

This subsystem's vulnerability is characterized by fluvial and costal characteristics (morphology), elevation above sea level, frequency of storms/rainfall events, sea level rise, as well as wave energy (Table 1). Exposure is proportional to the length of the coast and the number and density of beaches along the coast. Bays provide greater protection against storms, but they are also preferred human settlement areas. Estuaries host a wide variety of ecosystems and are sensitive to variations in sea level or changes in river flows [51].

Table 1. Exposure indicators of the hydro-geomorphological subsystem.

Factor	Indicator	Description	Impact on Vulnerability Increases (+)
Fluvial/costal characteristics	Beach coastline (km)	Kilometers of coastline, density of beaches	Higher beach coastline (+)
	Hydrographical characteristics (km)	River network, density of rivers	Higher river network (+)
Elevation above sea level	Low elevation coastal zones (km^2 or %)	Share of coastline with elevation up to 10 m	Higher share of Low Elevation Coastal Zones (LECZ) (+)
	Soil subsidence (m^2)	Surface of soil that is experimenting a decreasing	Higher area (+)
Frequency of storm/rainfall events	Rainfall intensity (mm or L/m^2)	Rainfall volume	Higher intensity/ frequency (+)
	Rainfall seasonality (mm or L/m^2)	Amount of rainfall per month	Impact of rainfall on high season (+)
	Frequency of storms (#)	Number of storms events in the last 10 years	Higher frequency of storms (+)
Sea-level rise	Sea-level rise (mm/yr)	Increasing in the level of the sea in *x* year	Higher Sea Level Rise (SLR) (+)
Waves	Wave regime (cm/yr or W/m^2/yr)	Changes in wave characteristics: height, period and energy	Higher wave intensity (+)

3.3. Exposure Indicators of Flood Characteristics

Coastal floods are caused by the combined action of tides, storm surges and wave run-up [52]. Flood-tides are related to a region's tidal range, wave energy, sediment supply and back barrier setting [53]. Astronomical tide plays an important role in high sea levels, but is predictable, in contrast to storm surges [52]. For some beaches and coastal areas, river discharge can influence flooding, especially when heavy rainfall coincides with extreme ocean conditions [12]. Indicators are shown in Table 2.

Table 2. Exposure indicators of flood characteristics.

Factor	Indicator	Description	Impact on Vulnerability Increases (+) Decreases (−)
Flood tide	Tide range (m)	Sea level variation due to tides within a day	Variable
	Tidal flooding (m)	Estimated range by tidal flooding; 2 scenarios with return periods of 50, 100 years (T50, T100)	Higher flooding (+)
Flood waves	Wave flooding (m)	Estimated range reached by wave flooding (2 scenarios T50, T100)	Higher flooding (+)
Velocity	Volume of flow (m^3/s)	Flow volume of rivers discharging into the coast (3 scenarios with return periods of T10, T50, T100)	More volume (+)
	Response time (scale)	Time that elapses between the moment of maximum rainfall and when the peak flow is reached	More time (−)
Inundation area	Area of fluvial flooding (km^2)	Surface area of potential fluvial flooding (3 scenarios with return periods of T10, T50, T100)	Larger area (+)
	Area of sea flooding (km^2)	Estimated flooded area according to a potential increase the elevation of the level of flooding in meters	Larger area (+)
Other flood characteristics	Level of flooding (m or %)	Level reached as a result of the joint action of the astronomical tide, storm surge and run-up generated by waves	Higher level of flooding (+)

3.4. Indicators for Socio-Economic and Political-Administrative Systems

Governance is increasingly recognized as a key factor for adaptive capacity [11]. The analysis of legal and regulatory contexts is an essential part of flood risk assessments [54], including administrative organization, legal frameworks, protected areas, contingency or crisis management plans [46,48,55–57] (Table 3).

Table 3. Indicators in the socio-economic and political-administrative context.

Factor	Indicator	Description
Legal and regulatory context	Regulatory context	Competences and responsibility on the coastal domain among the different levels of government
	Beaches at protected area	Beaches with risk of flood located in areas of environmental protection
Susceptibility including preparedness, coping and recovering	Contingency plans	Contingency plans designed to deal with flood risk situations and impacts on beaches
	Beach nourishment and recovering	Replenishing beach sediment in nourishment operations, recovering and cleaning
	Time recovery	Time needed to recover to a functional operation after coastal flood events
Other socio-economic and politico-administrative factors	Population	Number of people affected and number of inhabitants in potential flood zone
	Stock of affected capital	Losses and properties affected by floods

3.5. Resilience Indicators

Flood resilience describes the systemic ability to experience flooding with minimum damage and rapid recovery [46]. Aspects that can increase resilience include flood-proof infrastructure, dykes and dams, natural coastal morphology and habitat features. Factors such as sediment supply, wind action, or changes in the wave regime can help to prevent beach erosion and contribute to recovery [58–60]. Previous exposure to flooding may contribute to learning effects [61]. Resilience indicators are shown in Table 4.

Table 4. Resilience indicators.

Factor	Indicator	Description
Beach structure	Coast and beach profile	Coastal features and natural protection that have influence on flood resilience
	Length and width of the beach	Length, width and width variation during the year of beaches with risk of flood
	Sediment supply	Amount of sustained supply of sediment for the preservation and sustainability of the beaches
Previous exposure to flooding	Historical floods	Historical flood events and experience and knowledge gained in previous floods
Other resilience indicators	Flood protection	Existence of shelters and structural measures that physically prevent beach flooding
	Household disposable income	Household disposable income as a resilience factor

3.6. Uncertainties

Various uncertainties characterize any assessment of indicators. Data on precipitation, waves, surges or sea level rise is increasingly accurate, but it remains difficult to assess flooding probabilities, specifically under different scenarios of climate change [47]. Even more difficult is the assessment of demand side implication, i.e., as to how flooding events will affect tourist responses [8]. In acknowledging these uncertainties, an index for flood vulnerability assessment has been developed (Figure 1). The figure considers the various indicators, at the center of which are risks for tourism.

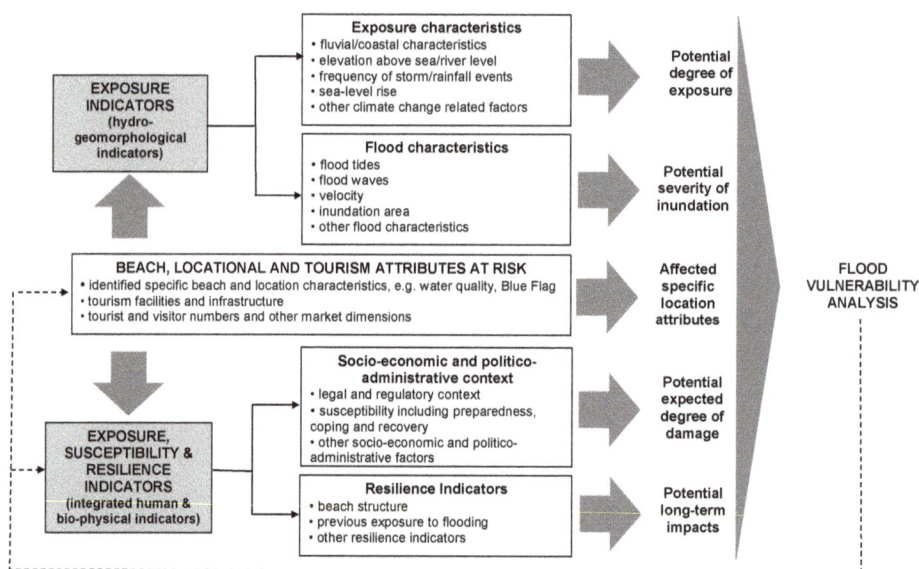

Figure 1. Flood vulnerability analysis for tourism.

3.7. Scaling and Weighting Indicators

In a final step, vulnerability assessments seek to integrate different indicators into a compound index, for which averages or weighted averages may be used [50]. Two main procedures can be distinguished to integrate indicators [42]: Raw values can be used, in which case indicators are presented independently; or values are transformed into homogeneous units. The first method provides data of the selected indicators directly, without manipulation, but is more subjective in its interpretation. The second method provides quantifiable and comparable indices as it standardizes the different measurement variables, but is strongly influenced by outlier observations and the unique value that is derived can hide divergences [50]. Here, weighting is also an issue, as indicators are often weighted based on expert judgment [44,49,62] and hence implying a degree of subjectivity [42,43,63]. In this analysis, the compound index is based on raw values, which has advantages of simplicity [64,65] and is considered sufficient given the tool's primary function of raising awareness.

4. Beach Vulnerability Assessment: Galicia

Beach tourism has considerable importance in Galicia with its 724 beaches in the North of Spain [66]. A total of 436 marine beaches and 68 river beaches are surveyed regarding water clarity, jellyfish, tar, floating materials, organic waste and other waste [67,68]. As of 2016 three-quarters (75.6%) of beaches were rated excellent for bathing, 16.5% good, 5.8% sufficient and 2.2% poor; 123 beaches are Blue Flag certified [69]. Two coastal areas, Rias Altas and Rias Baixas, receive most of the tourist arrivals (53.7%) and account for most overnight stays (63.2%). Tourism in coastal areas is concentrated in summer (64%), with most overnight stays in June (9.6%), July (18.7%), August (25%) and September (10.7%) [70,71].

4.1. Hydrogeomorphology—Fluvial/Coastal Characteristics

Galicia has an exceptionally long coastline (1720 km), corresponding to about 22% of the national total. A great part of the coast are cliffs (858.8 km), while beaches occupy over 180 km of the coast (17.2 km^2). The Atlantic coast has the highest density of beaches; 120 km from the border with Portugal to Cape Finisterre [72]. Figure 2 provides an overview of the coastline and its location within Spain.

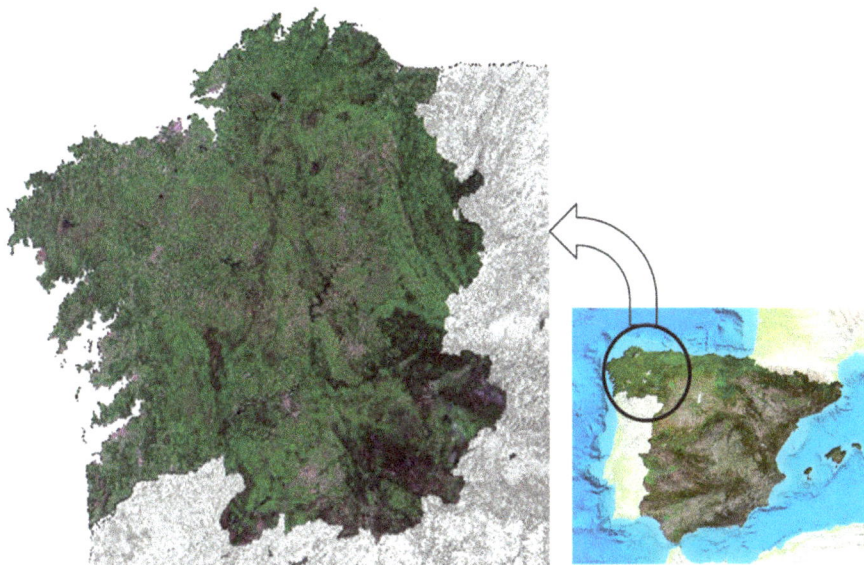

Figure 2. The region of Galicia in the North of Spain.

The entire region is prone to flooding for interrelated hydrographical and meteorological reasons. Rivers in Galicia are generally short and the region's river network is very dense. There are 8150 km of waterways representing 39.9% of Spain's total hydraulic network, in comparison to 5.9% of the country's surface area.

4.2. Hydrogeomorphology—Elevation above Sea Level

Low Elevation Coastal Zones (LECZ) are land areas including the coastline up to a 10-metre elevation. The extension of the LECZ area in Spain is 5953 km^2 (1.19% of the country). In Galicia, the Muros-Noia estuary is the lowest area [73]. The majority of the population in LECZ live in urban areas (84.9%). Throughout the country, the share of the population living in LECZ is expected to rise from 7.7% to 8.1% by 2100 [73].

4.3. Hydrogeomorphology—Frequency of Rainfall Events

Intense and frequent rainfall periods are often observed in Galicia [74] and are expected to increase substantially in autumn and winter, as well as in summer [75–77]. The average annual rainfall in Galicia is 1281 mm/yr [78], which can be compared to the Spanish average of about 500 mm/yr [79]. Within Galicia, there are notable differences in rainfall, ranging between 1014 mm in the North (A Coruña) to 1613 mm in the South (Pontevedra). Over the last 10 years, rainfall has fluctuated between 880–1800 mm/yr on the north coast and between 700–2050 mm/yr in the South, indicating considerable variability between years [80]. Over the year, rainfall is most intense in October-February; June-August are driest in both areas [80]. Figure 3 shows the geographical distribution of rainfall patterns in the main coastal tourist destinations [70,78].

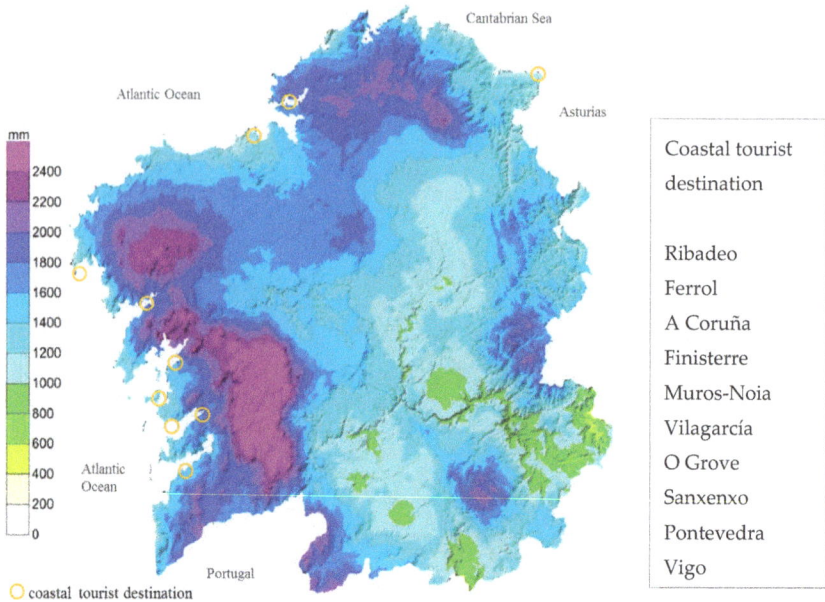

Figure 3. Map of cumulative rainfall in Galicia (2016). Source: MeteoGalicia [78].

4.4. Hydrogeomorphology—Frequency of Storms

The severity of a storm at the coast is a result of wave height and wave period [81]. Storms coming from the northwest cause the highest waves (42%). During the period 2004–2016, waves exceeded 10 m in 11 months and 8 m in 35 months. The maximum height reached by a wave was 27.8 m (storm "Christina" in 2014; [82]).

4.5. Hydrogeomorphology—Sea Level Rise

IPCC projections on sea level rise range from 26 cm to 82 cm by 2100 [7]. Projected sea level rise at the Galician coast averages 2–2.7 cm per decade [83]. Sea levels vary between summer and winter and trends of storm surges for the last 20 years are positive for the Atlantic, showing values around 0.5 mm/yr [84]. A sea level rise of 0.5 m could lead to the disappearance of about 22 km of beaches in the Basque Country and Cantabria—i.e., approximately 30% of the beach coastline in these regions [51].

4.6. Flood Characteristics—Tides, Waves, Velocity and Inundation Area

The magnitude of astronomical tides along the Spanish coast varies and oscillations can reach 4 m in the Cantabrian and Atlantic coasts, compared to a few centimeters in the Mediterranean [52]. The spring tidal range in Galicia varies between 0.2 m to more than 4.5 m [85]. Galicia is located in a climate zone with frequent low-pressure storms passing through in winter, as well as extra tropical cyclones as a result of rapid atmospheric pressure drops. Rapidly developing storms can bring heavy rains, wind and waves. Over the last 60 years, wave height has increased by 0.2 cm/yr along the Galician coast, with more intense changes during the winter (1.4 cm/yr) [52]. Large waves show an increase of up to 0.8 cm/yr, reflecting a more energetic sea. Average energy flow of the swell in the Cantabrian Sea and the Atlantic coast has grown at a rate of 0.07 W/m^2/yr [52].

Inundation as a result of flood risks for a return period of 10, 100 and 500 years identifies 210 Areas with Potential Significant Flood Risk (APSFR), including 42 coastal areas [86]. IHCantabria [72] analyzed flooded areas in relation to a potential increase in the level of flooding (considering astronomical tide, storm surge and wave run-up). Figure 4 represents the flooded area (km²) in 13 locations of the Galician coast under flooding level scenarios between 1 and 10 m. Results indicate that most vulnerable area is the Muros-Noia estuary, where a one-meter flooding could inundate 104.3 km².

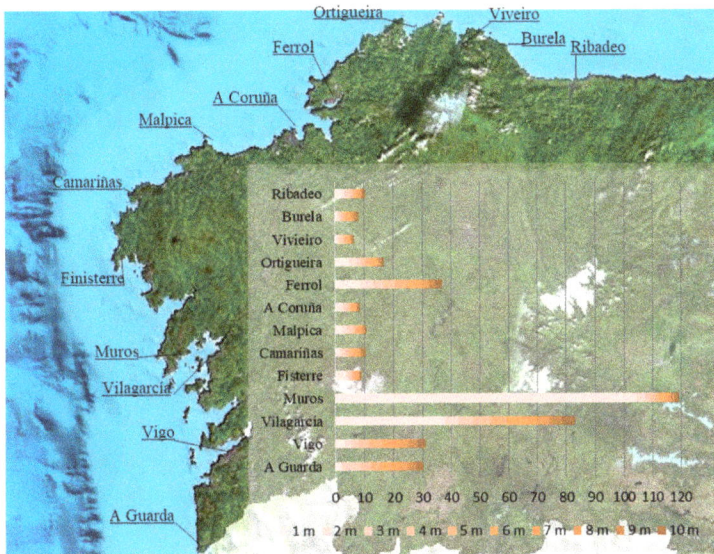

Figure 4. Inundation area (km²) in relation to flood level scenario (m). Source: Based on data supplied by IHCantabria [72].

4.7. Other Flood Characteristics

Recent years have seen more severe flooding events in Galicia, including beach berm erosion [52]. The 50-year flood level return period (FL50) analyses the potential flood risk due to storms and extreme events. Current FL50 along the coast of Galicia is between 3.44–3.91 m and 5.09–8.42 m for the dissipative, i.e., flat and shallow, beaches [48,72]. Table 5 shows an extrapolation of the long term FL50 trend for four major beach areas in Galicia (2020, 2030 and 2040 in comparison to 1960–1990). Table 6 shows FL50 projections in beaches for the end of century, averaged by provinces and for an interpolation of the trend line for 0.5 m, 0.85 m and 2.0 m sea level rise. The table also shows the FL50 percentage increase for 2100. Figure 5 illustrates the location of the four beach areas in Galicia.

Table 5. FL50 absolute and relative growth at coast and dissipative beaches in Galicia.

	Year	Vigo		Vilagarcia		A Coruña		Ribadeo	
		Coast	Dissipative Beach	Coast	Dissip. Beach	Coast	Dissip. Beach	Coast	Dissip. Beach
FL50	2020	0.05	0.06	0.01	0.14	0.04	0.07	0.05	0.09
Absolute	2030	0.11	0.13	0.02	0.32	0.08	0.16	0.11	0.19
growth (m)	2040	0.17	0.21	0.03	0.50	0.13	0.25	0.16	0.30
FL50	2020	1.3	1.10	0.20	1.70	0.96	1.26	1.35	1.69
Relative	2030	2.9	2.48	0.45	3.83	2.17	2.85	3.05	3.81
growth (%)	2040	4.6	3.85	0.68	5.96	3.38	4.43	4.75	5.93

Source: Based on data supplied by IHCantabria [72].

Table 6. FL50 projections for 2100 by extrapolating trends and for three SLR scenarios by provinces.

	Extrapolation of Trends	Scenario 1 SLR: 0.5 m	Scenario 2 SLR: 0.85 m	Scenario 3 SLR: 2 m
A Coruña	6.4	6.9 (+7.3%)	7.2 (+11.3%)	8.3 (+28.0%)
Lugo	5.6	6.1 (+9.3%)	6.3 (+13.3%)	7.4 (+32.5%)
Pontevedra	5.4	5.9 (+9.0%)	6.2 (+14.0%)	7.2 (+32.6%)

Source: Losada et al. [52] (pp. 77–78).

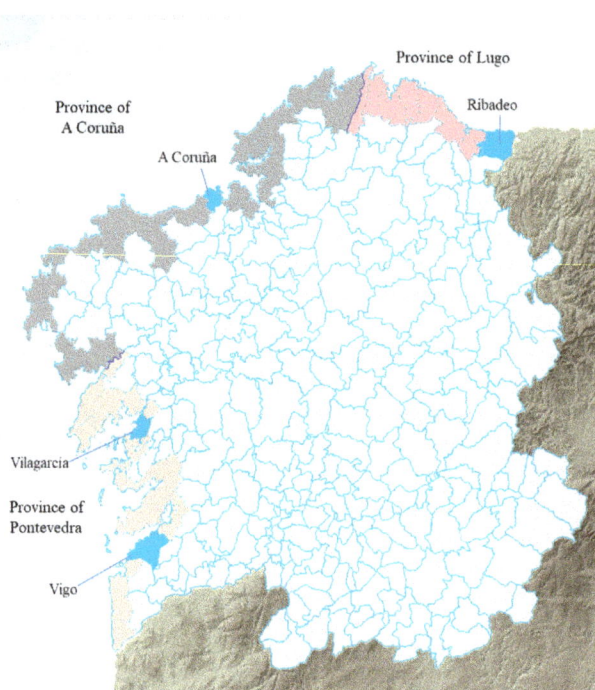

Figure 5. Location of the four communities studied.

4.8. Socio-Economic and Politico-Administrative Context: Legal and Regulatory Context

Spanish law for the protection and sustainable use of the coast distinguishes Public Maritime-Terrestrial Domains (PMTD) and Easement Zones (EZ). PMTD include inland waters, territorial sea, seashores and estuaries. The EZ comprises the strip of land needed to ensure public access to beaches extending 6–20 m landward from the coastal line; a zone of total protection extending landward from the coastline by 20–100 m; and an influence zone extending 500 m inland from the coastline [48]. Planning and building in this area is regulated by local authorities.

The state, regional and municipal administration have different levels of legal power in the Galician coast. The Spanish state is the owner of the maritime-terrestrial zone, beaches, territorial sea, natural resources of the economic zone and the continental shelf and responsible for ensuring its protection and conservation. As an Autonomous Community, Galicia has exclusive legal power over matters of land use planning, urban planning, coastal management, fishing in the estuaries, aquaculture, or port matters. It also has exclusive power in tourism promotion and management in its territory (Article 27.21 Statute of Autonomy).

"Landscape" and "Natural environment" are the most valued attributes when choosing Galicia as a destination, highlighted by a fifth of travelers (20.7%) as a reason for visitation [87]. This is also because Spain has established a wide range of protected areas, including conservation areas, protection areas for birds, biosphere reserves and others. In Galicia, 27 beaches under some form of environmental protection are formally recognized as Areas with Potential Significant Flood Risk (APSFR) [86].

4.9. Socio-Economic and Politico-Administrative Context: Susceptibility Including Preparedness, Coping and Recovery

Flood risk management in Spain is based on: (1) identification of Areas with Potential Significant Flood Risk, (2) danger maps for the entire Spanish coast and (3) management plans. At the national level, the General Direction for Sustainability of the Coast and Sea carries out the assessment of flood risks and also implements the European Flood Directive [88]. Regional plans include the Special Civil Protection Plan for Flood Risk in Galicia (Inungal, June 2016) and the Flood Risk Management Plan of Galicia-Costa 2015–2021 (January 2016), covering aspects of prevention-protection, preparation, recovery and evaluation.

Coastal erosion and the risks associated with flooding have led to the implementation of contingency plans and beach nourishment actions. Awareness and preparedness for floods has been growing. Weather forecasts on the arrival of storms are increasingly accurate and early warning systems in place. Storms causing floods mainly take place from October to January and much of the damage they cause can be addressed before arrival numbers peak (June-September). Coastal protection and beach recovery are also addressed in the Plan Litoral, launched in 2014. The plan addresses conservation, sustainability, storm impact mitigation and the protection, conservation and regeneration of beaches and dune systems [89]. There is also the PIMA-Adapta Plan, focused on adaptation to climate change.

4.10. Socio-Economic and Politico-Administrative Context: Other Socio-Economic and Politico-Administrative Factors

The population of the Spanish coastal municipalities increased at an annual rate of 1.9% in the first decade of the 21st century. The population density of the municipalities located on the coast of A Coruña (269 inhabitants/km^2) and Lugo (92 inhabitants/km^2) is below the average of Spain (435 inhabitants/km^2), but higher in Pontevedra (724 inhabitants/km^2) [90]. The total population in coastal municipalities is 1.1 million, of which about 11,500 people live in a flood zone [25]. Preferential Flow Zones (PFZs) are areas where serious damage to people and property may occur during flooding events. It is estimated that for a return period of 100 years, 26,800 people (0.53% of the population) live in PFZs in Galicia [25].

4.11. Resilience Indicators—Beach Structure

The coastline of Galicia is comprised of "hard" cliffs, formed by compact rocks that are resistant to erosion. The most important beaches are within the *rias* or cliff inlets. In the Cantabrian coast, estuaries with large intertidal zones and marshes in their environs are of special importance [51]. Dune ecosystems provide protection against flooding [48], though beach width and sediment supply are also relevant [91]. While there is sea level rise of 2.2–2.5 mm per year in Galicia, there are also cases of dune propagations exceeding 100 m per year [60,92,93].

4.12. Resilience Indicators—Previous Exposure to Flooding

During 1950–2010, eleven people died as a result of floods and 3875 houses were damaged in the provinces of A Coruña, Lugo and Pontevedra [94]. Table 7 provides an overview over historical flood events recorded within the period 1950–2010 in coastal municipalities. Tourism infrastructure, such as accommodation, has often been affected.

Table 7. Flood events in costal municipalities 1950–2011.

Province	Flood Cases	Examples of Tourism Infrastructure Affected in Flood Events
A Coruña	190	21/12/1995 — hotel facilities of Muxía
		20/10/2000 — beaches in Pobra do Caramiñal and Cedeira, hotel facilities in A Coruña
Pontevedra	146	21/12/1995 — seafronts and beaches in Baiona, Moaña and Marín
		20/10/2000 — flooding of beaches in Nigrán, Vilanova de Arousa, Portonovo, Marín and Baiona; hotel facilities in Baiona and A Guarda
		11/10/2001 — hotel facilities in Pontevedra and Sanxenxo
		04/09/2004 — hotel facilities in Vigo
		18/11/2006 — tourist facilities in Pontevedra, Sanxenxo, Vilagarcía de Arousa, Cangas and Marin.
Lugo	39	30/04/1998 — access road to the coast in Foz
		20/10/2000 — access road to Abrela beach in O Vicedo
		09/06/2010 — hotel facilities in Viveiro, access road to Catedrales beach

Source: from data of UNISDR [94].

4.13. Resilience Indicators—Other Resilience Indicators

Resilience of a society grows when the economic and social conditions of its inhabitants improve. The household disposable income in Galicia in 2015 was €25,614, slightly lower than the average household income in Spain (€26,092; [95]). The poverty risk rate in 2015, understood as the threshold below 60% of the median annual income per unit of consumption in the OECD, is 19.4% for Galicia as compared to Spain's 22.1%.

5. Tourism Vulnerability Assessment

The vulnerability assessment of 724 beaches in Galicia shows that 76 beaches are located in areas at risk of flooding [25] (see Appendix A for details). Of these beaches, 43 are at risk of coastal floods (waves, storm surges), 26 at risk of fluvial floods (river mouths) and seven at risk of both [86]. Based on this assessment, the vulnerability of these 76 beaches is evaluated with regard to tourism, considering three indicators: Level of visitation, tourism facilities and beach width. These indicators are assessed in terms of being at low, medium, or high risk and assigned a corresponding scoring on a scale from 1 to 3, corresponding to low-moderate-high vulnerabilities (Table 8). Note that "level of visitation", as the most important aspect, is weighted double. The result is an aggregated number, divided by the number of indicators, resulting in an average vulnerability (low-moderate-high). As outlined, this is an inherently subjective, expert-based approach to a vulnerability assessment [42].

Table 8. Assessment of beach vulnerability in regard to tourism.

Feature	Scoring 1–3
Level of visitation	Low to high
Tourism facilities	Depending on number and character
Beach width	more than 75 m; 15–75 m; less than 15 m

Figures 6 and 7 show the location of at risk beaches, in the context of their tourism vulnerability (low-moderate-high) in two regions, Rias Altas and Rias Baixas. Results indicate that the level of vulnerability of the different beaches varies considerably, indicating potential priorities for additional in-situ vulnerability assessments. While the maps do not suggest geographically focused risks, they highlight a total of 10 beaches at high risk and a significant number of other beaches at risk. The visualization of results in maps does not seem to indicate particular patters, such as specific coastlines being at lower risks than others, even though the south-western part of Rias Altas and the northern part of Rias Baixas appear less vulnerable. The maps also show that highly vulnerable beaches can be located close to those with low risks. For a detailed calculation of one beach, as an illustration of the method, see Appendix B.

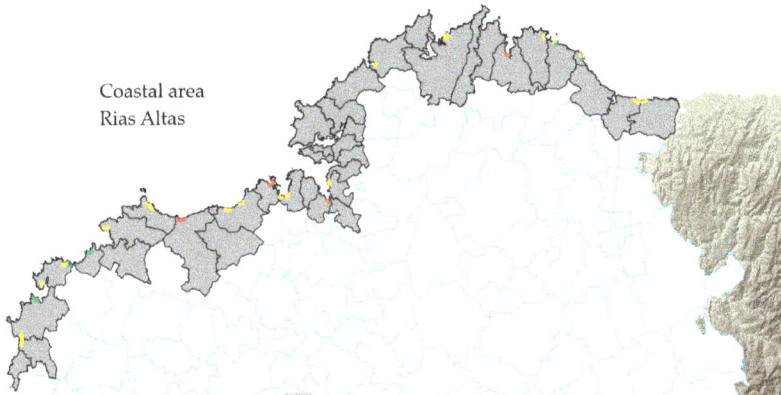

Figure 6. Tourism vulnerability of at risk beaches in Rias Altas. Green: Beaches with low tourism vulnerability; yellow: medium vulnerability; red: high vulnerability.

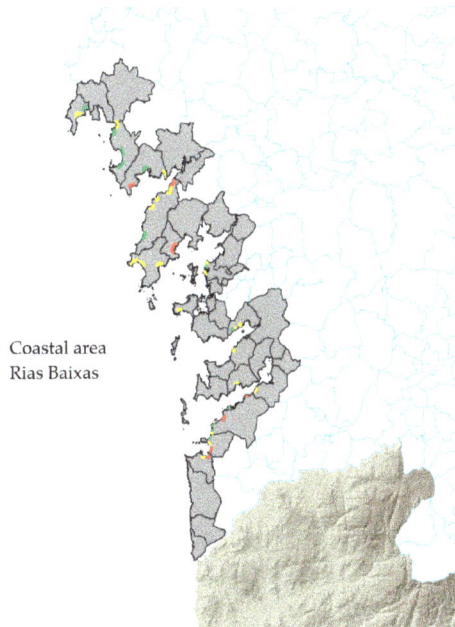

Figure 7. Tourism vulnerability of at risk beaches in Rias Baixas. Green: Beaches with low tourism vulnerability; yellow: medium vulnerability; red: high vulnerability.

6. Discussion and Conclusions

This paper developed a framework for the assessment of beach vulnerability to flood risks in coastal tourism regions. Using the Autonomous Spanish Community of Galicia as an example, 724 beaches have been assessed, of which 76 (11%) were identified as being at risk of flooding. A total of 43 beaches are at risk of flooding by waves, 26 in areas at risk of fluvial flooding in river mouths and 7 beaches are potentially exposed to both. In further analysis, the relative risk for tourism was evaluated for these 76 beaches. Results indicate that tourism vulnerabilities vary, depending on visitation levels, the existence of tourism facilities and beach width. While some at risk beaches have great importance for tourism, others are of little relevance.

An important aspect in the context of tourism and natural disaster risk analyses is their temporal dimension. Analysis shows that flooding events usually occur outside the tourism high season in July and August. However, under scenarios of climate change, it is expected that climate conditions in spring and autumn will become more suitable for tourism and tourism agencies already seek to increase arrivals in what is currently the shoulder season. This will imply a greater risk related to floods, as arrival peaks will become more closely aligned with flooding risks.

Results consequently have various implications for coastal and tourism management. For some beach destinations, the consideration of flooding risks may have to become part of destination planning and, potentially, adaptive measures related to crisis management. This will also require for climate change effects to be considered in planning strategies, such as the construction of roads or accommodation, which need to consider the likelihood of extreme events including flooding under scenarios of climate change. For investors, risk represents a cost, either in terms of insurance, more resilient constructions, reduced attractiveness for tourists, or cancellations in extreme situations, which may include loss of income in situations where infrastructure damage (e.g., roads) prohibits visitation. The severe consequences of flooding for destinations, have for instance been felt in the Caribbean in 2017, following a series of major hurricanes. The development of such frameworks and assessments as presented here can therefore play a significant role in improving the resilience of tourism at various scales [96–98].

More generally, coastal management strategies have to consider changed flood risks, also in relation to climate change, to protect important coastal economic, conservation and social assets, including those that are significant for tourism. Fundamental to the management of such assets is the development of beach vulnerability assessments that identify the most at-risk locations. Such information can enable evidence-based decision-making in the development of short and long-term adaptive strategies to reducing flood risk as well as the better allocation of economic resources in response to risk, increasing coastal destination resilience. Such information is valuable for a number of different stakeholders, including government, in determining resource allocation and priorities; insurance companies in their assessment of liabilities; businesses in relation to their own planning, adaptation and risk assessments; as well as those living in coastal areas (risks, property values). The index developed in this paper can also potentially be adapted for other coastal settings, particularly in Europe's coastal regions where similar data sets are likely to exist. Future work may improve the accuracy of the index, particularly when more data becomes available and as data and frameworks are contrasted to the impacts of actual events. Such developments could, for instance, include tourist demand and behavioral responses to flooding, which are insufficiently understood [5]. A further addition could be the inclusion of economic assessments of the potential direct, i.e., infrastructure damage and indirect, i.e., changes in tourist demand and behavior, costs of flooding into the assessment framework. However, a key issue in the development of such frameworks is the availability of existing indicators and data and the costs of development new ones.

Overall, it is acknowledged that the work presented in this paper is a pilot study that is necessarily based on simplifications. These include the choice of indicators, as well as the assumption that indicators are independent and sometimes extrapolated on the basis of existing trends, i.e., not considering trends in for instance rainfall or storm intensities. Future work should seek to address these issues in order to achieve more robust results.

Acknowledgments: There was no funding for this work.

Author Contributions: Diego R. Toubes conceived the work and collected and analyzed the data; Stefan Gössling interpreted data and drafted the paper; C. Michael Hall drew the conceptual framework, reviewed the methodology and carried out a substantive review and reformulation of the work; and Daniel Scott reviewed the methodology and carried out a substantive review and reformulation of the work.

Conflicts of Interest: The authors declare no conflict of interest.

Appendix A. Beaches Located at Potential Significant Flood Risk Areas in Galicia

Table A1. Coastal Flood Risk.

Beach Name	Municipality	Length	Width	Typology	Population Density	Tourism Facilities and Infrastructure	Water Quality 16	15	14	Blue Flag 16	15	14	Mid-Range of Tidal Flooding (m) T100	T500	Mid-Range of Wave Flooding (m) T100	T500
Covas	Viveiro	1500	10–400	village	High	promenade, camping, recreational area	G	G	G				124.15	126.1	11.86	11.68
O Torno		650	30–90	village	High	Seafront	E	na	na		✓					
Cubelas	Cervo	250	10–40	village	Medium	seafront, recreational area	na	P	P				34.73	35.73	23.96	25
Caosa		120	15–60	urban	Low		na	S	S							
Lago	Xove	700	25–80	rural	Medium	dunes, gardens and rest area	E	E	E							
O Portelo	Burela	460	25–110	village	High	Seafront	E	E	G	✓	✓		13.02	12.38	26.28	27.18
Penoural		109	35–70	village	Low	Seafront	P	P	P	✓						
Areal	A Pobra Caramiñal	1100	26	urban	High	promenade, yacht club, recreational areas	G	G	G				71.46	75.97	2.57	1.72
Corrubedo	Ribeira	5000	25–40	remote	Medium	natural park	E	E	E				59.13	57.87	22.11	24.59
O Prado		300	25	village	Medium	anchorage	S	P	P							
Suigrexa		420	20	rural	Medium		S	G	E							
Fontforrón	Porto do Son	100	7	rural	High		E	E	E				29	28.8	20.66	22.62
A Vila		170	15	village	High		S	G	E							
San Pedro	Carnota	250	160	village	Low	seafront	G	G	E				137.62	139.22	43.86	49.81
Ézaro	Dumbria	800	28	rural	Medium		E	P	P	✓	✓		26.79	32.56	67.53	89.95
Lires	Cee	123	20	remote	Medium		P	P	S				26.52	29.72	1.41	2.2
Langosteira	Fisterra	1970	26	rural	Medium		E	E	E	✓	✓		40.45	40.55	30.69	34.78
Espiñeirido	Muxia	250	15	rural	Low		E	E	E				25.94	26.02	59.19	62.3
A Cruz		70	100	rural	Medium		E	E	E							
Area da Vila		200	100	village	Medium	seafront, yacht club	P	P	P				10.24	9.65	11.59	13.59
Arou	Camariñas	130	50	rural	Medium	seafront	E	E	E				17.75	18.87	87.07	88.1
Camelle		150	105	village	Low	seafront	P	P	P							
Soesto	Laxe	850	30	remote	Low	surf	E	E	E				62.67	67.82	171.12	173.51
Playa Mayor	Malpica	378	60	urban	High	surf	S	S	P				15.27	15.77	18.64	18.14
Santa Cristina	Oleiros	1500	80	village	High		E	E	E			✓	123.35	123.18	1.35	2.35
Riazor	A Coruña	570	25	urban	High	promenade, recreational area	E	E	E	✓	✓	✓	31.6	31.95	68.47	72.07
Orzán		700	30	urban	High	promenade, recreational area	E	E	E	✓	✓	✓				
Matadero		80	20	urban	High	promenade	E	E	E	✓	✓	✓				
O Pedrido	Bergondo	500	40	urban	High	camping	S	G	G			✓	1619.52	1632.09	14.4	11.06

Table A1. *Cont.*

Beach Name	Municipality	Length	Width	Typology	Population Density	Tourism Facilities and Infrastructure	Water Quality 16	15	14	Blue Flag 16	15	14	Mid-Range of Tidal Flooding (m) T100	T500	Mid-Range of Wave Flooding (m) T100	T500
A Magdalena	Cedeira	1400	35	urban	High	recreational area	S	P	S				20.8	21.07	31.18	31.97
Santa Marta	Baiona	210	60-111	village	High	yacht club	G	G	G							
Ladeira		1650	28-68	village	High	dunes, camping	G	na	P	✓						
América		1300	50-100	village	High	recreational sport area	E	E	E		✓	✓	104.69	106.26	16.07	18.21
Panxón		1100	50-200	urban	High	promenade, yacht club	E	G	na		✓	✓				
Madorra	Nigrán	350	45	village	High		E	E	G							
Area Fofa		130	25	rural	Medium		E	G	E							
Patos		1400	25-80	village	High	seafront, sport area	E	E	E							
As Canas		150	25-80	rural	Low		S	S	S							
Muiños de Fortiñón	Vigo	120	60	rural	High		E	E	E	✓	✓	✓	45.69	45.2	20.66	23.2
O Portiño		130	30	village	Low		E	E	E							
A Sobreira		180	35	rural	Medium		E	E	E							
Arealonga	Redondela	180	35	rural	Medium		P	P	P							
A Punta		50	5-120	rural	High	seafront	E	E	E	✓	✓	✓	6.96	7.06	0.6	0.74
Bouzas	Vigo	450	25	village	Medium	seafront	E	E	E							
Portiño	O Grove	100	5-15	village	Medium	yacht club	G	S					27.69	33.6	3.48	1.67
O Bornal		400	20	village	Medium		E	E								
Terrón		500	15	rural	High		E	E								
As Brañas	Vilanova de Arousa	150	3	rural	Low		E	E					104.01	104.9	2.04	2.21
Con da Mina		200	10	village	Medium		E	E								
A Igrexa		250	6	rural	Low		E	E								

Table A2. Fluvial Flood Risk.

Beach Name	Municipality	Length	Width	Typology	Population Density	Tourism Facilities and Infrastructure	Water Quality 16	15	14	Blue Flag 16	15	14	Area of a Potential Fluvial Flooding (km²) T10	T100	T500	Volume of River Flow (m³/s) T10	T100	T500
Esteiro	Ribadeo	52	6-220	remote	medium	seafront	S	G	G				0.038	0.086	0.103	7.41	17.13	26.31
Arealonga	Barreiros	1000	45-170	village	medium	seafront	E	G	G				0.003	0.01	0.017	2.01	4.73	7.32
Covas *	Viveiro	1500	10-400	village	high	promenade, camping, ruiniform formations, recreational area	E	E	G	✓			0.003	0.006	0.009	7.66	17.15	25.99
Sardiñeiro	Fisterra	370	60	village	low	seafront	P	P	P				0.005	0.002	0.022	6	13.05	19.52

Table A2. *Cont.*

Beach Name	Municipality	Length	Width	Typology	Population Density	Tourism Facilities and Infrastructure	Water Quality 16	15	14	Blue Flag 16	15	14	Area of a Potential Fluvial Flooding (km²) T10	T100	T500	Volume of River Flow (m³/s) T10	T100	T500
Lires *	Cee	123	20	remote	medium		P	P	S				0.018	0.141	0.183	14.9	32.18	48
San Pedro *		250	160	village	low	seafront	G	G	E				0.012	0.016	0.019	4.66	8.99	12.71
San Mamede	Carnota	300	45	remote	low		E	E	E				0.009	0.02	0.037	6.7	14.48	21.6
Porto Cubelo		20	20	rural	low		E	E	E				0.008	0.017	0.022	2.51	5.55	8.38
A Magdalena *	Cedeira	1400	35	urban	high	recreational area	S	S	S				0.012	0.034	0.048	3.77	8.8	13.56
Arnela		100	20	urban	medium	seafront	na	P	P				0.002	0.004	0.007	3.03	6.23	9.08
Razo	Carballo	800	30	rural	high	dunes, marshland, fossil beach	G	S	S				0.037	0.067	0.083	14.6	31.02	45.9
A Concha	Ortigueira	750	30	village	medium	anchorage, cultural heritage	G	E	E	✓	✓	✓	0.016	0.038	0.051	15.5	22.7	28
Queiruga		1200	75	rural	medium		E	E	E				0.003	0.014	0.015	3.29	6.56	9.42
Pozo	Porto do Son	100	30	rural	medium		S	G	E				0.013	0.018	0.02	7.61	14.95	21.33
Coira		720	20	village	high	seafront	E	E	E				0.001	0.003	0.006	8.97	17.39	24.66
Ornanda		300	25	rural	high	1st category camping	E	E	E			✓	0.001	0.021	0.028	18.9	32.87	48.07
Bastiagueiro	Oleiros	500	100	village	high	seafront, recreational area, camping, bike path	E	E	E	✓	✓	✓	0.013	0.026	0.034	4.35	9.7	14.67
San Francisco		820	30	village	high	seafront, recreational area	E	E	E	✓	✓	✓	0.001	0.004	0.007	1.33	3.08	4.75
Ventin	Muros	160	30	rural	low	recreational area	E	E	E				0.008	0.012	0.012	2.3	5.05	7.58
Paramean		250	30	village	medium		G	G	G				0.007	0.013	0.015	11.4	22.12	31.44
Perbes	Miño	540	45	village	high	promenade, camping	E	E	E	✓	✓	✓	0.007	0.001	0.017	7.3	14.55	20.9
Coroso	Ribeira	1700	20	village	high	seafront, leisure port	E	E	E		✓	✓	0.043	0.069	0.089	8.95	20.34	31.01
Areal *	Pobra do Caramiñal	1100	26	urban	high	promenade, yacht club, recreational areas	G	G	G				0.13	0.232	0.27	20.4	42.22	61.76
Barrañan		1100	20	rural	high	seafront, dunes	E	E	E	✓	✓	✓	0.091	0.114	0.125	23.9	35.9	45
Alba-Sabón	Arteixo	850	30-80	remote	medium	seafront	E	E	E	✓	✓	✓	0.064	0.124	0.157	24.6	37.6	47.6
Samil	Vigo	1250	60	village	high	promenade, recreational area	E	G	S				0.043	0.866	1.034	90.5	138.5	175.5
Xunqueira	Moaña	500	50	village	high	promenade, recreational area	G	S	G				0.3	0.087	0.115	17.5	45.88	70.43
Santa Marta *	Baiona	210	60-111	village	high	yacht club	G	G	G				0.016	0.045	0.057	27	52.52	74.5
Arealonga *	Redondela	180	35	rural	medium		P	P	P				0.012	0.017	0.022	9.52	18.67	26.6
Loira	Marin	330	40	village	high	promenade, recreational area	G	S	S				0.008	0.016	0.022	22.9	34.8	43.9
Laño		585	15	village	low		S	S	P				0.001	0.004	0.006	9.14	17.95	25.59
Covelo	Poio	60	5	village	medium	promenade, anchorage	E	E	G				0.002	0.006	0.008	9.13	18.16	26.05
Chancelas peq.		140	8-30	village	medium	promenade	E	E	E				0.001	0.001	0.002	6.76	12.69	17.7

* Beaches at both risk, coastal-sea action and fluvial-river mouths. Sources: [66,69,86,99,100].

B. Calculation Procedure for Riazor beach, A Coruña Municipality

To illustrate the use of the index, beaches located in Areas of Potentially Significant Flood Risk (APSFR) are assessed against their tourism vulnerability. For this, the index considers the level of visitation, the characteristics of tourism facilities and infrastructures, as well as beach width. These are defined as:

- Level of visitation (*Lv*). Beaches score low to high (1–3), depending on use, with urban beaches being considered the most visited (score 3), those in the vicinity of a tourist destination being medium frequented (score 2) and remote beaches scoring 1. This indicator is double-weighted.
- Tourism facilities and infrastructure (*Tf*). Here, the highest score is given to camping, recreational areas or important cultural heritage, in addition to other infrastructures like promenade, seafront or marinas. A lower scoring (2) applies to beaches with less prominent infrastructure and the lowest scoring (1) to beaches without infrastructure.
- Beach width (*W*). The smaller the beach, the more vulnerable it is considered. The indicator score is 1 for beaches that are more than 75 m wide, score 2 for beaches between 15–75 m wide and score 3 for beaches with 15 m width or less.

The vulnerability index is calculated on the basis of the following equation:

$$Vt = \frac{(Lv \times 2) + Tf + W}{3}$$

To illustrate the procedure, the following section shows the calculation for Riazor beach, in the municipality of A Coruña (Rias Altas).

Indicator: Level of visitation (Lv)
- Typology (remote, rural, village or urban) [101]: urban
- Level of occupation: high
- Coastal tourist destination: yes
  The beach scores 3, given its significant importance for tourism.

Indicator: Tourism facilities and infrastructure (Tf)
- Attributes at risk: promenade, recreational area
- Blue flag certified
  The beach is scoring high again (3), as a result of its facilities and infrastructure.
  Assessment of the indicator: 3

Indicator: Width of the beach (W)
- Average width = 25 m.
  Here the beach scores 2.
  Based on the equation weighting the scores, Riazor beach has a vulnerability index of 3.7, i.e., it has a high vulnerability to flooding. (High vulnerability: $Vt \geq 3.7$; moderate: $2 \leq Vt < 3.7$ and low: $Vt < 2$).

References

1. Hall, C.M. Trends in ocean and coastal tourism: The end of the last frontier? *Ocean Coast. Manag.* **2001**, 44, 601–618. [CrossRef]
2. United Nations Environment Programme (UNEP). *Report on the Seminar "Coastal Tourism in the Mediterranean: Adapting to Climate Change", Cagliari, Italy, 8–10 June 2009*; United Nations Environment Programme (UNEP), République Française-Ministère de l'Écologie, du Développement et de l'Aménagement Durables, United Nations World Tourism Organization (UNWTO), Priority Actions Programme: Paris, France, 2009.
3. Hall, C.M. Global change, islands and sustainable development: Islands of sustainability or analogues of the challenge of sustainable development? In *Routledge International Handbook of Sustainable Development*; Redclift, M., Springett, D., Eds.; Routledge: Abingdon, UK, 2015; pp. 55–73.

4. Hoegh-Guldberg, O.; Beal, D.; Chaudhry, T.; Elhaj, H.; Abdullat, A.; Etessy, P.; Smits, M. *Reviving the Ocean Economy: The Case for Action–2015*; WWF International: Gland, Switzerland, 2015.

5. Scott, D.; Hall, C.M.; Gössling, S. *Tourism and Climate Change: Impacts, Adaptation and Mitigation*; Routledge: London, UK, 2012.

6. Intergovernmental Panel on Climate Change (IPCC). *Climate Change 2007: The Physical Science Basis*; Contribution of Working Group I to the Fourth Assessment Report of the Intergovernmental Panel on Climate Change; Cambridge University Press: Cambridge, UK, 2007.

7. Van Oldenborgh, G.J.; Collins, M.; Arblaster, J.; Christensen, J.H.; Marotzke, J.; Power, S.B.; Rummukainen, M.; Zhou, T. Annex I: Atlas of Global and Regional Climate Projections. In *Climate Change 2013: The Physical Science Basis. Contribution of Working Group I to the Fifth Assessment Report of the Intergovernmental Panel on Climate Change*; Stocker, T.F., Qin, D., Plattner, G.-K., Tignor, M., Allen, S.K., Boschung, J., Nauels, A., Xia, Y., Bex, V., Midgley, P.M., Eds.; Cambridge University Press: Cambridge, UK; New York, NY, USA, 2013.

8. Gössling, S.; Peeters, P.; Hall, C.M.; Dubois, G.; Ceron, J.P.; Lehmann, L.; Scott, D. Tourism and water use: Supply, demand, and security. An international review. *Tour. Manag.* **2012**, *33*, 1–15. [CrossRef]

9. Vitousek, S.; Barnard, P.L.; Fletcher, C.H.; Frazer, N.; Erikson, L.; Storlazzi, C.D. Doubling of coastal flooding frequency within decades due to sea-level rise. *Sci. Rep.* **2017**, *7*, 1399. [CrossRef] [PubMed]

10. Hallegatte, S.; Green, C.; Nichols, R.J.; Corfee-Morlot, J. Future flood losses in major coastal cities. *Nat. Clim. Chang.* **2013**, *3*, 802–806. [CrossRef]

11. Intergovernmental Panel on Climate Change (IPCC). Summary for policymakers. In *Climate Change 2014: Impacts, Adaptation, and Vulnerability. Part A: Global and Sectoral Aspects*; Contribution of Working Group II to the Fifth Assessment Report of the IPCC; Cambridge University Press: Cambridge, UK; New York, NY, USA, 2014; pp. 1–32.

12. Monbaliu, J.; Chen, Z.; Felts, D.; Ge, J.; Hissel, F.; Kappenberg, J.; Narayan, S.; Nicholls, R.J.; Ohle, N.; Schuster, D.; et al. Risk assessment of estuaries under climate change: Lessons from Western Europe. *Coast. Eng.* **2014**, *87*, 32–49. [CrossRef]

13. Government of Galicia. *Strategic Plan of Galicia, 2015–2020*; Government of Galicia: Galicia, Spain, January 2016. Available online: http://www.planestratexico.gal/es/descarga (accessed on 8 July 2016).

14. European Environment Agency (EEA). *Mapping the Impacts of Recent Natural Disasters and Technological Accidents in Europe*; Environmental Issue Report 35; EEA: Copenhagen, Denmark, 2004.

15. Risk Management Solutions. *Central Europe Flooding, Event Report*; Risk Management Solutions: Newark, CA, USA, August 2002. Available online: http://forms2.rms.com/rs/729-DJX-565/images/fl_2002_central_europe_flooding.pdf (accessed on 24 June 2016).

16. World Health Organization (WHO). *Floods in the WHO European Region: Health Effects and Their Prevention*; WHO Regional Office for Europe: Copenhagen, Denmark, 2013.

17. Olcina Cantos, J. Prevención de riesgos: Cambio climático, sequías e inundaciones. In *Fundación Nueva Cultura del Agua, Panel Científico-Técnico de Seguimiento de la Política de Aguas*; Convenio Universidad de Sevilla-Ministerio de Medio Ambiente: Seville, Spain, 2008.

18. Ministry of Agriculture and Fisheries, Food and Environment (MAPAMA). *Gestión de los Riesgos de Inundación*; Ministerio de Agricultura y Pesca: Alimentación y Medio Ambiente, Spain, 2016. Available online: http://www.mapama.gob.es/es/agua/temas/gestion-de-los-riesgos-de-inundacion/ (accessed on 24 June 2016).

19. United Nation Office for Disaster Risk Reduction (UNISDR). *Disaster & Risk Profile Spain. PreventionWeb, Project of UNISDR*; UNISDR: Geneva, Switzerland, 2016. Available online: http://www.preventionweb.net/countries/esp/data (accessed on 26 June 2016).

20. Frei, C.; Schöll, R.; Fukutome, S.; Schmidli, J.; Vidale, P.L. Future change of precipitation extremes in Europe: Intercomparison of scenarios from regional climate models. *J. Geophys. Res. Atmos.* **2006**, *111*. [CrossRef]

21. Feyen, L.; Dankers, R.; Bódis, K.; Salomon, P.; Barredo, J.I. Fluvial flood risk in Europe in present and future climates. *Clim. Chang.* **2012**, *12*, 47–62. [CrossRef]

22. Alfieri, L.; Burek, P.; Feyen, L.; Forzieri, G. Global warming increases the frequency of river floods in Europe. *Hydrol. Earth Syst. Sci.* **2015**, *19*, 2247–2260. [CrossRef]

23. Ciscar, J.C.; Iglesias, A.; Feyen, L.; Goodess, C.M.; Szabó, L.; Christensen, O.B.; Nicholls, R.; Amelung, B.; Watkiss, P.; Bosello, F.; et al. Climate change impacts in Europe. In *Final Report of the PESETA Research Project*; Office for Official Publications of the European Communities: Luxembourg, 2009.

24. Mojtahedi, S.M.H.; Oo, B.L. Coastal buildings and infrastructure flood risk analysis using multi-attribute decision-making. *J. Flood Risk Manag.* **2016**, *9*, 87–96. [CrossRef]
25. Aguas de Galicia. *Estudio Ambiental Estratégico. Plan de Xestión do Risco de Inundación da Demarcación Hidrográfica de Galicia-Costa (Ciclo 2015–2021)*; Aguas de Galicia: Santiago de Compostela, Spain, 2016.
26. United Nations World Tourism Organization (UNWTO). *Tourism Highlights, 2015 Edition*; UNWTO: Madrid, Spain, 2015. Available online: http://www.e-unwto.org/doi/pdf/10.18111/9789284416899 (accessed on 16 March 2016).
27. Instituto Nacional de Estadística (Spanish Statistic National Institute) (INE). Cuenta Satélite del Turismo de España. 2016. Available online: http://www.ine.es/dyngs/INEbase/es/operacion.htm?c=estadistica_C&cid=1254736169169&menu=ultiDatos&idp=1254735576863 (accessed on 9 February 2017).
28. Exceltur. Estudio de Impacto Económico del Turismo: IMPACTUR Galicia 2014. Available online: http://www.exceltur.org/wp-content/uploads/2016/02/IMPACTUR-Galicia-2014.pdf (accessed on 3 September 2016).
29. Gómez-Martín, M.B. Climate potential and tourist demand in Catalonia (Spain) during the summer season. *Clim. Res.* **2006**, *32*, 75–87. [CrossRef]
30. Rutty, M.; Scott, D. Will the Mediterranean become "too hot" for tourism? A reassessment. *Tour. Plan. Dev.* **2010**, *7*, 267–281. [CrossRef]
31. Olcina, C.; Vera-Rebollo, J.F. Climate change and tourism policy in Spain: Diagnosis in the Spanish Mediterranean coast. *Cuadernos de Turismo* **2016**, *38*, 323–571.
32. Amelung, B.; Nichols, S.; Viner, D. Implications of global climate change for tourism flows and seasonality. *J. Travel Res.* **2007**, *45*, 285–296. [CrossRef]
33. Hein, L.; Metzger, M.J.; Moreno, A. Potential impacts of climate change on tourism; a case study for Spain. *Curr. Opin. Environ. Sustain.* **2009**, *1*, 170–178. [CrossRef]
34. Moreno, A. *Turismo y Cambio Climático en España. Evaluación de la Vulnerabilidad del Turismo de Interior Frente a los Impactos del Cambio Climático. Ministerio de Agricultura*; Alimentación y Medio Ambiente: Madrid, Spain, 2010.
35. Vázquez, M.X.; Prada, A. Avaloración económica dos efectos do cambio Climático no Turismo. In *Evidencias e Impactos do Cambio Climático en Galicia*; Pérez, V., Fernández, M., Gómez, J., Eds.; Xunta de Galicia, Consellería de Medio Ambiente e Desenvolvemento Sostible: Santiago de Compostela, Spain, 2009; pp. 675–699.
36. Bujosa, A.; Rosselló, J. Climate change and summer mass tourism: The case of Spanish domestic tourism. *Clim. Chang.* **2013**, *117*, 363–375. [CrossRef]
37. Barrios, S.; Ibañez, J.N. Time is of the essence: Adaptation of tourism demand to climate change in Europe. *Clim. Chang.* **2015**, *132*, 645–660. [CrossRef]
38. Connor, R.F.; Hiroki, K. Development of a method for assessing flood vulnerability. *Water Sci. Technol.* **2005**, *51*, 61–67. [PubMed]
39. Balica, S.F.; Douben, N.; Wright, N.G. Flood vulnerability indices at varying spatial scales. *Water Sci. Technol.* **2005**, *60*, 2571–2580. [CrossRef] [PubMed]
40. Balica, S.F.; Wright, N.G. Reducing the complexity of Flood Vulnerability Index. *Environ. Hazards* **2010**, *9*, 321–339. [CrossRef]
41. Adger, W.N.; Hughes, T.P.; Folke, C.; Carpenter, S.R.; Rockstrom, J. Social-ecological resilience to coastal disasters. *Science* **2005**, *309*, 1036–1039. [CrossRef] [PubMed]
42. Kaly, U.; Briguglio, L.; McLeod, H.; Schmall, S.; Pratt, C.; Pal, R. Environmental Vulnerability Index (EVI) to summarise national environmental vulnerability profiles. *SOPAC Technol. Rep.* **1999**, *275*, 73.
43. McLaughlin, S.; Cooper, J.A.G. A multi-scale coastal vulnerability index: A tool for coastal managers? *Environ. Hazards* **2010**, *9*, 233–248. [CrossRef]
44. Balica, S.F.; Wright, N.G.; van der Meulen, F. A flood vulnerability index for coastal cities and its use in assessing climate change impacts. *Nat. Hazards* **2012**, *64*, 73–105. [CrossRef]
45. Kaján, E.; Saarinen, J. Tourism, climate change and adaptation: A review. *Curr. Issues Tour.* **2013**, *16*, 167–195. [CrossRef]
46. Nicholls, R.; Zanuttigh, B.; Vanderlinden, J.P.; Weisse, R.; Silva, R.; Hanson, S.; Narayan, S.; Hoggart, S.; Thompson, R.C.; de Vries, W.; et al. Developing a holistic approach to assessing and managing coastal flood risk. In *Coastal Risk Management in a Changing Climate*; Zanuttigh, B., Nicholls, R.J., Vanderlinden, J.P., Thompson, R.C., Burcharth, H.F., Eds.; Butterworth-Heinemann: Oxford, UK, 2015; pp. 9–53.

47. Shumann, A.H. Introduction-hydrological aspects of risk management. In *Flood Risk Assessment and Management. How to Specify Hydrological Loads, Their Consequences and Uncertainties*; Shumann, A.H., Ed.; Springer Science + Business Media: Dordtrecht, The Netherlands, 2011; pp. 1–11.
48. Williams, A.; Micallef, A. *Beach Management: Principles & Practice*; Earthscan: London, UK, 2009.
49. Peduzzi, P.; Dao, H.; Herold, C.; Rochette, D.; Sanahuja, H. *Feasibility Study Report—On Global Risk and Vulnerability Index-Trends per Year (GRAVITY)*; United Nations Development Programme Emergency Response Division UNDP/ERD: Geneva, Switzerland, 2001. Available online: http://www.grid.unep.ch/products/3_Reports/ew_gravity1.pdf (accessed on 24 January 2017).
50. Briguglio, L. *Methodological and Practical Considerations for Constructing Socio130 Economic Indicators to Evaluate Disaster Risk*; Programme on Information and Indicators for Risk Management, IADB-ECLAC-IDEA; Institute of Environmental Studies, University of Colombia: Manizales, Colombia, 2003.
51. Cendrero, A.; Sánchez-Arcilla, A.; Zazo, C. Impactos sobre las zonas costeras. In *Evaluación Preliminar de los Impactos en España por Efecto del Cambio Climático*; Universidad de Castilla-La Mancha, Ministerio de Medio Ambiente: Toledo, Spain, 2005; pp. 469–524.
52. Losada, I.; Izaguirre, C.; Diaz, P. *Cambio Climático en la Costa Española*; Oficina Española de Cambio Climático, Ministerio de Agricultura, Alimentación y Medio Ambiente: Madrid, Spain, 2014.
53. Davis, R.A., Jr.; FitzGerald, D.M. *Beaches and Coasts*; Blackwell: Oxford, UK, 2004.
54. Penning-Rowsell, E.C.; Parker, D.J.; de Vries, W.S.; Zanuttigh, B.; Simmonds, D.; Trifonova, E.; Hissel, F.; Monbaliu, J.; Lendzion, J.; Ohnle, N.; et al. Innovation in coastal risk management: An exploratory analysis of risk governance issues at eight THESEUS study sites. *Coast. Eng.* **2014**, *87*, 210–217. [CrossRef]
55. Craig-Smith, S.J.; Tapper, R.; Font, X. The coastal and marine environment. In *Tourism and Global Environmental Change: Ecological, Economic, Social and Political Interrelationships*; Gössling, S., Hall, C.M., Eds.; Routledge: London, UK, 2006; pp. 107–127.
56. Faulkner, B.; Vikulov, S. Katherine, washed out one day, back on track the next: A post-mortem of a tourism disaster. *Tour. Manag.* **2001**, *22*, 331–344. [CrossRef]
57. Wilks, J.; Moore, S. *Tourism Risk Management for the Asia Pacific Region: An Authoritative Guide for Managing Crises and Disasters*; Commonwealth of Australia: Canberra, Australia, 2003.
58. French, P.W. *Coastal Defenses. Process, Problems and Solutions*; Routledge: London, UK, 2001.
59. Lorenzo, F.; Alonso, A.; Pagés, J.L. Erosion and accretion of beach and spit systems in Northwest Spain: A response to human activity. *J. Coast. Res.* **2007**, *23*, 834–845. [CrossRef]
60. Pethick, J.; Crooks, S. Development of a coastal vulnerability index: A geomorphological perspective. *Environ. Conserv.* **2000**, *27*, 359–367. [CrossRef]
61. Gotangco, C.K.; See, J.; Dalupang, J.P.; Ortiz, M.; Porio, E.; Narisma, G.; Yulo-Loizaga, A.; Dator-Bercilla, J. Quantifying resilience to flooding among households and local government units using system dynamics: A case study in Metro Manila. *J. Flood Risk Manag.* **2016**, *9*, 196–207. [CrossRef]
62. Penning-Rowsell, E.C.; Ashley, R.; Evans, E.P.; Hall, J.W. *Foresight: Future Flooding: Scientific Summary, Vol 1: Future Risks and Their Drivers*; Department of Trade and Industry: London, UK, 2004.
63. Cendrero, A.; Fischer, D.W. A procedure for assessing the environmental quality of coastal areas for planning and management. *J. Coast. Res.* **1997**, *13*, 732–744.
64. Briguglio, L. The Vulnerability Index and Small Island developing states: A review of conceptual and methodological issues. In Proceedings of the AIMS Regional Preparatory Meeting on the Ten Year Review of the Barbados Programme of Action, Praia, Cape Verde, 1–5 September 2003.
65. Farrugia, N. Conceptual Issues in Constructing Composite Indices, Occasional Paper on Islands and Small States, 2/2007. In Proceedings of the International Conference on Small States and Economic Resilience, Valletta, Malta, 23–25 April 2007.
66. Galicia Tourism. Rías y Playas en Galicia. 2016. Available online: http://www.turismo.gal/localizador-de-recursos?filtro=%23STP:15&langId=es_ES (accessed on 3 August 2016).
67. Servicio Gallego de Salud [Galician Health Service] SERGAS. Censo de las Playas de Galicia. Programa de Vigilancia Sanitaria 2016. Available online: http://www.sergas.es/Saude-publica/Censo-praias-Galicia (accessed on 8 August 2016).
68. European Environment Agency (EEA). *European Bathing Water Quality in 2015*; EEA Technical report 9/2016; Publications Office of the European Union: Luxembourg, 2016.

69. Association of Environmental and Consumer Education (ADEAC). Bandera Azul 2016. Available online: http://www.adeac.es/bandera_azul (accessed on 14 October 2016).

70. Instituto Nacional de Estadística (Spanish Statistic National Institute) (INE) Encuesta de Ocupación Hotelera 2015. Available online: http://www.ine.es/jaxi/menu.do?type=pcaxis&path=%2Ft11%2Fe162eoh&file=inebase&L=0 (accessed on 24 July 2016).

71. Galicia Tourism. Area of Research and Study of Tourism of Galicia. 2016. Available online: http://www.turismo.gal/aei/portal/index.php (accessed on 20 September 2016).

72. Instituto de Hidráulica Ambiental de Cantabria (Institute of Environmental Hydraulics of Cantabria) (IHCantabria). Cambio Climático en la Costa Española, Proyecto C3E. Available online: http://www.c3e.ihcantabria.com/ (accessed on 4 March 2017).

73. Center for International Earth Science Information Network (CIESIN). *Urban-Rural Population and Land Area Estimates Version 2. Palisades*; Columbia University, NASA Socioeconomic Data and Applications Center (SEDAC): Washington, DC, USA, 2017. Available online: http://sedac.ciesin.columbia.edu/data/set/lecz-urban-rural-population-land-area-estimates-v2 (accessed on 13 March 2017).

74. Cabalar Fuentes, M. Los temporales de lluvia y viento en Galicia. Propuesta de clasificación y análisis de tendencias (1961–2001). *Investig. Geogr.* **2005**, *36*, 103–118. [CrossRef]

75. Cruz, R.; Lago, A.; Lage, A.; Rial, M.E.; Díaz-Fierros, F.; Salsón, S. Evolución recente do clima de Galicia. Tendencias observadas en variables meteorolóxicas. In *Evidencias e Impactos do Cambio Climático en Galicia*; Pérez, V., Fernández, M., Gómez, J., Eds.; Xunta de Galicia, Consellería de Medio Ambiente e Desenvolvemento Sostible: Santiago de Compostela, Spain, 2009; pp. 19–59.

76. Martínez de la Torre, A.; Míguez Macho, G. Modelización dun Escenario de futuro Cambio Climático en Galicia. In *Evidencias e Impactos do Cambio Climático en Galicia*; Pérez, V., Fernández, M., Gómez, J., Eds.; Xunta de Galicia, Consellería de Medio Ambiente e Desenvolvemento Sostible: Santiago de Compostela, Spain, 2009; pp. 543–556.

77. Xunta de Galicia. *Galicia Cambio Climático, Primer Informe. Consellería de Medio Ambiente y Ordenación del Territorio*; Government of Galicia, Santiago de Compostela: Galicia, Spain, 2012. Available online: http://cambioclimatico.xunta.gal/novas (accessed on 29 October 2016).

78. MeteoGalicia. Climatological Reports. Galician Meteorological Agency. Consellería de Medio Ambiente e Ordenación do Territorio da Xunta de Galicia. 2017. Available online: http://www.meteogalicia.gal/observacion/informesclima/informesIndex.action?request_locale=gl (accessed on 20 February 2017).

79. Agencia Española de Meteorología (Spanish Meteorological Agency) (AEMET). *Resúmenes Climatológicos*; AEMET: Madrid, Spain, 2016. Available online: http://www.aemet.es/es/serviciosclimaticos/vigilancia_clima/resumenes?w=0&datos=2 (accessed on 22 December 2016).

80. MeteoGalicia. Network Weather Stations. Available online: http://www.meteogalicia.gal/observacion/rede/redeIndex.action (accessed on 18 March 2017).

81. Puertos del Estado. *Informe Climático, Extremos Máximos de Oleaje Por Direcciones*; Banco de Datos Oceanográficos de Puertos del Estado; Ministerio de Fomento: Madrid, Spain, 2017.

82. Puertos del Estado. Banco de datos Oceanográficos. 2017. Available online: http://www.puertos.es/es-es/oceanografia/Paginas/portus.aspx (accessed on 16 February 2017).

83. Rosón, G.; Cabanas, J.M.; Fernández Pérez, F.; Herrera Cortijo, J.L.; Ruiz-Villarreal, M.; Castro, C.G.; Piedracoba, S.; Álvarez-Salgado, X.A. Evidencias do Cambio Climático na hidrografía e a dinámica das Rías e da plataforma Galega. In *Evidencias e Impactos do Cambio Climático en Galicia*; Pérez, V., Fernández, M., Gómez, J., Eds.; Xunta de Galicia, Consellería de Medio Ambiente e Desenvolvemento Sostible: Santiago de Compostela, Spain, 2009; pp. 287–303.

84. Cid, A.; Castanedo, S.; Abascal, A.J.; Menéndez, M.; Medina, R. A high resolution hindcast of the meteorological sea level component for Southern Europe: The GOS dataset. *Clim. Dyn.* **2014**, *43*, 2167–2184. [CrossRef]

85. MeteoGalicia. Tides and Moons. In *Galician Meteorological Agency*; MeteoGalicia: Santiago de Compostela, Spain, 2017. Available online: http://www.meteogalicia.gal/web/prediccion/maritima/mareasIndex.action (accessed on 23 March 2017).

86. Aguas de Galicia. Anexo 1 Caracterización das ARPSIS. In *Plan de Xestión do Risco de Inundación da Demarcación Hidrográfica de Galicia-Costa (ciclo 2015–2021)*; ARPSIS: Santiago de Compostela, Spain, January 2016.

87. Department of Culture and Tourism. Enquisa de destino 2009. In *Análise Estatística Sobre o Turismo en Galicia*; Secretaría Xeral para o Turismo: Xunta de Galicia, Spain, 2010.
88. Ministry of Agriculture and Fisheries, Food and Environment (MAPAMA). *Aplicación de la Directiva de Inundaciones y del R.D. 903/2010 en la Costa Española. Costas y Medio Marino*; Ministerio de Agricultura y Pesca, Alimentación y Medio Ambiente: Madrid, Spain, 2017. Available online: http://www.mapama.gob.es/es/costas/temas/proteccion-costa/directiva-inundaciones/ (accessed on 5 March 2017).
89. Ministry of Agriculture and Fisheries, Food and Environment (MAPAMA). *Actuaciones para la Protección de la Costa. Ministerio de Agricultura y Pesca*; Alimentación y Medio Ambiente: Madrid, Spain, 2017. Available online: http://www.mapama.gob.es/es/costas/temas/proteccion-costa/directiva-inundaciones/ (accessed on 6 March 2017).
90. BBVA Foundation. La población en España: 1900–2009. Fundación BBVA, 2010. Available online: http://www.fbbva.es/TLFU/dat/cuadernos_FBBVA_51espana_web.pdf (accessed on 12 March 2017).
91. Pompe, J.J.; Rinehart, J.R. Beach quality and the enhancement of recreational property values. *J. Leis. Res.* **1995**, *27*, 143–154.
92. Pérez-Alberti, A.; Vázquez-Paz, M. Caracterización y dinámica de sistemas dunares coseros en Galicia. In *Las dunas en España*; Sanjaume, E., Gracia, F.J., Eds.; Sociedad Española de Geomorfología: Zaragoza, Spain, 2011; pp. 161–187.
93. Willis, C.M.; Griggs, G.B. Reductions in fluvial sediment discharge by coastal dams in California and implications for beach sustainability. *J. Geol.* **2003**, *111*, 167–182. [CrossRef]
94. United Nations Office for Disaster Risk Reduction (UNISDR). DesInventar, Disaster Information Management System. Disaster & Risk Profile Spain. 2017. Available online: http://www.desinventar.net/DesInventar/profiletab.jsp?countrycode=esp&continue=y (accessed on 5 March 2017).
95. Instituto Nacional de Estadística (Spanish Statistic National Institute) (INE). Gasto total y Gastos Medios por Comunidad Autónoma de Residencia. 2017. Available online: http://www.ine.es/jaxiT3/Datos.htm?t=10721 (accessed on 3 April 2017).
96. Hall, C.M.; Prayag, G.; Amore, A. *Tourism and Resilience. Dividual, Organisational and Destination Perspectives*; Channel View: Bristol, UK, 2018.
97. Orchiston, C. Tourism business preparedness, resilience and disaster planning in a region of high seismic risk: The case of the Southern Alps, New Zealand. *Curr. Issues Tour.* **2013**, *16*, 477–494. [CrossRef]
98. Jopp, R.; DeLacy, T.; Mair, J. Developing a framework for regional destination adaptation to climate change. *Curr. Issues Tour.* **2010**, *13*, 591–605. [CrossRef]
99. Ministry of Agriculture and Fisheries, Food and Environment (MAPAMA). Guía de Playas. Ministerio de Agricultura y Pesca, Alimentación y Medio Ambiente, Spain, 2016. Available online: http://www.mapama.gob.es/es/costas/servicios/guia-playas/default.aspx (accessed on 17 July 2016).
100. European Environment Agency (EEA). *State of Bathing Waters*; European Environment Agency: Copenhagen, Denmark, 2016. Available online: http://www.eea.europa.eu/themes/water/interactive/bathing/state-of-bathing-waters#parent-fieldnametitle (accessed on 30 June 2016).
101. Williams, A. Definitions and typologies of coastal tourism beach destinations. In *Disappearing Destinations. Climate Change and Future Challenges for Coastal Tourism*; Jones, A., Phillips, M., Eds.; CAB International: Wallingford, UK, 2011; pp. 47–65.

environments

MDPI

Article

Testing Extended Accounts in Scheduled Conservation of Open Woodlands with Permanent Livestock Grazing: Dehesa de la Luz Estate Case Study, Arroyo de la Luz, Spain

Pablo Campos [1,*], Bruno Mesa [1], Alejandro Álvarez [1], Francisco M. Castaño [2] and Fernando Pulido [2]

[1]   Spanish National Research Council (CSIC), Instituto de Políticas y Bienes Públicos (IPP), C/Albasanz, 26–28, E-28037 Madrid, Spain; bruno.mesa@cchs.csic.es (B.M.); alejandro.alvarez@cchs.csic.es (A.Á.)
[2]   Institute for Dehesa Research (INDEHESA), Universidad de Extremadura, 06071 Badajoz, Spain; franmcmartin@hotmail.com (F.M.C.); nando@unex.es (F.P.)
*   Correspondence: pablo.campos@csic.es; Tel.: +34-91-602-2535

Received: 22 September 2017; Accepted: 10 November 2017; Published: 15 November 2017

Abstract: Standard Economic Accounts for Agriculture and Forestry do not measure the ecosystem services and intermediate products embedded in the final products recorded, and omit the private non-commercial intermediate products and self-consumption of private amenities. These limitations of the standard accounts are addressed by the extended Agroforestry Accounting System, which is being tested at the publicly-owned Dehesa de la Luz agroforestry estate. The extended accounts simulate conservation forestry of holm oak and cork oak for the current as well as successive rotation cycles during which scheduled conservation of the cultural woodland landscape of the Dehesa de la Luz is carried out, improving the natural physical growth of the firewood and cork. The estimated results for 2014 reveal that private ecosystem services make up 50% of the firewood and grazing products consumed; the private environmental income accounts for 13% of the total private income; and the private environmental asset represents 53% of the total opening capital. The net value added is more than 2.3 times the amount estimated using the standard accounts. The landowner donates intermediate products of non-commercial services at a value of 85 €/ha, which are used to enhance the supply of public products.

Keywords: Agroforestry Accounting System; standard accounts; private ecosystem services consumed; private intermediate services; private environmental income and asset; private profitability rate; public products

---

1. Introduction

The sustainable management of the Spanish dehesa is important to rural development at the local, national, and European Union levels due to its environmental and economic value [1–4]. Open woodlands in five autonomous communities in West and Central Spain predominate over an area of 6,151,318 ha (Table 1 and Figure 1) [5]. Open holm oak woodland accounts for 73% of Spanish open woodland in the five main dehesa regions. In the absence of statistics from the government regarding public dehesas, we estimated the extent of publicly-owned Mediterranean open woodlands where the fraction of tree cover is between 5% and 75%. These public open woodland formations occupy 738,615 ha and represent 12% of the estimated total area of open woodland formations in the Spanish dehesa area (Table 1). Most of these open woodlands do not form part of dehesa estates [6]. This agroforestry system is defined as an anthropogenic land use system based mainly on extensive livestock grazing in the Mediterranean woodlands, shrublands, and grasslands, where more than 20%

of the area of the estate is occupied by broadleaved species with a canopy cover fraction of between 5% and 60% [7] (p. 7). Spanish dehesa agroforestry estates cover a total area of 3,606,154 ha and the open woodlands within them account for 2,203,002 ha (Table 2 and Figure S9) [7,8]. The natural conditions and medieval process of land appropriation have led to the concentration of most of the dehesa areas within large dehesa estates. For example, the 4575 dehesa estates of 200 hectares or more represent 64% of the total dehesa area, with an average estate size of 502 ha. The remaining 107,812 dehesa estates with less than 200 ha have an average estate size of 12 ha (Table 2).

Table 1. Open woodland area in the five autonomous communities in West and Central Spain (ha).

Tree Species	Andalucía	Castilla-La Mancha	Castilla y León	Extremadura	Madrid	Total
Holm oak	1,302,901	1,019,286	676,305	1,353,119	119,848	4,471,460
Cork	207,101	21,724	6753	138,334	190	374,102
Other oaks	27,158	156,562	662,997	91,069	20,033	957,819
Others [1]	118,637	105,871	103,116	7593	12,721	347,937
Open woodlands [2]	1,655,796	1,303,443	1,449,171	1,590,115	152,792	6,151,318

Notes: [1] Others includes Spanish juniper, wild olive, narrow-leaved ash, and carob tree. [2] Open woodlands are between $CCF_{trees} \geq 5\%$ and $CCF_{trees} \leq 75\%$, where CCF (canopy cover fraction) is the stand area covered by the tree canopies. Includes the stand ages of polewood and old growth only. Source: Own elaboration based on Reference [5].

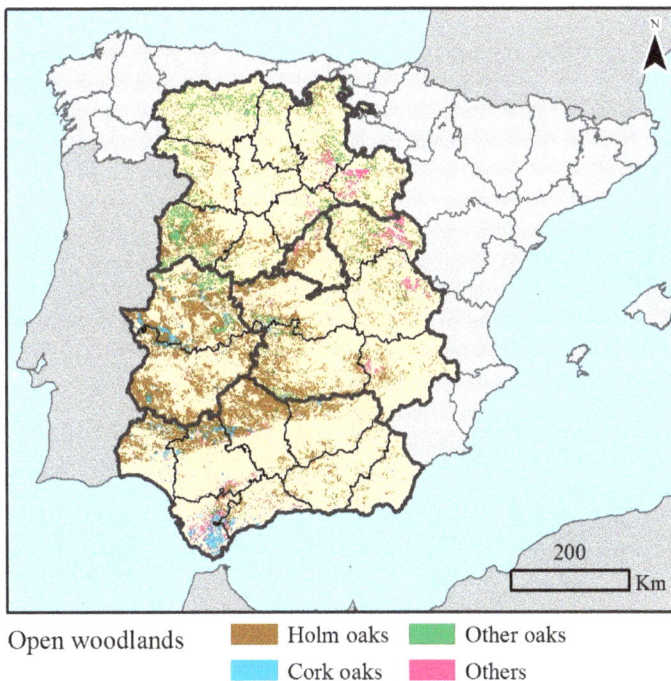

Figure 1. Map of open woodlands in the five autonomous communities in West and Central Spain by tree species. Source: Own elaboration based on Reference [5].

In the Iberian Peninsula, stakeholders have warned of the economic and environmental consequences of the decline of Spanish dehesas [6,7,9–11]. This concern is also present in the Portuguese montados, where woodlands cover 1,066,000 ha (holm oak: 329,000 ha and cork oak: 737,000 ha), mainly in the Alentejo region [12,13].

Table 2. Numbered classification of dehesa estates according to surface area in the five Spanish autonomous communities in West and Central Spain.

Dehesa Estates Size Class (ha)	Dehesa Estates Number		Area of Dehesa States			
			Open Woodland		Total	
	N°	%	ha	%	ha	%
0 < ha ≤ 10	87,395	78	102,611	5	152,867	4
10 < ha ≤ 50	12,015	11	183,203	8	287,939	8
50 < ha ≤ 100	4612	4	209,429	10	330,672	9
100 < ha ≤ 150	2322	2	177,758	8	285,042	8
150 < ha ≤ 200	1468	1	161,912	7	253,716	7
200 < ha ≤ 300	1698	2	265,382	12	416,935	12
300 < ha ≤ 500	1521	1	373,223	17	582,026	16
500 < ha ≤ 1.000	979	1	394,791	18	658,528	18
ha > 1.000	377	0	334,693	15	638,429	18
Total	112,387	100	2,203,002	100	3,606,154	100
0 < ha ≤ 10	87,395	78	102,611	5	152,867	4
10 ≤ ha ≤ 200	20,417	18	732,302	33	1,157,369	32
ha > 200	4575	4	1,368,089	62	2,295,918	64

Source: Modified from Reference [7] (Table 23, p. 46).

The deficiency or complete lack of natural regeneration has been identified as the key problem in current dehesa open woodland management. Current grazing levels reveal that, in general, both landowners and governments overlook the question of compatibility with woodland regeneration and that today, after several centuries of inadequate grazing management, the dehesas are suffering an ongoing process of increasing natural death rates due to diseases and ageing of the trees. The regeneration of trees in Spanish dehesas is either null or scarce in 46–70% of plots and normal or abundant in 28–45% according to data from the Third National Forest Inventory for five autonomous communities with dehesas [5,7,8]. The data which reflect this general lack of regeneration are accentuated if the analysis is restricted to the tree species which most frequently form part of the livestock farming estates (holm oak, cork oak, and Pyrenean oak), with a lack of woody seedlings in 82%, 96%, and 65% of each species, respectively, in the inventoried plots [10].

The Economic Accounts for Agriculture and Forestry (EAA/EAF) is the government statistical office regulation for accounting the final products and net value added from the agriculture and forestry sector market [14]. The omission of economic statistics for dehesas from standard national EAA/EAF (hereinafter standard accounts) prevents us from determining the contribution of these agroforestry estates to the provincial and regional economies of the autonomous regions in which they are most widespread (Tables 1 and 2, Figure 1 and Figure S9). The EAA/EAF estimates the aggregated sales of products classified as agricultural and forestry products produced in the territory at the regional or national scale without distinction between types of enterprise. Thus, agroforestry estates are not a separate part of the agricultural and forestry product statistics. For example, in the case of livestock production, no distinction is made between whether it is produced in a grazing system or in an industrial feedlot. The only economic information available on dehesas is that which has been published in scientific articles relating to a small number of large dehesas in the Spanish communities of Andalucia and Extremadura. Data from the scarce scientific publications regarding testing of the Agroforestry Accounting System (AAS) (hereinafter extended accounts) in a group of large private dehesas coincide with those from studies conducted using the same extended accounts in Mediterranean ranches in California [15,16] (see Supplementary Materials (SM) 1–6).

We follow the reference [17] and define products (or outputs) as goods (tangible products) and services (intangible products) produced in the accounting period in the estate for current or future consumption by people. The products measured in the Dehesa de la Luz case study are: cork and firewood natural growths, firewood harvest, acorns, grass, stored water flow, intermediate and final private services, manufactured gross capital formation of plantation and dry-stone wall, livestock products, private amenity, recreation, landscape, livestock biodiversity, and carbon. We measure the environmental assets and/or manufactured capitals of the abovementioned products and the environmental asset of hunting.

The objective of the management at Dehesa de la Luz, undertaken by the public owner and scheduled in this study, is to reach the highest potential consumption of the goods and services produced, subject to the condition that the net worth (see SM 2) of environmental assets are not diminished at the closing of the accounting period. Achieving this objective requires the continued future presence of private forestry activities, animals, and services simultaneously, both in space and/or sequentially over time. In this study, our purpose is to examine the assumed hypothesis of economic rationale of the public landowner and the leasehold family livestock owners with regard to the supply of private manufactured intermediate products of services (hereinafter intermediate services). The framwork of our case study are as follows: the owners receive a normal manufactured net operating margin (operating benefit) from their manufactured capital investments in forestry conservation, livestock, and infrastructures, with which they produce intermediate services (ISS) to be used up as own intermediate consumption of services (SSo) for public and private activities. These services contribute to the economic activities of Dehesa de la Luz as inputs to the final products of public recreation, landscape (including human-made historical-cultural legacy), and threatened livestock biodiversity activities as well as private amenity activities. Thus, we hypothesize that the intermediate services which we attribute to private activities explain the investment rationale of the public landowner, as well as the employment and investment by leasehold family livestock owners who decide to continue with their private activities despite incurring net manufactured monetary operating margins at basic prices (after including subsidies and taxes linked to the production process) below the normal margins of the market for alternative investment options (see SM 4).

The objectives of this study are to test the monetary extended accounts for the year 2014 of the individual private activities analyzed and for Dehesa de la Luz as a whole, the ecosystem services, the intermediate services, the environmental income, the environmental assets, the total private income and its factorial distribution, and the total private profitability rates [18,19]. With regard to previous applications of the extended accounts, the novelty of the present study is that it attempts to illustrate, through the real case study of Dehesa de la Luz, the estimation of intermediate service values hidden and omitted by the standard accounts, and to assign them to the activities which produce them as well as to the public and private activities which consume them as own intermediate consumption (inputs).

2. Materials and Methods

The case study of Dehesa de la Luz is presented in this section, which deals with the modeling of physical growth functions for tree volume, firewood, cork, and acorns (Section 2.1) and the specific valuation criteria applied for each of the private products estimated in 2014 (Section 2.2). The previously published development of extended accounts is not presented in the main text of this study [4,16,19–21], although we summarize the key accounting concepts in the Supplementary Materials (see SM 1–4, 8–9).

Primary data have been collected through field work at Dehesa de la Luz by following ad hoc protocols in forestry, livestock, and infrastructure products and costs. Reference is made to appropriate published data where such data have been used.

2.1. Dehesa de la Luz Public Ownership Case Study

The town council of Arroyo de la Luz (Cáceres province, Spain) (Figures 2 and 3), as public owner of Dehesa de la Luz, aims to establish conservation-orientated management of forest resources and threatened autochthonous livestock breeds based on scientific information as well as to improve the offer of public products in Dehesa de la Luz. Determining the total private income and total private capital of the individual activities at Dehesa de la Luz allows us to estimate their economic profitability rates and, where appropriate, justify the compensation received from the government in reciprocity for the contribution to the intermediate services which are re-employed as input in the supply of public products at Dehesa de la Luz. The public landowner mitigates the manufactured (man-made) investment risk and favors the supply of public products, leasing livestock grazing for most of the cattle rearing activity (with the exception of pure breeds in danger of extinction), as well as pruning firewood from holm oaks by self-employed family labor. Small game hunting, with no investment by either the landowner or hunters, is practiced by members of the local hunting society [19,22].

In this Dehesa case study, we use 'number of trees' (and omit tree biomass) because the holm oak and cork oak are not commercial wood species and acorn and periodical cork harvests are the main products, with the silviculture undertaken being similar to that of fruit tree management (cork is a periodically harvested product from the same cork oak trees). In addition, we want to highlight the quantity of plantation trees vs natural regeneration. The inventories performed in 2014 over a total area of 978 ha of Dehesa de la Luz reveal that 93% of the area is occupied by holm oaks (*Quercus ilex* L.), with a small number of cork oaks (*Quercus suber* L.) dispersed among the former (Table 3).

Sixty percent of the trees originate from natural regeneration and the remaining 40% are young trees from recent plantations. Fifty three percent of the latter trees are cork oaks planted in 1993 and the other 47% are holm oaks planted in 1993 and 2014 (Figure 4). Seven percent of the non-wooded area includes parts which are occupied by paths, roads, water courses, and pools as well as infrastructures currently used for livestock management.

Over more than 50 years, the holm oaks and cork oaks from natural regeneration in Dehesa de la Luz have diminished by 17% (see SM 5). The diameter distribution of the adult trees reveals marked ageing of the woodland. This situation led the public owner, having recently regained ownership of the trees, to initiate the recovery of the holm and cork oaks, impoverished by excessive pruning carried out by the local private owners of the trees over more than a century (a local private societal enterprise ("Sociedad Forestal") bought the holm oaks and the cork oaks in the 1880s and managed them up until the 1990s, when the regional government of Extremadura bought the trees and donated them to de municipality of Arroyo de la Luz, which previously had the ownership of grass and agricultural uses) (Figure 4). This involved the mixed plantation of holm oaks and cork oaks in 1993 and the plantation (densification) of holm oaks in 2014 (Figure 5).

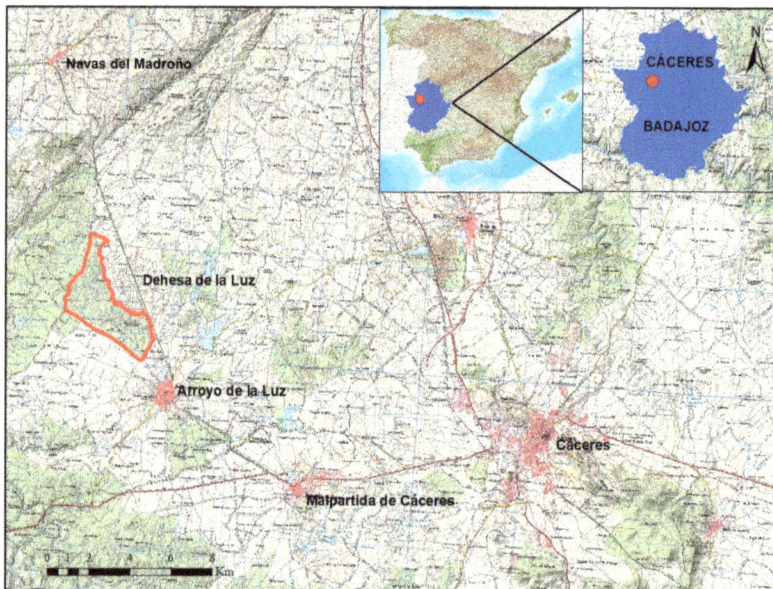

Figure 2. Location map of Dehesa de la Luz estate.

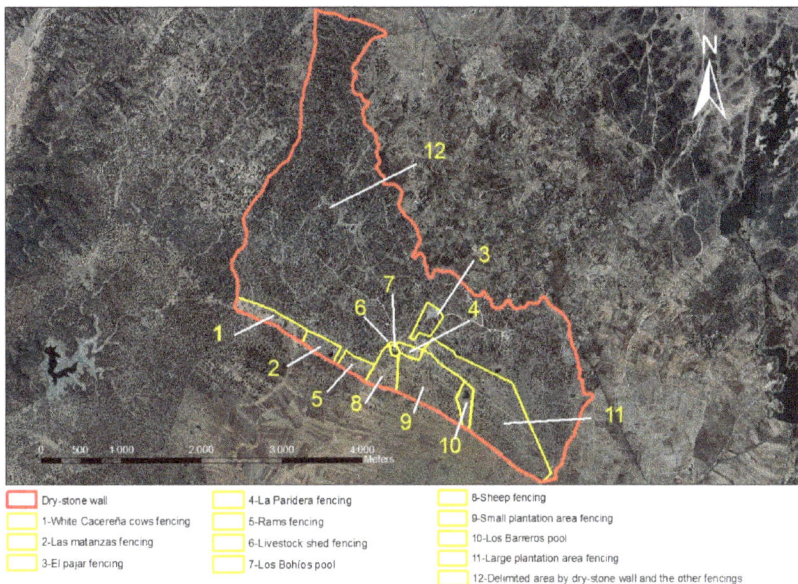

Figure 3. Orthophoto of Dehesa de la Luz.

Table 3. Open woodlands and other land use areas in Dehesa de la Luz estate (2014).

Class	N [1]	Canopy Cover Fraction (%)	Area [2] (ha)	(%)
1. Wooded area	47,968	19	909	93
1.1 Natural regeneration	29,007	17	756	77
Holm oaks	28,248	17	756	77
Cork oaks	759	0		
1.2 Plantation	18,961	27	153	16
Holm oaks	8895	21		
Cork oaks	10,066	6	153	16
2. Non-wooded area			69	7
Pools			11	1
Other			58	6
3. Total			978	100

Notes: [1] Number of trees; [2] Area assigned to the main species (that which has the greatest number of trees).

The distribution of the trees by diameter class allows us to verify that the holm oaks from natural regeneration, of more than 25 cm, make up more than a third of the total. If all the holm oaks are considered, including those which were planted, 68% have a diameter at breast height (Db) of more than 25 cm (Table 4).

Thus, according to this diameter distribution, most of the holm oaks present in Dehesa de la Luz are more than 60 years old and the natural regeneration is insufficient to replace the existing woodland. In the case of cork oaks, 86% of those originating from natural regeneration have a diameter of more than 25 cm, the opposite being the case for the total population, since 95% have a diameter of less than 25 cm due to the quantity of cork oaks distributed throughout the estate and the density of the plantation carried out in 1993 (Table 4).

Table 4. Density per species, origin, and diameter class in Dehesa de a Luz (2014).

Diameter Class (cm)	Holm Oaks (Trees)			Cork Oaks (Trees)			Total (Trees)
	Natural Regeneration	Plantation	Total	Natural Regeneration [1]	Plantation	Total	
5–25	5959	5758	11,717	82	10,018	10,100	21,817
30–50	8563		8563	224	48	272	8835
55–75	9309		9309	199		199	9508
80–100	6011		6011	66		66	6077
105–125	1437		1437	12		12	1449
130–150	106		106	3		3	109
155–175				1		1	1
Total	31,385	5758	37,143	587	10,066	10,653	47,796

Notes: [1] Does not include 172 trees without branches, those which on which the diameter is impossible to measure, and the lesser non-inventoriable trees.

Figure 4. Holm oak pruning in Dehesa de la Luz. Photograph: Daniel González.

Figure 5. Old holm oaks and young holm densification in Dehesa de la Luz estate. Photograph: Daniel González.

In the inventory conducted at the opening of the 2014 accounting period, there was a greater presence of bovine livestock belonging to family livestock owners, making up 77% of the census

(Table S6). Bovine rearing involves producing calves for sale after weaning at an age of between five and seven months. They graze during the whole year in the leased enclosures, and there are jointly owned Limousin studs for mating with the cows. The bovine belonging to family livestock owners are a cross with foreign breeds: Charolaise, Friesian, and Limousin [22], although there are also some pure Retinta cows.

The landowner has autochthonous livestock species, including black Merina sheep, white Cacereña cows, and Cordobes donkeys, although there are also pure breed foreign species such as Rambouillet Merina sheep and Hispano-bretón mares (Figures 6, 7 and S7). Of all the livestock belonging to the public landowner, the Rambouillet Merina sheep are the most numerous, comprising 70% of their livestock. Regarding the type of livestock on the estate, bovine make up the largest percentage (79%) of the total, followed by ovine (18%) (Table S6).

Figure 6. Endangered Cacereña White cow rearing in Dehesa de la Luz. Photograph: Daniel González.

Figure 7. Endangered Black Merina Sheep. Photograph: Fernando Pulido.

Recreational hunting for small game is leased by the public landowner to the local hunting society. However, in practice, no resource rent is paid for hunting, as the hunters deem the resource to have a reduced market value due to the scarcity of game species captured. Although the hunting product value may be modest, it has been incorporated by discounting its expected resource rent from the estimated market price of the land.

At Dehesa de la Luz there are legacy-cultural values such as the presence of archaeological sites dating from pre-Roman times and the medieval era, as well as contemporary cultural-historical constructions such as Roman-Visigothic tombs (Figures S10 and S11), the 18-km dry-stone wall which encloses the whole estate occupied by the Dehesa de la Luz (Figure 8), and a stone shepherds' hut (Figure S8). The Ermita de la Virgen de la Luz sanctuary is also a noteworthy construction situated within the estate, although the public economic services associated with it have not been addressed in our valuation on this occasion (Figure S12).

Figure 8. Historical cultural legacy dry-stone walling. Photograph: Daniel Gónzalez.

2.2. Modeling Natural Growth of Trees and Extracted Products

2.2.1. Forest Stand Inventories

The models for holm oak and cork oak production functions are estimated for the full cycle of the woodland on the basis of the existing adult trees in 2014 from natural regeneration, young trees planted in 1993, and the densification in 2014.

Modeling the production functions starts with an inventory of 34 plots, a stem count of the scarce number of adult cork oaks dispersed among the holm oaks and in 20 reforested plots in the area occupied by the 1993 plantation.

The site is divided into six forest stands, allowing a detailed analysis based on the physical and geographic characteristics of the woodland (see SM 5). Using this management division, mortality between the years 1956 and 2010 was analyzed through orthophotos and geographical information system (GIS) software, which allowed us to determine the existence of trees in each year and the potential occurrence of regeneration (see SM 5). This field data provides the basis for modeling the future conservation forestry schedule. Based on the estimated tree volume growth, firewood pruning and cork stripping rotations, and the mortality and commercial cycles of the trees, it is possible to schedule the conservation forestry for the expected future growth and extracted products at Dehesa de la Luz. The schedule is designed in accordance with the estimated area, location, and year of intervention as well as the type of activity or treatment to be applied (see SM 5). Natural growth and extractions are estimated by physical functions fitted to the environmental characteristics and woodland management of the Dehesa de la Luz case study [19] (see SM 5).

2.2.2. Holm Oak: Tree Volume, Firewood Natural Growth, and Acorn Yield Functions

Calculating the full production cycle of the holm oak involves using the functions for estimated age (Ae) and volume (V) to calculate physical natural growth (ng), based on the measurements carried out in the inventories, the age functions developed by Reference [23], and the official databases of the National Forest Inventory (NFI). This cycle is assumed to correspond to the point at which the power function of the growth based on estimated volume tends towards asymptotic curve, estimating the point of tangency between the linear (ng_{linear}) and power (ng_{power}) functions in order to select the forestry cycles for assisted holm oak landscape regeneration (see SM 5):

$$ng = 29.5437 \cdot (Db)^{0.8156} \cdot \left(\frac{1}{Ae + 72.9785} \right) \tag{1}$$

$$ng_{linear} = -0.0002 \cdot V + 1.930 \tag{2}$$

$$ng_{power} = 2.905 \cdot V^{-0.084} \tag{3}$$

where Db is the breast height diameter in cm, Ae is the estimated age in years, and V is the volume in dm^3.

The models for annual holm oak firewood product ($P_{firewood}$) are estimated in kg, based on the measurement of a pilot pruning of 30 holm oaks representative of the diameter classes recorded in the estate inventory. This model serves to calculate firewood growth according to the models developed to estimate the total volume of holm oak based on the measurements taken and the functions used in the second National Forest Inventory (NFI) in the province of Caceres [24]. Based on this estimate, it is possible to determine the time period necessary to replace extracted firewood between two consecutive pruning operations without exceeding the accumulated growth since the previous pruning. This period is the minimum rotation between two consecutive pruning operations (see SM 5). Only holm oak firewood is considered, as pruning is not performed on cork oaks.

$$P_{firewood} = 0.6661 \cdot Db^{1.3314} \tag{4}$$

The function for acorn production from adult holm oaks originating from natural regeneration ($P_{acornnr}$) is estimated in kg by modeling the count of cupules on the floor at the end of the 'montanera' (Iberian pig fattening period) in the months of December and January, over three consecutive seasons (2013–2104 to 2015–2016). To estimate the acorn production function for young, planted holm oaks, the acorn yield model developed by Reference [25] is applied.

$$P_{acornnr} \equiv F(Cca, Da, Wa) \tag{5}$$

where Cca is the tree canopy cover area, Da is the average density of acorns per square meter, and Wa is the average weight of the acorns.

2.2.3. Cork Oak: Tree Volume, Cork Natural Growth, and Acorn Yield Functions

Given the average age of cork oaks and the scarce number of inventoried adult trees, it is not possible to obtain an acceptable production cycle using the algorithms applied to the holm oak; hence, a maximum cycle for cork stripping of 150 years is used [26].

The estimate of the cork production function considers the inventories conducted stem by stem and the areas with planted trees. The model used to estimate cork yield (Pc) is taken from Reference [27]. The results obtained using this model are contrasted with the data from the last cork stripping in 2010 (these extractions being performed the same year for all the cork oaks dispersed throughout the estate), obtaining similar yield results. It is assumed that cork growth is linear during the period between stripping, with the debarking rotation (td) applied in Dehesa de la Luz being 10 years.

The cork oak acorn yield function uses a different model for the plantation cork oaks and the natural regeneration adult cork oaks. In the case of the young cork oaks, the fitted model for cork oak acorn yield published by Reference [28] is used. To calculate acorn yield in adult cork oaks, a coefficient is estimated which relates the mean yield obtained in young holm oaks and cork oaks.

2.2.4. Carbon Uptake

The carbon stored through the sink effect of the woodland is calculated using models developed by References [29,30], based on volume and growth measurements performed in the inventories (see SM 5). These models allow aboveground, large root, and fine root carbon to be measured both in holm oaks and cork oaks.

2.3. Private Activity Economic Valuation Criteria

The economic valuation uses the extended accounts. The novelties incorporated are described below, referring to conceptual aspects divulged in publications [4,16,19–22] as well as to Supplementary Materials concerning the methodological details and some of the extended accounts application methods employed at Dehesa de la Luz.

Concepts and equations for products and costs are described in Supplementary Materials sections 1. The reader should consult the tables and Supplementary Materials for detailed explanations on accounting variables. Measuring the total private economic value of an agroforestry estate in a consistent manner using social income theory may be an impossible task [18,21]. We conducted exhaustive data collection at the estate itself to value the multiple private economic values currently consumed. Some public services have been estimated according to their public landowner production cost. The latter price is the EAA/EAF valuation criterion for non-market goods.

2.3.1. Forestry Activities

Private forestry activities in this study are classified into manufactured (human-made) conservation forestry activity (CF), cork, firewood, and grazing (grass and acorns) activities. The CF products include intermediate services and the final product of gross formation of fixed manufactured capital (GFCF) from tree plantations, the replacement of failed plantations, and densification (Figure 4). The cork and firewood products only incur ordinary costs of raw materials, services, and work in progress used in the course of extractions, and their products are natural growth, cork stripping, and holm oak firewood. Manufactured CF also enhances the natural growth of cork and firewood as well as acorn yield, increasing the value of these environmental assets.

The natural growths of cork and firewood over the period are final products classified as gross production-in-progress formation in the supply side of the production account and are registered as entries to the production-in-progress environmental assets of the capital account for the same period. The current inventories of holm oaks and cork oaks are fixed environmental assets of biological resources, valued according to the discounted expected future resource rents of cork and firewood from harvesting rotations beyond the current one. Future trees (not yet existent) which will replace those of current cycles also generate fixed cork and firewood environmental assets classified as land. The sum of the three types of environmental assets of natural growths of cork and firewood comprises the total value of their environmental assets. These valuation approaches avoid double accounting when measuring total environmental incomes and assets of cork and firewood (see SM 6 and [21]).

Grazing only incurs the ordinary cost of ploughing (Figure S6) (see the development of estimates for full production cycles of holm oak and cork oak in SM 6).

2.3.2. Water Storage Activity

The main function provided by the pools, wells, and springs is that of supplying water as an intermediate raw material product (hereinafter intermediate water) for livestock drinking troughs, although a secondary use for some pools in certain months is the rearing of tench (Figure 9). The value of the intermediate water is the ordinary cost of production (maintenance cost and ordinary consumption of fixed capital), plus the normal return from immobilized manufactured fixed capital (pools). Water pools are valued at their market replacement production cost, corrected by a factor which takes into account the state of conservation of each individual pool. The physical intermediate water and consumption were not measured.

Figure 9. Pool made with compacted soil. Photograph: Daniel González.

2.3.3. Livestock Activity

Regarding field data collection for livestock, little difficulty is involved in the physical inventory at the start of the accounting period, entries, withdrawals, and valuation of commercial products. Self-employed family labor is valued by the residual method (see SM 8) [4,16,22,31,32] if there is a positive net operating margin for livestock. If the latter is negative, that is, a monetary loss for the family livestock activity, we assume that there is a positive trade-off against a self-consumed intermediate service (ISSnca) by family livestock owners (see SM 8). This ISSnca is considered an input of own intermediate consumption of service (SSo) of the family livestock private amenity [16]. In this case study, we did not measured the total product of amenity activity, but rather their SSo. Thus, the potential environmental income for livestock is not measured.

2.3.4. Intermediate Services of Infrastructure Activities

Service activities include fencing and other infrastructures, footpaths for the public visitors, and the dry-stone perimeter wall, given its public service function as a cultural landscape with historical constructions.

The same valuation criteria as those used for the water services are followed for the livestock infrastructure services. The main function of the fencing, access gates, livestock infrastructures, and main gates is that of livestock management. Sanitation management and livestock foodstuff storage require the use of infrastructures (sheds, tanks, stables and portable troughs). The fencing and the dry-stone perimeter wall produce commercial and non-commercial intermediate services which are consumed by the livestock activity and the public landscape activity, respectively. Infrastructure service activity in 2014 also saw the final product of the dry-stone perimeter wall improvements (manufactured gross fixed capital formation) valued at restoration cost.

The roads, paths, and bridges for the free public access are mainly used for public recreational activity in Dehesa de la Luz. Hence, their construction is suitable for vehicle and pedestrian access. There is a public right of way for access to the Ermita de la Luz sanctuary.

The dry-stone perimeter wall serves the same purpose as the fencing as well as providing a public service given its cultural-historical interest (Figure 8). The concept of cultural-historical value of a fixed-capital manufactured asset refers precisely to its condition as an ancient man-made construction and as such it is assumed that citizens wish to contribute to its maintenance costs in return for using the services provided by its existence in its current state of conservation. This cultural-historical asset has survived to the present day in a partially complete state as regards the historic construction, with broken parts of the wall having been replaced with stone and construction materials. The cost incurred includes maintenance work, investment in restoration, and consumption of fixed capital of post-2004 restoration works at replacement cost. The value per cubic meter of stone wall is assumed to be the market value of its restoration. The market price of construction weighted by a correction factor that takes into account the current state of conservation is used. The capital value of the dry-stone wall is divided among the livestock activity, considering the equivalent linear meters of wire fencing, and the remaining capital value is attributed to the cultural-historical service provided by the dry-stone perimeter wall. The value of the intermediate service is estimated using the same criteria as those for the livestock and public recreational infrastructure, although the capital in this case is estimated according to the cost of the quasi-restoration of stone work weighted by a correction coefficient of 0.6.

2.3.5. Private Amenity Activity

The family livestock owners' private amenity product is valued in accordance with their production cost, thus obtaining a null net operating margin. However, the value of the environmental asset of the private amenity embedded in the market price of the land is estimated based on available published information [4,16,33].

2.4. Public Activity Valuation Criteria

We measured the imputed market value of the product, cost, and change of net worth associated with greenhouse effect carbon (environmental asset revaluation in this case). Other public activities are final services and these are not valued at simulated market price, but at public landowner production cost. Firstly, we registered the conservation forestry, livestock, dry-stone wall, and roads that produce intermediate services to be used as inputs (own intermediate consumption) by free access public recreation, option value of landscape services, and existence value of the threatened livestock biodiversity service. Secondly, we registered the respective public activities as inputs of own intermediate consumption of services (SSo). Finally, as standard accounts criterion apply, we assumed the value of public services to be equal to their SSo production cost.

Public profitability denotes the ratio between the benefits (capital income) and the immobilized capital (average annual capital invested in the economic activity) of public activities. To estimate public benefit, we needed to measure the public products at simulated market prices [4,21]. To determine this latter value, we needed to employ several non-market valuation techniques based on consumer preferences.

2.4.1. Public Recreation, Landscape, and Threatened Livestock Biodiversity Activities

In this case study of Dehesa de la Luz, we omitted the valuation of public services produced by the simulated market price criterion which consumers are willing to accept to finance the private costs of the landowners and the livestock owners, as well as the direct costs to the government for the management of public activities. Due to the omission of the public willingness-to-pay criterion, it is not possible to determine the true product values of the public activities considered in Dehesa de la Luz. The valuation of the free access for public recreation (Figure 10), landscape, and threatened livestock biodiversity is conducted using the valuation criterion of private 'own' services costs incurred by the public landowner, family livestock owners, and family foresters.

Figure 10. Main free access entrance gate at Dehesa de la Luz. Photograph: Daniel Gónzalez.

2.4.2. Carbon Activity

The only exception to the valuation of public services at production cost is that of carbon, which is valued at simulated market price. The carbon service involves estimating the fixation and environmental consumption service of 2014 carbon dioxide emissions from the firewood consumption and their revaluation in future cycles of the woodland, which is consistent with the standard economic-environmental accounting criteria (SEEA-CF) and the extended accounts valuation criterion [4,32,34].

We estimated the environmental income from carbon stored in the trees by the variation in capital values between the opening and closing of the 2014 accounting period. This variation in net carbon assets is equivalent to the sum of the carbon environmental net operating margin and the environmental gains. The margin is calculated as the difference, over the period, between the values for carbon fixation from natural growth of firewood and cork, and the equivalent emissions, estimated from the firewood extracted in pruning operations and natural mortality in the woodland in 2014. The environmental asset gain is estimated by the revaluation of the carbon environmental asset, adjusted by the deduction of the expected fixation value at the opening of the period. The total environmental asset of carbon is recorded as fixed capital land (FClce). This environmental asset has two components: first, the carbon fixed by trees in the current production cycle, and second, the carbon that is expected to be fixed or emitted in successive production cycles. These production cycles were simulated according to silvicultural models.

2.5. Private Ecosystem Services

Ecosystem services are classified in this study according to the International Classification of Ecosystem Services (CICES) [35] and defined as 'the contributions of ecosystems to benefits (products:

goods and services) used in economic and other human activity' [17] (p. 19, para. 2.23). Ecosystem services can be intermediate or final, depending on the classification of products in which they are embedded. The SEEA-EEA technical guidelines clarify the latter criterion: 'There is common misunderstanding of the role of classifications with regard to the distinction between final and intermediate ecosystem services. Put simply, it is not the case that ecosystem services must be neatly classified between those that contribute directly to economic and social beneficiaries and those that support the ongoing functioning of ecosystems. For example, when water is extracted from a lake it would be considered final if the beneficiary was a household but intermediate if consumed by wild deer' [36] (p. 53, para. 5.33).

The private ecosystem services refer to the embedded contributions as natural production factor inputs to the values of the total products consumed from the landowner's private forestry activities at Dehesa de la Luz in 2014. The absence of cork stripping explains the null value for ecosystem services of cork consumed (the natural growth of cork and firewood are not consumed) in 2014, as their contribution is taken into account in the accounting period in which they are extracted (consumed) (for methodological details see [19]).

2.6. Intermediate Products of Services

If the operating benefits of manufactured investments in conservation forestry, livestock, and infrastructures for public services (recreation and cultural legacy) according to standard accounts are lower than the normal in an alternative investment, then business-as-usual investment theory states that the aforementioned activities are not competitive investments. Our hypothesis provides a solution for this unexplained occurrence which does not fit into currently accepted investment theory. We assume that the land and livestock owners obtain non-commercial intermediate services, which entail that the owners receive competitive operating benefits (manufactured net operating margin at basic prices). Our extended accounts measure the hidden donated and self-consumed intermediate services that are omitted in the standard accounts measurements of intermediate services.

In this research, our extended agroforestry accounts incorporate the intermediate products of services (intermediate services) in a manner which is consistent with the SEEA-CF [34] and SEEA-EEA [17] methodologies; although, in the latter, an ongoing approach to establish a standard for institutional sectors of the ecosystem accounts has not been agreed on. Our extended Agroforestry Accounting System (extended accounts) adopts a novel development to the SEEA-EEA model B accounts [17] (p. 134, Section 6.3.2 and p. 144, Annex A6), [21] (p. 28, SM. Eq. (3.1)), [4] (p. 50). In this model B, the ecosystem is considered as a factor of the production function of the individual products and the ecosystem is not an additional institutional sector, as treated in model A of the accounts [37] (p. 13), [17] (p. 17, para. 2.13).

Agroforestry ecosystems potentially produce environmental intermediate products of raw materials and services (although we term the latter 'intermediate services' for simplicity) with the absence of human labor and manufactured capital inputs in their production function, and more generally, they supply multiple manufactured intermediate products. The latter necessarily incorporate the values of human labor and manufactured capital contributions which, along with the contribution from the natural environment, can potentially provide embedded values of ecosystem services estimated by the residual valuation method. We underline the fact that the estimated intermediate services for conservation forestry, livestock, and infrastructures activities are manufactured (human-made) intermediate services.

Intermediate products are goods and services produced on the agroforestry estate that are used during the same period in which they are generated as own intermediate consumptions (inputs) by the same activity that produces them (intra-consumption) or by other activities (inter-consumption) on the same estate for the generation of the final products of the period. The classification and valuation of the intermediate products and the individual intermediate consumptions coincide by definition and, where all the products of the estate on which they are produced and consumed are considered, their entire aggregate value also coincides.

Intermediate products are valued at the prices of formal markets or, in the absence of formal transactions, simulated markets. It is assumed in the imputed prices of the individual donated and self-consumed intermediate products that they correspond to the normal opportunity cost of the immobilized manufactured investment in the production of the individual intermediate product. Opportunity cost is defined by the total ordinary cost plus a normal manufactured net operating margin (see SM 4 for details of its calculation).

The assumed hypothesis of a continuing investment by the landowner in the activities of conservation forestry, livestock, and infrastructures with recreation and legacy services must reflect the achievement of a normal profit (manufactured net operating margin). Establishing sufficient evidence of obtaining a persistent profit margin in the medium term from manufactured margin at producer prices below the norm for the mentioned activities is justified by the omission in the standard accounts of the non-commercial intermediate services for the individual private activities previously mentioned.

The intermediate services are classified into commercial and non-commercial categories. The latter are noted as 'compensated' by the government (these are conventional operating and capital subsidies), donated by the public landowner, and the self-consumption of private amenities by the leasehold family livestock owners (these are used as input in the supply of self-consumed private amenity products). The public landowner aims to encourage the intermediate services to promote the supply of public products, accepting a lower private monetary manufactured net operating margin against a benefit in the form of non-commercial intermediate services for donations. The family livestock owners accept a lower private monetary manufactured net operating margin against a benefit in the form of non-commercial intermediate services for private amenity self-consumption. The family livestock owner benefit from his investment in livestock rearing is characterized by the acceptance of lower or zero compensation from self-employed family labor and, occasionally, a negative private monetary manufactured net operating margin from manufactured investment (excluding the land).

3. Results

3.1. Physical Assets and Yields of Forestry

3.1.1. Open Woodlands Condition and Expected Future Improvement Trends

Table 5 presents the scheduling for full cycles of conservation forestry for the proposed cultural landscape valued as the final environmental asset, indicating future interventions and the rotation period applied (see SM 5). Regarding this future horizon, if the proposed future plantations and interventions continue as scheduled (Table 5), the product, growth, and other parameters representative of the forest species present in the estate will increase.

The estimated average age of the adult holm oaks and cork oaks at Dehesa de la Luz in 2014 was 165 ($\pm$4.2) and 109 ($\pm$3.6) years, respectively. In 2014, the density of the naturally regenerated holm oaks and cork oaks was more than double that of the planted trees, reaching a similar density by 2100 (Table 6). The canopy cover (CCF) of the estate circa 2014 was 19%, increasing to 31% by 2100 with the planned conservation forestry schedule. The conservation forestry cycles estimated for holm oak and cork oak are 225 and 150 years, respectively. For the pruning of holm oak firewood, a rotation period of 41 years has been established, which is compatible with the growth of the holm oaks, and which will be reduced to 27 years once the currently existing aged trees have gone. In the case of cork stripping, the current rotation period of 10 years is maintained.

Table 5. Schedule of the future assisted regeneration of holm oaks and cork oaks at Dehesa de la Luz.

Forest Stand	Plot (ha)	Production Pruning Next (year)	Production Pruning Period [1] (years)	Densification Next (year)	Densification Period (years)	Formative Pruning Next (year)	Formative Pruning Period (years)	Replacing Failed Plants Next (year)	Replacing Failed Plants Period (years)	Debarking Next (year)	Debarking Period (years)	Regeneration Felling Next (year)	Regeneration Felling Period (years)	Grazing Delimitation Next (year)	Grazing Delimitation Period (years)
1	21	2029	41	2120	110	2121	110	2015	110	2020	10				
1	18.8	2028	41	2120	110	2121	110	2015	110	2020	10				
1	24.7	2027	41	2120	110	2121	110	2015	110	2020	10				
1	26	2026	41	2120	110	2121	110	2015	110	2020	10				
2	23.5	2025	41	2014	35	2015	35	2019	35	2020	10				
2	21.1	2030	41	2014	35	2015	35	2019	35	2020	10				
2	21.3	2031	41	2014	35	2015	35	2019	35	2020	10				
2	23.2	2045	41	2022	35	2023	35	2027	35	2020	10				
3	22.1	2046	41	2022	35	2023	35	2027	35	2020	10				
4	24.8	2039	41	2018	205	2019	205	2023	205	2020	10				
4	28.8	2040	41	2026	210	2027	210	2031	210	2020	10				
4	23.7	2041	41	2026	210	2027	210	2031	210	2020	10				
4	22.8	2042	41	2026	210	2027	210	2031	210	2020	10				
4	22.2	2043	41	2030	210	2031	210	2035	210	2020	10				
4	16.4	2047	41	2022	210	2023	210	2027	210	2020	10				
4	19.2	2048	41	2022	210	2023	210	2027	210	2020	10				
4	19.1	2049	41	2030	210	2031	210	2035	210	2020	10				
4	21.2	2050	41	2030	210	2031	210	2035	210	2020	10				
4	21.8	2051	41	2030	210	2031	210	2039	210	2020	10				
4	27.6	2052	41	2034	210	2035	210	2039	210	2020	10				
4	20.8	2055	41	2034	210	2035	210	2039	210	2020	10				
4	20.5	2056	41	2034	210	2035	210	2039	210	2020	10				
5	23.8	2020	41	2034	210	2035	210	2031	210	2020	10				
5	20	2023	41	2026	225	2027	225	2043	225	2020	10				
5	20.4	2024	41	2026	225	2027	225	2043	225	2020	10				
5	22.9	2044	41	2038	225	2039	225	2043	225	2020	10				
5	19.2	2053	41	2038	225	2039	225	2043	225	2020	10				
5	21.8	2054	41	2038	225	2039	225	2047	225	2020	10				
5	27.2	2016	41	2038	225	2039	225	2047	225	2020	10				
5	17.7	2017	41	2042	225	2043	225	2047	225	2020	10				
5	16.5	2018	41	2042	225	2043	225	2047	225	2020	10				
5	24.3	2019	41	2042	225	2043	225	2051	225	2020	10				
5	26.1	2021	41	2042	225	2043	225	2051	225	2020	10				
5	26.1	2022	41	2046	225	2047	225			2020	10				
R	24.6	2032	41			2159	150			2023	10	2144	150	2144–2164	130
R	23	2033	41			2159	150			2023	10	2144	150	2144–2164	130
R	21.8	2034	41			2159	150			2023	10	2144	150	2144–2164	130
R	12.5	2035	41			2159	150			2023	10	2144	150	2144–2164	130
R	7.7	2035	41			2159	150			2023	10	2144	150	2144–2164	130
R	16.9	2036	41			2159	150			2023	10	2144	150	2144–2164	130
R	21	2037	41			2159	150			2023	10	2144	150	2144–2164	130
R	25.2	2038	41			2159	150			2023	10	2144	150	2144–2164	130

Notes: [1] The pruning period is 41 years until all the trees from natural regeneration have been replaced by planted trees, when the pruning period becomes 27 years. Soil tilling is carried out over the area where pruning was performed the previous year.

Table 6. Projection of the future condition and supply of the main products of holm oaks and cork oaks at Dehesa de la Luz.

Class	Average Age (years)			Density (Trees)			Acorn Production (t)			Cork Growth (t)			Firewood Growth (m³)			Canopy Cover Fraction (%)		
	Year 2014	Year 2050	Year 2100	Year 2014	Year 2050	Year 2100	Year 2014	Year 2050	Year 2100	Year 2014	Year 2050	Year 2100	Year 2014	Year 2050	Year 2100	Year 2014	Year 2050	Year 2100
1. Natural regeneration	163	200	250	32,144	29,681	27,054	147.4	154.0	156.2	1.9	2.7	3.0	63.0	52.1	36.3	17	19	19
Holm oak	165	201	251	31,385	29,202	26,646	146.2	152.7	154.9				63.0	52.1	36.3	17	18	18
Cork oak	109	145	195	759	479	408	1.2	1.3	1.3	1.9	2.7	3.0				0	0	1
2. Plantation	19	40	81	15,824	26,475	26,384	9.3	56.5	195.5	4.4	10.4	23.1	2.4	19.5	46.3	2	5	12
Holm oak	15	30	69	5758	17,244	18,198	4.2	35.8	158.1				2.4	19.5	46.3	1	3	7
Cork oak	21	57	107	10,066	9231	8186	5.1	20.6	37.4	4.4	10.4	23.1				1	3	5
3. Total	117	124	166	47,968	56,156	53,438	156.6	210.5	351.7	6.3	13.1	26.1	65.4	71.6	82.6	19	24	31

3.1.2. Physical Natural Growth and Extractions of Firewood, Cork, Acorn, and Grass

The pruning of holm oaks in 2014 took place over an area of 19 ha. The annual growth of firewood accounts for 45% of the firewood extracted, which is due, in the first place, to the fact that extractions carried out in 2014 were larger than the accumulated growth since the last pruning. This was firstly because the holm oak firewood was extracted beyond the maximum cycle established in this study, and secondly due to dead holm oak firewood being extracted (estimated at 37% of the amount of green firewood extracted and making up 23% of the total in 2014). Table 7 shows the values for the growth and extraction of firewood, valued at stumpage price per ton.

The grazing price of acorn and grass (including browse) was estimated to be 0.035 €/forage unit (FU) at Dehesa de la Luz in 2014 [19,22]. Grass and acorn make up 87% and 13%, respectively, of the total grazing value (Table 7). The acorn yield per tree, obtained using the cupules count model, is below that expected for holm oaks of that diameter due to the ageing of adult trees and the excessive pruning that has taken place in the past.

Grazing (including acorn, grass and browse) are the main forestry activity raw material at Dehesa de la Luz. The value of cork growth at Dehesa de la Luz is 6% that of grazing.

Table 7. Annual products of wood, cork, acorns, and grass at Dehesa de la Luz (2014).

Class	Unit (u)	Yield (u/100 Trees)	Quantity (u)	Price (€/u)	Value (€)
Firewood extraction	t	19.4	147.0	3.7	538.0
Annual firewood growth	t	0.2	65.7	2.1	139.8
Annual cork growth	kg	58.4	6325.7	0.3	2198.8
Grazing consumption	100 FU *		8234.9	3.5	28,723.6
Grass and browse	*100 FU*		*7131.5*	*3.5*	*24,875.0*
Acorn	*100 FU*	*2.4*	*1103.4*	*3.5*	*3848.6*

* FU: Physical forge unit represents a kilogram of barley with humidity of 14.1% which provides a content of 2.723 kcal/kg DM (dry matter) of metabolisable energy.

3.1.3. Carbon Uptake

The value of carbon fixation by holm oaks is almost four times that of cork oaks, while the carbon emissions from holm oaks are more than 10 times greater due to the quantity of firewood extracted in 2014. Due to the quantity of carbon emissions from the holm oaks, the net fixation value is negative, whereas in the case of the cork oaks it remains positive (Table S9).

3.1.4. Livestock Grazing

The 2014 accounting period total metabolic energy requirements of the landowner's and family's livestock that feed on the Dehesa de la Luz estate is estimated to be 1013.7 FU/ha. Eighty three percent of these energy requirements are provided by grazing, while the remaining 17% comes from the provision of supplementary foodstuff. In the case of family livestock owners, the accounting period total physical energy requirements of the livestock are estimated at 794 FU/ha.

Eighty two percent of these family's livestock energy requirements are met by grazing and the remaining 18% corresponds to supplementary foodstuff. For the landowner's livestock, the requirements are estimated at 219.4 FU/ha, with 88% of that coming from grazing and 12% supplementary foodstuff (Table 8).

Regarding the different livestock, bovine consume 796 FU/ha, of which 81% is grazed and the other 19% is supplemented. Equine consumption is estimated at 39 FU/ha, of which 94% corresponds to grazing and 6% to supplementary foodstuff (Table 8).

The total price of the feed consumed is estimated at 0.074 €/FU, that of the family livestock owners being double that of the landowner (Table 9) [19,22]. There is no marked difference in the prices of supplementary foodstuff (Table 9).

Table 8. Livestock grazing and supplementary foodstuff consumption in Dehesa de la Luz (2014: FU/ha).

| Class | Forage Units (FU) | | | | |
| | Grazing | | | Supplements | Total |
	Grass and Browse	Acorn	Total		
1. Family livestock owners	561.8	86.9	648.7	145.5	794.3
1.1 Bovine	547.7	84.7	632.5	143.8	776.2
1.2 Equine	14.1	2.2	16.3	1.8	18.1
2. Landowner	167.6	25.9	193.5	25.9	219.4
2.1 Ovine	140.9	21.8	162.6	15.7	178.3
Rambouillet Merina	129.0	20.0	148.9	14.3	163.3
Black Merina	11.9	1.8	13.7	1.3	15.0
2.2 Bovine	8.8	1.4	10.2	9.6	19.8
2.3 Equine	17.9	2.8	20.7	0.6	21.3
Total	729.4	112.8	842.2	171.5	1013.7

Table 9. Price of grazing and supplementary foodstuff by owner and livestock type at Dehesa de la Luz (2014: €/100 FU).

Class	Grazing	Supplements	Total
1. Family livestock owners	4.2	27.2	8.4
1.1 Bovine	4.0	27.2	8.3
1.2 Equine	12.1	29.5	13.8
2. Landowner	1.1	22.4	3.6
2.1 Ovine	0.7	26.4	3.0
Rambouillet Merina	0.7	26.4	3.0
Black Merina	0.7	26.4	3.0
2.2 Bovine	0.0	15.7	7.6
2.3 Equine	4.5	25.0	5.1
Total	3.5	26.5	7.4

3.2. Selected Physical Capital and Product Indicators per Livestock Type

The number of calves born to each reproductive female is higher for the landowner than for the family livestock owners. In contrast, the number of equine births is greater among the family livestock owners than for the landowner. The fertility rate of the two ovine breeds differs moderately, the figure being 0.7 for the Rambouillet Merina and 0.8 in the case of the Black Merina sheep (Table S7).

The sale of calves per reproductive female is greater in the case of the landowner than for the family livestock owners. The ratio of calf sales to births is 78% in the case of the landowner and 71% for the family livestock owners. Concerning ovine livestock, the ratio of sales to births is 43% for the Rambouillet Merina and 38% for the black Merina. The equine livestock belonging to the landowner had a sales-to-births ratio of 33%, while in the case of the family livestock owners, no sales of foals were made during the accounting period.

Table S7 shows the average prices used per livestock type and owner for the different livestock product valuations. In the case of calf sales, it can be seen that the landowner's price is higher than that of the family livestock owners.

3.3. Selected Economic Indicators of Private Activities at Dehesa de la Luz

Table 10 presents the main accounting identities used in the estimation of income, total capital, and private yield rates in the case study of Dehesa de la Luz [4,16,18,20,32].

Table 10. Intermediate services, ecosystem services, immobilized capital, incomes, and profitability rates for selected identities of the extended accounts.

Class	Identities
Intermediate services (ISS)	ISS = ISSc + ISSnc
Ecosystem services consumed (ES)	ES = TPc − ICmo − LCo − CFCo − NOMmo
Net value added (NVA)	NVA = TP − IC − CFC
Net operating margin (NOM)	NOM = TP − TC
Labor cost (LC)	LC = LCe + LCse
Opening capital (Co)	Co = WPo + FCo
Capital revaluation (Cr)	Cr = Cc + Cw − Co − Ce
Capital gains (CG)	CG = Cr − Cd + Cad
Capital income (CI)	CI= NOM + CG
Total income (TI)	TI = NVA + CG
Environmental income (EI)	EI = TI − LC − CIm
Resource rent (RR)	RR = ENOM + WPeu − NGe
Immobilized capital (IMC)	IMC = Co + WC
Operating profitability (o)	o =NOM/IMC
Capital gain profitability (g)	g = CG/IMC
Current profitability (r)	r = CI/IMC

Abbreviations: ISSc: commercial intermediate services. ISSnc: non-commercial intermediate services. TPc: total product consumption. ICmo: ordinary manufactured intermediate consumption. LCo: ordinary labor cost. CFCo: ordinary consumption of fixed capital. NOMmo: ordinary manufactured net operating margin. TP: total product. IC: intermediate consumption. CFC: consumption of fixed captial. TC: Total cost. LCe: labor cost employees. LCse: labor cost self-employed. WPo: opening work in progress. FCo: opening fixed capital. Cc: closing capital. Cw: capital withdrawals. Ce: capital entries. Cd: capital destructions. Cad: capital adjustments. CIM: manufactured capital income. NOMe: environmental net operating margin. WPeu: environmental work in progress used. NGe: environmental natural growth. WC: working capital.

3.3.1. Net Value Added

Estimating the private net value added for both the public landowner and the leasehold family livestock owners at Dehesa de la Luz is extremely complex and somewhat controversial in the academic sphere as well as in national accounting offices (Table 11). Although no landowner private amenity product is consumed, the service activity is that which contributes most to the total net value added (NVA) at Dehesa de la Luz, as is the case at large private dehesas [4,32]. This is due to the allocation of the intermediate product of infrastructure services at a value of 3% of the immobilized capital (Table 11 and Table S10). In 2014, there was large own investment in the restoration of the dry-stone wall.

The next most important contributors to the NVA after services are the forestry activity (Table 11 and Table S2) and livestock (Table 11 and Table S8). As for water activity, this relates to the intermediate product of services and the amortization of livestock drinking water infrastructures (Table 11). The labor income is concentrated in livestock rearing (31 €/ha), firewood (13 €/ha), and conservation forestry (12 €/ha)

The incomes from employee labor and self-employed family labor account for similar quantities (Table 11).

Table 11. Owners and government production account for Dehesa de la Luz (2014: €/ha).

Class	Forestry	Water	Livestock	Services	Amenity	Owners	Recreation	Landscape	Carbon	Biodiversity	Public	Total
	1	2	3	4	5	6 = Σ 1 a 5	7	8	9	10	11 = Σ 7 a 10	12 = 6 + 11
1. Total product (TP)	128	19	300	185	18	650	15	119	3	5	142	792
1.1 Intermediate product (IP)	56	19	57	99		230						230
Intermediate raw materials (IRM)	34	19				53						53
Intermediate services (ISS)	22		57	99		178						178
Commercial (ISSc)	22			20		42						42
Non-commercial (ISSnc)	1		57	79		136						136
Compensated (ISSncc)			34			34						34
Donated (ISSncd)	1		5	79		85						85
Amenity (ISSnca)			18			18						18
1.2 Final product (FP)	72		244	86	18	420	15	119	3	5	142	562
Sales (FPs)	17		74			91						91
Gross fixed capital formation (GFCF)	52		28	86		167						167
Gross work in progress formation (GWCPF)	2		142			144						144
Autoconsumption (FPa)					18	18						18
Public goods and services (PCS)							15	119	3	5	142	142
2. Total cost (TC)	101	4	292	107	18	521	15	119	2	5	142	663
2.1. Intermediate consumption (IC)	69		260	86	18	433	15	119	2	5	142	575
Raw materials (RM)	35		100			135						135
Bought (RMb)	31		52			83						83
Own (RMo)	4		48			53						53
Services (SS)	33		30	86	18	167	15	119	2	5	142	309
Bought (SSb)	33		10	86		129						129
Own (SSo)	0		20		18	38	15	119	2	5	140	178
Environmental (SSe)											2	2
Work in progress used (WPu)	1		131			131						131
2.2 Labor cost (LC)	25		31			57						57
Employees (LCe)	12		18			30						30
Self-employed (LCse)	13		14			27						27
2.3 Consumption of fixed capital (CFC)	7	4		20		32						32
3. Net operating margin (NOM)	27	15	8	79		129					0	129
4. Gross valued added (GVA)	59	19	40	99		217					0	217
5. Net valued added (NVA)	52	15	40	79		185					0	185

3.3.2. Intermediate Services

The intermediate services are estimated at 178 €/ha for the total area of Dehesa de la Luz in 2014. These intermediate services are produced by conservation forestry, livestock, and infrastructure service activities in the following proportions: 12%, 32%, and 56%, respectively (Table 12).

The non-commercial intermediate services (ISSnc) contribute 136 €/ha. Non-commercial intermediate services compensated (ISSIncc) by the government to the owners of the land and livestock make up 34 €/ha. The intermediate services donated (ISSncd) by the public owner to recreational visitors and society as a whole (public landscape conservation services and conservation of biological and cultural diversities) add up to 85 €/ha (Tables 11 and 12). The cultural diversity service attributed to the dry-stone wall contributes 75% of the ISSncd. The family livestock owners consume amenity intermediate services (ISSnca) to a value of 18 €/ha. The ISSc values of the conservation forestry are below the aggregate values of their intermediate raw material products of grazing and firewood (Table 11). The ISSnc of livestock rearing are higher than the ISSc of conservation forestry. The ISSnc of the infrastructure services exceed the combined value of the conservation forestry and the livestock rearing.

Table 12. Intermediate product of services by activity for Dehesa de la Luz (2014: €/ha).

Class	Commercial	Non-Commercial				Total
		Compensated	Donated	Amenity	Total	
Conservation forestry	22		1		1	22
Livestock		34	5	18	57	57
Family livestock owners		14		18	32	32
Bovine		14		17	32	32
Equine				0	0	0
Landowner		19	5		25	25
Ovine		16	1		17	17
Rambouillet Merina		14			14	14
Black Merina		1	1		3	3
Bovine		2	4		6	6
Equine		2			2	2
Infrastructures services	20		79		79	99
Fencing	11		64		64	75
Other infrastructures	9					9
Paths			15		15	15
Private	42	34	85	18	136	178

3.3.3. Ecosystem Services

Grazing makes up 98% of the total ecosystem services consumed, and the remaining 2% corresponds to firewood (Table 13). If the ecosystem services of grazing and firewood are compared with the net environmental operating margin, it can be appreciated that the latter coincides with the service of grazing, and is inferior to the firewood service due to the fact that extraction exceeds growth [19].

Table 13. Landowner ecosystem services consumed at Dehesa de la Luz (2014: €/ha).

Class	Total Product Consumption	Ordinary Manufactured Intermediate Consumption	Ordinary Labor Cost	Ordinary Consumption of Fixed Capital	Ordinary Manufactured Net Operating Margin	Ecosystem Services	
	TPc	ICmo	LCo	CFCo	NOMmo	ES	%
Provisioning	200	274	44	4	−147	25	100.0
Cork		0				0	0.0
Firewood	21	8	13		0	1	2.2
Grazing	29	5			0	24	97.8
Water	19	0		4	15	0	0.0
Livestock	131	260	31		−162	na	na
Regulating	22	55	12	7	−52	0	0.0
Conservation forestry	22	55	12	7	−52	na	na
Cultural	117	104		20	−8	0	0.0
Amenity	18	18				0	0.0
Infrastructure services	99	86		20	−8	na	na
Total	339	433	57	32	−207	25	100.0

na: Not applicable.

3.3.4. Total Capital

The environmental asset of the dehesas generally makes up more than 80% of the total capital [32]. Given the scarce crop land and the fact that livestock management activities only take place up to the offspring weaning stage, the need for investment in machinery is limited. As for livestock rearing infrastructures (such as pools and wire fenced enclosures) and the residential dwellings for families of individual owners as well as the managers of the institutional owners (public and private), investment is mainly undertaken by large dehesa owners [32,38].

According to our hypothesis of non-commercial intermediate products, the dry-stone wall provides a non-commercial intermediate cultural-historical service. It is donated by the landowner to promote the final public products consumed by open access recreational visitors and society as a whole through its preservation. The valuation of the dry-stone wall in accordance with the cost of restoration is 1721 €/ha (Table 14 and Table S11). The priority condition of providing a cultural-historical public service underlies its substantial contribution to the private capital of the estate, only surpassed by the environmental asset contribution of the private amenity service (2518 €/ha) (Table 14). Paradoxically, this environmental asset does not present the consumption of its private amenity service due to the fact that the owner is an institution.

The past trend towards the depreciation of the raw materials environmental asset, namely, grazing, firewood and acorns, the latter due to the decline in acorn yield resulting from the ageing of holm oaks at Dehesa de la Luz, explains the fact that in 2014, they only accounted for 35% (1396 €/ha) of the total environmental asset (4007 €/ha) (Table 14). The investment in infrastructures (3148 €/ha) and livestock (444 €/ha) accounted for 47% (3592 €/ha) of the total opening capital value (7599 €/ha) invested in Dehesa de la Luz in 2014 (Table 14). The remaining 53% corresponds to the contribution of the environmental asset.

Table 14. Private capital balance account of Dehesa de la Luz (2014: €/ha).

Class	1. Opening Capital	2. Capital Entries				3. Capital Withdrawals						4. Revaluation	5. Closing Capital
		2.1 Bought	2.2 Own	2.3 Other	2.4 Total	3.1 Used	3.2 Sales	3.3 Destructions	3.4 Reclassifications	3.5 Other	3.6 Total		
	(Co)	(Ceb)	(Ceo)	(Ceot)	(Ce)	(Cwu)	(Cws)	(Cwd)	(Cwrc)	(Cwo)	(Cw)	(Cr)	(Cc)
1. Capital (C = WP + FC)	7599		311		311	131	2	1	2	0	137	−38	7735
2. Work in progress (WP)	161		144		144	131			2		133	1	172
2.1 Cork (WPc)	23		2		2				2		2	1	24
2.2 Firewood (WPf)	7		0		0	1			0		1	0	6
2.3 Non-breeding livestock (WPnb)	131		142		142	131					131		142
3. Fixed capital (FC)	7438		167		167		2	1		0	3	−39	7563
3.1 Land (FCl)	3275											−19	3255
Commercial (FClc)	756											1	757
Cork (FClco)	22											1	23
Firewood (FClf)	1												1
Grass and browse (FClg)	622											0	622
Acorn (FCla)	18											1	19
Hunting (FClh)	93												93
Environmental (FCe)	2518											−20	2498
Amenity (FCea)	2518											−20	2498
3.2 Biological resources (FCbr)	1016		28		28		2	1			3	25	1065
Cork (FCbrc)	528											16	544
Firewood (FCbrf)	6									0	0	0	6
Acorn (FCbra)	168											1	169
Breeding and draught livestock (FCbrb)	313		28		28		2	1			3	8	346
3.3 Plantations (FCp)	76		52		52							−14	114
3.4 Infrastructure (FCco)	2578		86		86							−26	2639
3.5 Pools (FCp)	494											−5	489

3.3.5. Capital Income

Private activities at Dehesa de la Luz, before taking into account the non-commercial intermediate services (ISSnc), generate negative capital income at the producer's price (CIpp) of −18 €/ha in 2014. After including the SSIncc, we estimate a basic price capital income (RCpb) of 16 €/ha. If we add to the latter the SSIncd and SSInca, we obtain a social private price capital income of 118 €/ha (Table S12).

The environmental income (EI) accounts for 19% of the total social capital income of Dehesa de la Luz, and the remaining 81% corresponds to the total manufactured capital income (CIm). Concerning forestry activity, the measurement of capital income at the producer's price, basic price, and social price reveals positive values of 35, 35, and 36 €/ha, respectively (Table S12). Livestock activity generates capital income for the three types estimated at −42, −8, and 15 €/ha, respectively (Table S12). The infrastructure services contribute to the different capital income with −5, −5, and 73 €/ha respectively. The environmental income from the forestry activity is 43 €/ha, which is more than its total capital income as the manufactured capital income is negative. The contribution of the private amenity services to the environmental income is negative, with a value of −20 €/ha due to the negative variation in the price of the land in 2014.

The total manufactured capital income, imputed as normal in its net operating margin component (except for the family livestock owners and loggers which are residually measured at basic prices), mainly comes from the infrastructure services, contributing 73 €/ha. The forestry and water service activities contribute −7 and 14 €/ha, respectively, to the manufactured capital income (Table S12).

3.3.6. Total Income

The net value added is the operating income which, determines the value of the long-term horizon total income, although the capital balance account revaluation/depreciation in the accounting period must be estimated in order to add them to the net value added, thus obtaining the true total income for short- and medium-term horizons. In the absence of extraordinary destructions, the variation in the price of the land is one of the main causes of the revaluation/depreciation of the environmental asset. In 2014, a total capital depreciation of −38 €/ha occurred (Table S13). The capital gains (in fact losses in 2014), which are estimated based on adjusted depreciation for accounting purposes to avoid double accounting and destructions due to livestock mortality, are subtracted, giving a negative capital gain of −11 €/ha. Quantities close to the NVA and total income are obtained with this limited loss of capital for 2014 (Table S14).

In 2014, a comparison of the AAS and EAA/EAF methodologies revealed marked differences if we consider that the estimated amount of NVA is 2.3 times greater with the AAS than with the EAA/EAF measurement (Table S14).

3.3.7. Profitability

Our definitions of the current and real profitability differ in that the first substitutes the variation in the current price of the land in that accounting period for the real average rate of variation (net rate of inflation) in the price of grazing land in Spain over the period of 1994–2014 [16,32]. The results for the current and real operating profitability rates coincide due to the effect of the variation in prices of the land and manufactured capital. These price variations only affect the rate of capital income (Table 15).

The operating and total current private profitability rates at social prices are positive, while the current capital gain is negative. The total real profitability of 2.3%, the real capital gain of which is positive for all the activities as a whole, reflects an overall rate in which the individual products display markedly different results (Table 10).

Table 15. Private profitability rates for Dehesa de la Luz (2014: %).

Class	Current Profitability			Real Profitability		
	Operating	Capital Gain	Total	Operating	Capital Gain	Total
1. Forestry	1.8	0.6	2.4	1.8	0.6	2.4
1.1 Cork	0.4	2.6	3.0	0.4	2.6	3.0
1.2 Firewood	1.2	0.4	1.6	1.2	0.4	1.6
Silviculture	1.2	0.4	1.7	1.2	0.4	1.7
Pruning	0.0	0.0	0.0	0.0	0.0	0.0
1.3 Grazing	3.0	0.2	3.2	3.0	0.2	3.2
Grass and browse	3.3	0.0	3.3	3.3	0.0	3.3
Acorn	2.1	0.8	2.9	2.1	0.8	2.9
1.4 Conservation forestry	0.0	−6.4	−6.4	0.0	−6.4	−6.4
2.Water	3.0	−0.2	2.8	3.0	−0.2	2.8
3. Livestock	1.8	1.4	3.3	1.8	1.4	3.3
3.1 Family livestock owners	1.1	2.1	3.1	1.1	2.1	3.1
Bovine	1.1	2.1	3.2	1.1	2.1	3.2
Equine	0.1	−0.6	−0.6	0.1	−0.6	−0.6
3.2 Landowner	5.7	−1.7	4.0	5.7	−1.7	4.0
Ovine	5.0	−2.1	2.9	5.0	−2.1	2.9
Rambouillet Merina	2.2	−1.7	0.6	2.2	−1.7	0.6
Black Merina	22.4	−4.8	17.6	22.4	−4.8	17.6
Bovine	16.9	0.1	17.0	16.9	0.1	17.0
Equine	−2.9	−2.0	−5.0	−2.9	−2.0	−5.0
4. Infrastructures services	3.0	−0.2	2.8	3.0	−0.2	2.8
Fencing	3.0	−0.2	2.8	3.0	−0.2	2.8
Other infrastructures	3.0	2.5	5.5	3.0	2.5	5.5
Paths	3.0	−1.0	2.0	3.0	−1.0	2.0
5. Amenity	0.0	−0.8	−0.8	0.0	1.4	1.4
Total	1.7	−0.1	1.5	1.7	0.6	2.3

3.4. Public Activities

3.4.1. Carbon Environmental Income and Asset Values at Simulated Market Price

There is no bought intermediate consumption of services (SSb) in the case of carbon activity. However, the environmental intermediate consumption of the service of carbon emissions from the firewood extracted (SSe) is registered. The price of the carbon natural growth and emission is valued at European trade prices for greenhouse effect carbon [39]. This European industrial market price for greenhouse effect carbon, applied to the annual growth of the holm oaks and cork oaks, works out at a market value of around 3 and 2 €/ha, respectively, in 2014, resulting in an almost null environmental net operating margin (Table 11 and Table S9).

The balance capital account for carbon shows a significant gross revaluation of 16 €/ha, caused by the discounted future growth of recent plantations, since the current adult trees present a negative net discounted value (Table S9). The environmental income and total income of carbon activity coincide at 14 €/ha in 2014 and represent a third of the estimated private environmental income at Dehesa de la Luz (Table S13).

3.4.2. Public Recreation, Landscape, and Threatened Livestock Biodiversity Products

Since information was not available on the public consumers' willingness-to-pay for the final products of public recreation, landscape, and threatened biodiversity activities consumed in 2014, in accordance with the EAA/EAF standard criterion, the value of the final products was estimated by the private own (self) manufactured intermediate consumption of services (SSo). Therefore, their net value added was estimated as null (Table 11). The private intermediate self-consumption of services by the private activities accounts for 21% of the private SSo, and the other 79% of private SSo corresponds to public activities. The landscape activity accounts for 85% of the private intermediate consumption of the public activities (Table 16). In other words, the final product of public activities is concentrated in the landscape activity at Dehesa de la Luz (Table 11). This result is mainly due to the landowner donated private non-commercial intermediate services (ISSncd) associated with the provision of the dry-stone wall public service (Table 12).

Table 16. Intermediate consumption of services at Dehesa de la Luz (2014: €/ha).

Class	Commercial	Non-Commercial				Total
		Compensated	Donated	Amenity	Total	
Private	20			18	18	38
Conservation forestry	0					0
Livestock	20					20
Family livestock owners	13					13
Bovine	13					13
Equine	0					0
Landowner	7					7
Ovine	6					6
Rambouillet Merina	6					6
Black Merina	1					1
Bovine	0					0
Equine	0					0
Amenity				18	18	18
Public	22	34	85		118	140
Recreation			15		15	15
Landscape	22	34	64		98	119
Biodiversity			5		5	5
Total	42	34	85	18	136	178

4. Discussion

4.1. Lack of Landowner Concern over Conservation Forestry Investment

The current state of the regeneration of dehesas is mainly the result of poor livestock grazing management, which has hampered the regeneration of trees. However, sustained grazing can be compatible not only with natural regeneration but also with plantations, as long as the individual trees in plantations are protected against controlled animal browsing. To achieve successful natural regeneration and plantation of trees in plots, it is necessary to establish appropriate areas of forest in the process of regeneration and to schedule the rotation of regeneration plots in dehesa open woodlands, based on the biological lifecycles of the trees.

The lack of investment in conservation forestry by a group of large private dehesa estates in Andalusia is worthy of mention [32]. It is unusual for owners to make investments for the benefit of future generations without receiving compensation from the government, given that competitive profitability results are mainly generated by amenities, and these are not affected in the short or medium term by the current rate of decline in raw material extractions of firewood, cork, acorns, and grass from dehesa woodlands. In this regard, the historical variations in the price of the land should also be taken into account. The private owner prefers to invest in land and livestock, which contribute in the short/medium term to avoiding negative monetary profitability along with medium to high private amenity or public profitability [4,16,32,38]. The manufactured investment in plantations today will only provide monetary capital income decades from now, which may be the main reason for the lack of woodland renewal. The high level of uncertainty with regard to the realization of future profits also underlies the uncertainty regarding the change of net worth in the present for these future yields (see SM 2). However, the landowner who, at some point in the future, harvests the products of these historical plantations will be the beneficiary of greater monetary operating margins since the historical costs of the conservation forestry will have been amortized.

Spanish dehesa woodland landscape conservation was mainly undertaken in the 1980s and 90s as a result of government compensation, co-financed by the European Union under the regulations of the programme for setting aside agricultural land of the Common Agricultural Policy (CAP) [40]. This is the case of Dehesa de la Luz, where a programme of plantation and tree densification was applied in holm oak and cork oak woodlands in 1993 and 2014, respectively. In this particular case, the landowner was totally compensated by the Extremadura government in return for commercial intermediate services associated with the conservation forestry activity carried out since 1993.

Environmental groups sometimes question the compensations (subsidies) provided under the Common Agricultural Policy (CAP) for livestock grazing which is not subject to the use of appropriate practices for the regeneration of the dehesa open woodland. The fact is that animal grazing (both livestock and game species) is essential to the existence of dehesa open woodland. Although public payments for livestock farming are not subject to auditing of compatibility with the natural regeneration of the trees, it is generally accepted that livestock provide the grazing with which the owner potentially constructs or destroys the dehesa open woodlands, in the latter case through inadequate management.

4.2. From Spanish Dehesa Private Low Commercial Operating to High Total Profitability Rates

Sufficient cash flow is important for large private family owners when their household livelihoods depend to a large degree on the monetary income from the dehesas. For large private owners of dehesas, dependence on government compensation is limited [16,41]. As for small and medium sized landowners, who were outside the scope of this research, we would imagine that for many of them, obtaining a positive net cash flow is a requirement for dehesa management. In these cases, it is the residual remuneration from self-employed family labor and income from the land and livestock that guide the landowners. The small leasehold livestock owners of Dehesa de la Luz, however, accept moderate or even null compensation for self-employed family labor and investment in livestock in return for self-consumption of amenities.

The operating profitability rate and the current gain rate should be important in the medium and long term and, to a lesser extent, in the short term due to volatility in the annual physical yield of grazing and the annual variation in the prices of land. The rationale that distinguishes investment in the dehesa from non-agrarian investment (e.g., public or industrial financial capital such as shares in a publicly traded company) is that land and livestock owners can benefit from self-consumed amenity services apart from the monetary benefit. For these reasons, private industrial landowners (capitalists) who, not being a natural person, cannot consume amenities, incur a potential loss because the market price of the dehesa land does include the private amenity discounted as a component of the price of the land [16,42]. Therefore, these private landowners tend to sell their estates to obtain greater monetary profitability from their investment in other forms of capital. In the case of public dehesas, the option of selling them is restricted by institutional and cultural settings. The loss to the public landowner of potential margin due to the absence of self-consumption of private amenities could, in this case, be counteracted by a greater supply of public products based on the provision of intermediate services of conservation forestry, threatened livestock, and historical-legacy service activities.

Published information on the private profitability of dehesas and montados is limited to the results for a group of large estates in Extremadura, Andalucía, and Alentejo. The profitability of dehesas has been estimated using extended accounts, showing moderate private commercial operating profitability both at producer's prices and basic prices. The results show −3% to 4% of the private commercial (excluding private amenities) operating profitability rates. Results from testing extended accounts reveal that the large dehesa estates obtain more highly competitive private real total profitability rates between 5% and 7% after taking into consideration the private amenities and real capital gains, mainly stemming from land revaluations that were not anticipated at the opening of the accounting period [1,4,15,16,32,43,44]. Other publications have applied standard accounts to agroforestry system estates [20,41,45–47]. Reference [41] defines a concept of 'profitability rate' which is estimated from the standard net operating surplus (which includes self-employed labor compensation) and the total capital. The standard accounts net operating surplus (NOS) could overestimate the standard capital investment profitability rate. The NOS includes the implicit remuneration of unpaid family labor (LCse). Thus, NOS is an operating income which includes the operating benefit of the investment (the extended accounts net operating margin (NOM)) plus the remuneration of the LCse. The investment theory consistent with the theory of capital profitability only includes the NOM. Therefore, by definition, NOM is less than or equal to NOS. The latter results when LCse = 0.

The high profitability rate reported by Reference [41] for the group of large dehesas studied ("grupo 4") is still more surprising if it is borne in mind that the extended broadened extend the standard accounts measurements to include the natural growth of cork and firewood, non-commercial intermediate services, private amenities, and capital gains. These authors estimate a commercial operating profitability of 6% on the total capital (which could be approximately 4% if our estimate of the residual compensation for imputed self-employed labor is excluded from the net operating surplus). It is debatable whether this result is significantly linked to the extraction of natural resources. In the absence of measurement errors, it is likely that it is significantly influenced by production processes based on the purchase of foodstuff for semi-industrial livestock production, especially of Iberian pigs and their crossbreeds. Reference [41] applies the EAA/EAF, which, as we know, do not estimate the natural growth of cork and firewood, and which limit the valuation of the products of these raw materials to the extractions of the period. Furthermore, since natural growth of firewood is omitted in the EAA/EAF, work-in-progress products extracted are not recorded as a cost in the period according to their value at the opening of the period, thus in the standard accounts double accounting of these values is avoided. The prices of grazing leases for the agrosilvopastoral systems of large family farms in Andalusia are estimated in Reference [31]. The available references [4,6,15,16,20–32,43,44,46,47] correspond to applications by other authors of EAA/EAF and AAS in dehesa estates that report commercial profitability rates for cork, firewood, grazing, and livestock both at producer's prices and at basic prices, which are notably lower than those of Reference [41].

The private operating profitability of public dehesas tends to be lower than that of private dehesas due to the absence of the self-consumption of private amenities. When the management by public owners is oriented towards increasing the supply of intermediate services in order to promote public activities, it can result in reduced commercial operating profitability at producer's prices [47]. This is the case of Dehesa de la Luz, with a substantial public landowner donation of non-commercial intermediate services to promote the supply of public products consumed by free access public recreation and society as whole when the services come from landscape, threatened biodiversity, and dry-stone wall legacy cultural services Tables 11, 12 and 16.

4.3. Comparison of Results of Standard versus Extended Accounts

The Economic Accounts for Agriculture and Forestry are produced by the government at the national level and include the economic activities of all the national agricultural estates. Thus, the aggregate result from all the individual estates corresponds to the total agroforestry 'estate' for the nation as a whole, since it includes all the national agricultural products. Therefore, the standard accounts is also applicable at the scale of the individual agroforestry estate, without the need for any conceptual change. The only change is instrumental and refers to the standard accounts part of the intermediate products, which is omitted, and another part, which is usually traded, is considered a final product of 'intra-consumption' when used as an input in the same estate. The extended accounts (AAS) measure all raw materials and services produced (intermediate products) and consumed (own intermediate consumption) by the estate activities in the accounting period.

We need to estimate the total product and total costs of single activities or products at the estate scale. The reason for this is that we need to estimate the benefit of a single product in order to estimate individual environmental assets and capitals. As an example, we can consider the acorn production consumed by livestock grazing on the estate. The discounted future benefit (resource rent) from the acorns gives the value of its environmental asset. The acorns are an intermediate product of forestry activity (raw material) and, during the same accounting period, are also an intermediate consumption of own raw material by livestock activity at the same estate.

In the case of Dehesa de la Luz, divergences between the standard versus extended accounts measurements of the private net value added of activities managed under the responsibility of the public landowner (either directly or delegated through leaseholds for family livestock rearing) are due to differences in the concept of economic products and the fact that the standard accounts do not include the natural growth of cork and firewood or the work-in-progress use of firewood from pruning operations and dead holm oaks. Both accounting methodologies coincide with the value of products according to market price or, where this is not available, the production cost. Another point of coincidence is that they both estimate the amortization (consumption of fixed capital) according to the lowest replacement cost of the manufactured fixed capital replaced.

With regard to the application of the extended accounts in Dehesa de la Luz, the private service activity includes the non-commercial intermediate service of the dry-stone wall. This intermediate service of restoration is not acknowledged in the standard accounts, which only include the livestock fencing service provided by the dry-stone wall. Hence, the most economical replacement cost for this service is to substitute it with a wire fence. In our case we accept this cost of replacement of the amortization, which is attributed to own intermediate consumption of services of the livestock rearing activity, while the additional cost of restoration, over and above the cost of the wire fence, is considered a non-commercial intermediate service of the dry-stone wall donated by the landowner to maintain the cultural landscape at Dehesa de la Luz.

Apart from the deficiencies described above, another problem with the standard accounts relates to the 'timing' of the measurement of net value added for cork and firewood, since the only criterion applied is that of extraction, whereas natural growth is omitted in the valuation of the product over the accounting period. However, as this problem of the 'timing' of the net value added measurement is not an issue with the extended accounts, this methodology for net value added measurement is more consistent with economic theory.

Comparing the standard and extended accounts, the standard accounts value the net value added at the producer's price whereas the extended accounts calculate it at the social price (producer's prices plus non-commercial intermediate services). The extended accounts private net value added is more than 2.3 times the value estimated using the standard accounts.

4.4. Private Incomes and Capital Sensitivity to Discount Rate Changes

The normal discount rate of 3% applied in Dehesa de la Luz to the future resource rents (see SM 3) from firewood, cork, acorn, and grass raw materials (intermediate products) gives their individual environmental asset, which are consistent with market prices for the land declared by the landowners of the Andalusian dehesa estates. Our choice of discount rate coincides with the rates applied in the net present value method used by the Spanish government for the valuation of estates, that is, applying the rate of return on 30-year public debt for the three years prior to the valuation [48,49]. The discount rates applied in the valuation of woodlands in the United Kingdom [50,51] are also similar to our rate and to those of the Spanish government. The manufactured capital invested is not affected by variations in the discount rate, but the manufactured capital income, environmental income, and environmental asset are affected [52]. In this case, the values of the intermediate infrastructure services were imputed, applying normal 3% rates of return. The natural growth of cork and firewood, both current and future, were estimated in accordance with the net present value of their discounted resource rent. From a baseline discount rate of 3%, reducing this rate by half would increase the environmental asset of Dehesa de la Luz by 72%, and increasing the discount rate by 50% would lead to a decrease in the environmental asset of −17% (Table 17).

Table 17. Private income and capital sensitivity to discount rate changes at Dehesa de la Luz.

Class	Manufactured Opening Capital (€/ha)	Manufactured Capital Income (€/ha)	Manufactured Working Capital (€/ha)	Environmental Income (€/ha)	Environmental Asset (€/ha)	Index Respect Environmental Asset (%)
Discount rate to 1.5%	3591.7	37.8	94.0	101.1	6876.7	172
Discount rate to 3.0%	3591.7	77.4	89.0	22.8	4007.3	100
Discount rate to 4.5%	3591.7	117.1	87.2	16.8	3324.3	83

4.5. Strengths and Weaknesses of Testing Extended Accounts in Dehesa de la Luz

This study shows the versatility of the Agroforestry Accounting System, applied in this case to individually quantify of the intermediate services, the total income, and factorial distribution corresponding, among other indicators, to private activities at Dehesa de la Luz. These extended accounts provide the owners with a suitable tool for decision-making with regard to conservation forestry and its relationships with other private and public economic activities at Dehesa de la Luz. The lack of social price valuation of public services in this case study makes it impossible to present a relevant comparison of commercial vs non-commercial values in this study. Nevertheless, private non-commercial intermediate services have been measured and these represent the main products consumed which are supplied by private activities at Dehesa de la Luz.

The scale of the extended accounts testing at Dehesa de la Luz provides a high degree of robustness to the quantification and valuation of the products and private costs, taking into account the observed economic rationale of the public landowner, leasehold family livestock owners, and loggers. The availability of detailed inventories of the woodland, livestock, man-hours employed, as well as the consumption of raw material and services per type of activity allows the physical yields and economic results to be assigned, thus minimizing individual product measurement bias. In this situation, the estimates of ecosystem services, intermediate services, environmental income, environmental asset, labor income, net operating margin, net value added, capital gains, change of net worth, capital income, and total private income are feasible and consistent with the theory of economic market valuation, both real and simulated. However, the fact that these results are subject to the author's choice of discount rate and its future variations creates an unknown level of uncertainty. This is inherent to all economic activity, which includes changes of net worth in estimates of net operating margin and total capital income.

The hypothesis that intermediate services donated by the public landowner are embedded in the value of the public services for which public consumers are prepared to pay is somewhat controversial, as the consumption of public products at Dehesa de la Luz, in accordance with the consumer's willingness-to-pay, has not been valued. This weakness in the extended accounts is similar to that of the standard accounts with regard to the manufactured gross fixed capital formation, which it also values according to production cost.

The weakness which we believe to be most important in the application of the extended methodology to conservation forestry is that it does not incorporate future variations in the public environmental services of carbon and water trade-off in the context of surplus demand for irrigation water in the lower Tagus river basin. Improvements regarding the densification and natural growth of young holm oaks and cork oaks, in contrast to the alternative land use option of treeless grazing, will lead to a decrease in surface water run-off beyond Dehesa de la Luz to the pool and dams. This competition between the environmental services of carbon and the surface water yield regulated in the pools of Arroyo de la Luz and collected in the reservoirs of the lower Tagus basin is a critical issue which has not been addressed in this study and will be a prioritized aspect of future research [53].

5. Concluding Remarks

5.1. Dehesa de la Luz Open Woodland Cultural Landscape Conservation

The long-term conservation of the cultural landscape of dehesa open woodland is not viable without animal grazing, although agricultural activity may be absent as currently occurs, at least to a certain degree, in dehesa systems of large estates [4,32,38]. A dehesa is shaped by livestock management and of continual investment in forestry over long natural production cycles, so that the natural landscape matures and reaches a fragile balance between the conservation and consumption of its natural resources.

5.2. Critical Ecosystem Capital and Economic Data Lag Failures

A general limitation of the concept of ecosystem services consumption and which, therefore, is present in the Dehesa de la Luz case study, is the lack of certainty of the signs of its physical and monetary variation over the same accounting period. The estimated ecosystem services and environmental net operating margin values for Dehesa de la Luz are proximate, apparently indicating that consumptions do not significantly exceed the accumulated natural growth of firewood and grass yield in the accounting period. In fact, firewood extraction was greater than natural growth in 2014. Therefore, these economic results do not necessarily mean that ecological decline is absent in Dehesa de la Luz, due to the short-sightedness of the market that omits non-catastrophic physical environmental decline, if indeed it exists, from both the short- and medium-term revaluation of the market price of the land. This limitation is inherent to the economic valuation of the natural environment in general. Environmental conservation is a preference expressed in social choice, subject to the restraint of tolerable cost to current generations in order to guarantee physical capital above the critical thresholds of irreversible loss of such capital. It could be said that the environmental asset value tells us the importance which actual people give to the future consumption of nature services, although the value of the current consumption does not provide us with unequivocal information regarding the variation in the biological condition of the environmental asset. This ecological condition may not be explicit in the economic value until critical thresholds of the ecological integrity of the cultural ecosystem are reached.

5.3. Socially Tolerable Government Cost for Improving Dehesa Public Services

The public service of the cultural landscape of Dehesa de la Luz is favored by investment in conservation forestry, and this public benefit is one of the most important factors justifying public payment to land and livestock owners. However, future commercial yields of firewood and grazing (acorns, browse, and grass) do register in the market, although these private yields are considered sub-products with no cost, since the costs of plantation and densification are assigned to the conservation forestry activity.

In this study, we show the legitimacy of potential payment to private landowners for losing monetary income when this is valued according to the value of non-commercial intermediate services consumed in the production of public services. Based on the results obtained in this study, the demands for compensation for lost monetary income by landowners can be legitimated since the cost to the landowner of promoting the production of public environmental services is identified. However, the social legitimacy of the payment of lost monetary income has not been considered in our study. To address the social legitimacy of government compensation to land and livestock owners, it is necessary to collate the variation in the compensated production of public services valued according to the willingness-to-pay of active and passive consumers with the public expenditure incurred.

The social legitimacy of public compensation for the private non-commercial intermediate services used as inputs for the renovation of dehesa landscape, autochthonous livestock breeds, and unique constructions of public interest (dry-stone wall) is not covered in this study of Dehesa de la Luz, as the valuation of public services in accordance with the consumer's willingness-to-pay and the direct cost of government administration of public activities were omitted. We assumed, however, that public consumers are at least prepared to pay the cost associated with private intermediate services ascribed as inputs to the production of public free access recreation, landscape, and livestock biodiversity activities (Table 11).

5.4. Private Amenity versus Public Services Trade-Off in Spanish Dehesas

The current, predominantly environmental service economy associated with Spanish dehesas is illustrated by the large contribution of private amenity services in large private dehesas, and to a lesser extent evidenced by the ecosystem services embedded in the firewood, cork, and grazing products consumed [4,16]. Dehesa public landowners face the challenge of counteracting the loss of private amenity services by incorporating new public products in greater amounts than those offered by private dehesas, in which much of the public free access use is lost. Private dehesas limit or completely avoid the consumption of certain public services, particularly recreational use, since the private landowner has the right to prohibit entry to the estate. Although the public dehesa owner also has this right, public recreational use is frequently favored where there is an effective demand.

The conceptual impossibility of self-consumption of private amenity services by the public landowner, as with institutional property (private non-family, non-profit entities, and public institutions) has a significant influence on the differences in the composition of the final private product of public dehesas. This high importance of the private amenity is the main factor underlying the modest ecosystem service and net environmental margin values measured in Dehesa de la Luz, in comparison to those estimated in large private family dehesas in Andalusia [4,16,52].

At Dehesa de la Luz, the public owner promotes free access to visitors for recreational use. The public owner could charge visitors to Dehesa de la Luz, either collecting money or payment in kind for at least part of this use (in this study we did not estimated the public recreational value). In this situation, we are not able to compare whether the public property would generate a recreational net operating margin that exceeds the loss of the private amenity net operating margin. However, the public property of Dehesa de la Luz does not lose private amenity environmental gains which we assume to be represented by unpredicted future variation in the price of land at the opening of the accounting period. The environmental asset of the amenity, like that of any other capital stock, represents the current discounted value of future resource rent and not those of past accounting periods.

5.5. Spanish Dehesas Public and Private Governance Concerns

Finally, the conservation of the dehesa cultural landscape is dependent on the continuation of livestock grazing, investment in conservation forestry, and government public service activities. The challenge facing both public and private social interests in dehesa open woodland landscape regeneration is to reach an equitable and inclusive agreement on the distribution of conservation payments among consumers, government (in representation of society as a whole), and landowners for the supply of intermediate services.

Supplementary Materials: The Supplementary Materials are available online at www.mdpi.com/2076-3298/4/4/82/s1.

Acknowledgments: This study received funding as well as information from the Town council of Arroyo de la Luz in the framework of the Dehesa de la Luz study agreement with the Spanish National Research Council (CSIC) and the University of Extremadura (UEX). We would like to thank the Town council for their support, in particular Santos Jorna, Isabel Molano, Pedro Solana, José Tapia, Manuel García and Daniel González. We also thank Mercedes Bertomeu, Manuel Bertomeu and Gerardo Moreno of the UEX for their contributions to the design and protocol of the woodland inventory sampling and Eloy Almazán for his collaboration in editing the maps. We thank the following for their help in the field work: Lorenzo Castaño, Paulino Ramos, Teodoro Sanguino, Juan Antonio Lucas, Eusebio Bermejo, Daniel Reguero, Raquel Plata, family livestock owners and loggers. The authors take full responsibility for any deficiencies which readers may find in this study.

Author Contributions: Pablo Campos is the main author and responsible for designing the contents, writing the main text and Supplementary Materials of the article, the methods and accounting valuation criteria. Bruno Mesa is co-responsible for designing the technical methodology for modelling the full-cycle silviculture and is co-writer on parts 2.2 and 3.1 of the main text and parts SM 5 and SM 6 of the Supplementary Materials as well as coordinator of the data processing through designing the Excel application which is used to estimate the products and valuation of the complete cycle in the silvicultural models for holm oak and cork oak. Alejandro Álvarez is responsible for data processing through the desing of Excel spreadsheets for the extraction of wood, livestock rearing and infrastructure services and is also co-writer of parts 3.1 and 3.2 of the main text and SM 8 of the Supplementary Materials. He is co-responsible with Bruno Mesa and Pablo Campos for editing the add-on information used in this study. Francisco M. Castaño co-responsible, along with Bruno Mesa, for designing the modelling approach for the complete cycle silviculture and is responsible for the forest inventory field data collection and survey for estimating pruning and chopping of wood , and is also responsible for the design and development of the GIS cartography used. Fernando Pulido is co-responsible for the development of the sampling protocols for the woodland inventory, the extractions of wood and qualitative modelling of the conservation forestry.

Conflicts of Interest: The authors declare no conflict of interest.

References

1. Campos, P.; Ovando, P.; Montero, G. Does private income support sustainable agroforestry in Spanish dehesa? *Land Use Policy* **2008**, *25*, 510–522. [CrossRef]
2. Díaz, M.; Campos, P.; Pulido, J.P. The Spanish dehesas: A diversity in land-use and wildlife. In *Farming and Birds in Europe*; Pain, D.J., Pienkowski, W., Eds.; Academic Press: London, UK, 1997; pp. 178–209, ISBN 9780125442800.
3. Rodríguez-Estévez, V.; Sánchez-Rodríguez, M.; Arce, C.; García, A.R.; Perea, J.M.; Gómez-Castro, A.G. Consumption of Acorns by Finishing Iberian Pigs and Their Function in the Conservation of the Dehesa Agroecosystem. In *Agroforestry for Biodiversity and Ecosystem Services—Science and Practice*; Kaonga, M.L., Ed.; InTech: Rijeka, Croatia, 2012; pp. 1–22, ISBN 9789535104933.
4. Ovando, P.; Campos, P.; Oviedo, J.L.; Caparrós, A. Ecosystem accounting for measuring total income in private and public agroforestry farms. *For. Policy Econ.* **2016**, *71*, 43–51. [CrossRef]
5. Dirección General de Conservación de la Naturaleza. *Mapa Forestal de España 1:50.000*; Ministerio de Medio Ambiente: Madrid, Spain, 2007.
6. Campos, P.; Carranza, J.; Miguel Coleto, J.; Diaz, M.; Diéguez, E.; Escudero, A.; Ezquerra, F.J.; Lopez, L.; Fernandez, P.; Montero, G.; et al. *Libro Verde de la Dehesa*; Universidad de Extremadura: Plasencia, Spain, 2010; pp. 1–48.
7. Ministerio de Agricultura, Pesca y Alimentación. *Diagnóstico de las Dehesas Ibéricas Mediterráneas*; Secretaría General de Agricultura y Alimentación, Dirección General de Desarrollo Rural: Madrid, Spain, 2008; Tomo 1; Unpublished.
8. Ministerio de Agricultura, Pesca y Alimentación. *Diagnóstico de las Dehesas Ibéricas Mediterráneas*; Secretaría General de Agricultura y Alimentación, Dirección General de Desarrollo Rural: Madrid, Spain, 2008; Tomo 2; Unpublished.
9. Alejano, R.; Domingo, J.M.; Fernández, M.; Alaejos, J.; Calzado, A.; Carevic, F.; Del Campo, A.; Domínguez, L.; Fernández de Villarán, R.; Flores, E.; et al. *Manual Para la Gestión Sostenible de las Dehesas Andaluzas*; Foro para la Defensa y Conservación de la Dehesa "Encinal" and Universidad de Huelva: Huelva, Spain, 2011; p. 465, ISBN 9788461540020. Available online: http://rabida.uhu.es/dspace/handle/10272/6641 (accessed on 14 September 2017).
10. Plieninger, T.; Rolo, V.; Moreno, G. Large-scale patterns of *Quercus ilex*, *Quercus suber*, and *Quercus pyrenaica* regeneration in Central-Western Spain. *Ecosystems* **2010**, *13*, 644–660. [CrossRef]
11. Senado. *Informe de la Ponencia de Estudio Sobre la Protección del Ecosistema de la Dehesa*; Boletín Oficial de las Cortes Generales: Madrid, Spain, 2010; pp. 1–27. Available online: http://www.senado.es/legis9/publicaciones/pdf/senado/bocg/I0553.PDF (accessed on 14 September 2017).
12. Instituto da Conservação da Natureza e das Florestas. *IFN6—Áreas dos Usos do Solo e das Espécies Florestais de Portugal Continental. Preliminary Results*; ICNF: Lisboa, Portugal, 2013; p. 35. Available online: http://www.icnf.pt/portal/florestas/ifn/resource/ficheiros/ifn/ifn6-res-prelimv1-1 (accessed on 14 September 2017).
13. Lauw, A.; Ferreira, A.G.; Gomes, A.A.; Moreira, A.C.; Fonseca, A.; Belo, A.; Azul, A.M.; Mira, A.; Murilhas, A.; Pinheiro, A.C.; et al. *Livro Verde dos Montados*; ICAAM: Évora, Portugal, 2013; pp. 1–61.
14. European Communities. *Manual on the Economic Accounts for Agriculture and Forestry EEA/EAF 97 (Rev. 1.1)*; EC, EUROSTAT: Luxembourg, 2000. Available online: http://ec.europa.eu/eurostat/documents/3859598/5854389/KS-27-00-782-EN.PDF/e79eb663-b744-46c1-b41e-0902be421beb (accessed on 14 September 2017).

15. Oviedo, J.L.; Ovando, P.; Forero, L.; Huntsinger, L.; Álvarez, A.; Mesa, B.; Campos, P. The private economy of dehesas and ranches: Case studies. In *Mediterranean Oak Woodland Working Landscapes. Dehesas of Spain and Ranchlands of California*; Campos, P., Huntsinger, L., Oviedo, J.L., Starrs, P.F., Díaz, M., Standiford, R., Montero, G., Eds.; Springer: Dordrecht, The Netherlands, 2013; pp. 389–424, ISBN 9789400767065.

16. Oviedo, J.L.; Huntsinger, L.; Campos, P. The Contribution of Amenities to Landowner Income: Case of Spanish and Californian Hardwood. *Rangel. Ecol. Manag.* **2017**, *70*, 518–528. [CrossRef]

17. United Nations; European Commission; Food and Agriculture Organization of the United Nations; Organization for Economic Co-operation and Development; World Bank Group. *System of Environmental Economic Accounting 2012—Experimental Ecosystem Accounting [SEEA-EEA]*; United Nations: New York, NY, USA, 2014; p. 198, ISBN 9789211615753. Available online: http://ec.europa.eu/eurostat/documents/3859598/6925551/KS-05-14-103-EN-N.pdf (accessed on 14 September 2017).

18. Campos, P. Cuentas agroforestales: Retos de la medición de la renta total social de los montes de Andalucía. In *Economía y Selviculturas de Los Montes de Andalucía*; Campos, P., Díaz-Balteiro, L., Eds.; Memorias científicas de RECAMAN; Editorial CSIC: Madrid, Spain, 2015; Volume 1, memoria 1.1; pp. 18–152, ISBN 9788400100407.

19. Campos, P.; Mesa, B.; Castaño, F.M.; Álvarez, A.; Pulido, F.J. *Economía de la Actividad Forestal Privada del Propietario de la Dehesa de la Luz*; Working Paper 2017-03; Instituto de Políticas y Bienes Públicos (IPP) CSIC: Madrid, Spain, 2017. Available online: http://ipp.csic.es/sites/default/files/content/workpaper/2017/2017_03_ippwp_camposmesacastanoalvarezpulido.pdf (accessed on 14 September 2017).

20. Campos, P.; Daly, H.; Oviedo, J.L.; Ovando, P.; Chebil, A. Accounting for single and aggregated forest incomes: Application to public cork oak forests of Jerez in Spain and Iteimia in Tunisia. *Ecol. Econ.* **2008**, *65*, 76–86. [CrossRef]

21. Campos, P.; Caparrós, A.; Oviedo, J.L.; Ovando, P.; Álvarez-Farizo1, B.; Díaz-Balteiro, L.; Carranza, J.; Beguería, S.; Díaz, M.; Herruzo, A.C.; et al. *Bridging the Gap between National and Ecosystem Accounting*; Working Paper, 2017-04; Instituto de Políticas y Bienes Públicos (IPP) CSIC: Madrid, Spain, 2017. Available online: http://ipp.csic.es/sites/default/files/content/workpaper/2017/2017_04_ippwp_campos_etal.pdf (accessed on 20 September 2017).

22. Campos, P.; Álvarez, A.; Mesa, B.; González, D. Economía de las vacas, ovejas y yeguas de la Dehesa de la Luz. In *La Dehesa de la Luz en la Vida de Los Arroyanos*; Campos, P., Pulido, F., Eds.; Ayuntamiento de Arroyo de la Luz. Editorial Luz y Progreso: Arroyo de la Luz, Spain, 2015; pp. 157–218, ISBN 9788460682127.

23. Plieninger, T.; Pulido, F.J.; Konold, W. Effects of land-use history on size structure of holm oak stands in Spanish dehesas: Implications for conservations and restoration. *Environ. Conserv.* **2003**, *30*, 61–70. [CrossRef]

24. Ministerio de Agricultura, Pesca y Alimentación. *Segundo Inventario Forestal Nacional, 1986–1997: Extremadura: Provincia de Cáceres*; Instituto Nacional para la Conservación de la Naturaleza: Madrid, Spain, 1994; ISBN 8485496604.

25. Fernández-Rebollo, P.; Carbonero-Muñoz, M.D. *Control y Seguimiento de los Programas Agroambientales Para el Fomento de la Dehesa en Andalucía. Technical Report*; Consejería de Agricultura y Pesca, Junta de Andalucía: Seville, Spain, 2008; Unpublished.

26. Montero, G.; Cañellas, I. *Manual de Reforestación y Cultivo del Alcornoque*, 2nd ed.; INIA-Mundi-Prensa: Madrid, Spain, 1999; p. 103; ISBN 9788484761211.

27. Montero, G.; Torres, E.; Cañellas, I.; Ortega, C. Modelos para la estimación de la producción de corcho en alcornocales. *Investigación Agraria Sistemas y Recursos Forestales* **1996**, *5*, 97–127.

28. Montero, G.; Pasalodos-Tato, M.; López-Senespleda, E.; Ruiz-Peinado, R.; Bravo-Oviedo, A.; Madrigal, G.; Onrubia, R. Modelos de selvicultura y producción de madera, frutos y fijación de carbono de los sistemas forestales de Andalucía. In *Economía y Selviculturas de Los Montes de Andalucía*; Campos, P., Díaz-Balteiro, L., Eds.; Memorias Científicas de RECAMAN; Editorial CSIC: Madrid, Spain, 2015; Volume 1, memoria 1.2; pp. 153–396, ISBN 9788400100407.

29. Montero, G.; Ruiz-Peinado, R.; Muñoz, M. *Producción de Biomasa y Fijación de CO_2 Por Los Bosques Españoles*; Monografía INIA, Serie Forestal: Madrid, Spain, 2005; p. 270, ISBN 9788474985122.

30. Rolo, V.; Moreno, G. Interspecific competition induces asymmetrical rooting profile adjustments in shrub encroached open oak woodlands. *Trees Struct. Funct.* **2012**, *26*, 997–1006. [CrossRef]

31. Campos, P.; Ovando, P.; Mesa, B.M.; Oviedo, J.L. Environmental income of livestock grazing on privately owned silvopastoral farms in Andalusia, Spain. *Land Degrad. Dev.* **2016**. [CrossRef]

32. Ovando, P.; Campos, P.; Mesa, B.; Álvarez, A.; Fernández, C.; Oviedo, J.L.; Caparrós, A.; Álvarez-Farizo, B. Renta y capital de estudios de caso de fincas agroforestales de Andalucía. In *Renta Total y Capital de las Fincas Agroforestales de Andalucía*; Campos, P., Ovando, P., Eds.; Memorias científicas de RECAMAN; Editorial CSIC: Madrid, Spain, 2015; Volume 4, memoria 4.2; pp. 156–445, ISBN 9788400100445.

33. Ministerio de Agricultura y Pesca, Alimentación y Medio Ambiente. *Encuesta de Precios de la Tierra*; Ministerio de Agricultura y Pesca, Alimentación y Medio Ambiente: Madrid, Spain, 2014. Available online: http://www.mapama.gob.es/es/estadistica/temas/estadisticas-agrarias/encuestadepreciosdelatierra2014_tcm7-407561.pdf (accessed on 15 September 2017).

34. United Nations; European Union; Food and Agriculture Organization of the United Nations; International Monetary Fund; Organization for Economic Cooperation and Development; World Bank. *System of Environmental–Economic Accounting 2012—Central Framework [SEEA-CF]*; United Nations: New York, NY, USA, 2014; p. 378, ISBN 879211615630. Available online: https://unstats.un.org/unsd/envaccounting/seeaRev/SEEA_CF_Final_en.pdf (accessed on 14 September 2017).

35. Haines-Young, R.; Potschin, M. *CICES V4.3—Revised Report Prepared Following Consultation on CICES Version 4: Technical report to the European Environment Agency*; United Nations: New York, NY, USA, 2013. Available online: http://unstats.un.org/unsd/envaccounting/seearev/GCComments/CICES_Report.pdf (accessed on 25 October 2017).

36. United Nations Environmental Program; United Nations Statistics Division; Convention on Biological Diversity. *SEEA Experimental Ecosystem Accounting: Technical Recommendations, Consultation Draft*; United Nations: New York, NY, USA, 2015. Available online: https://unstats.un.org/unsd/envaccounting/workshops/ES_Classification_2016/SEEA%20EEA%20Tech%20Rec%20Consultation%20Draft%208.1%20Dec2015%20final.pdf (accessed on 25 October 2017).

37. Obst, C.; Hein, L.; Edens, B. National Accounting and the Valuation of Ecosystem, Assets and Their Services. *Environ. Resour. Econ.* **2016**, *64*, 1–23. [CrossRef]

38. Oviedo, J.L.; Campos, P.; Caparrós, A. Valoración de servicios ambientales privados de propietarios de fincas agroforestales de Andalucía. In *Renta Total y Capital de Las Fincas Agroforestales de Andalucía*; Campos, P., Ovando, P., Eds.; Memorias científicas de RECAMAN; Editorial CSIC: Madrid, Spain, 2015; Volume 4, memoria 4.1; pp. 8–155, ISBN 9788400100445.

39. Sistema Electrónico de Negociación de Derechos de Emisión de Dióxido de Carbono (SENDECO2). Available online: http://www.sendeco2.com/es/precios-co2 (accessed on 14 September 2017).

40. Ovando, P.; Campos, P.; Montero, G. Forestaciones con encina y alcornoque en el área de la dehesa en el marco del Reglamento (CE) 2080/92 (1993–2000). *Revista Española de Estudios Agrosociales y Pesqueros* **2007**, *214*, 173–186.

41. Gaspar, P.; Mesias, F.J.; Escribano, M.; Rodriguez de Ledesma, A.; Pulido, F. Economic and management characterization of dehesa farms: Implications for their sustainability. *Agrofor. Syst.* **2007**, *71*, 151–162. [CrossRef]

42. Campos, P.; Oviedo, J.L.; Caparrós, A.; Huntsinger, L.; Coelho, I. Contingent valuation of woodland owners private amenities in Spain, Portugal and California. *Rangel. Ecol. Manag.* **2009**, *62*, 240–252. [CrossRef]

43. Campos, P.; Riera, P. Rentabilidad social de los bosques. Análisis aplicado a las dehesas y los montados ibéricos. *Información Comercial Española* **1996**, *751*, 47–62.

44. Campos, P.; Rodríguez, Y.; Caparrós, A. Towards the Dehesa total income accounting: Theory and operative Monfragüe study cases. *Forest Syst.* **2001**, *10*, 43–67.

45. Campos, P. *Evolución y Perspectivas de la Dehesa Extremeña*; Editorial de la Universidad Complutense: Madrid, Spain, 1984; p. 498.

46. Campos, P.; Sesmero, J. Análisis económico de un grupo de dehesas de Extremadura (1983–1984). In *Conservación y desarrollo de Las Dehesas Portuguesa y Española*; Campos, P., Martín, M., Eds.; Ministerio de Agricultura, Pesca y Alimentación: Madrid, Spain, 1987; pp. 487–534, ISBN 8474795338.

47. Rodríguez, Y.; Campos, P.; Ovando, P. Commercial economy in a public Dehesa in Monfragüe Shire. In *Sustainability of Agro-silvo-pastoral Systems. Dehesas & Montados*; Schnabel, S., Gonçalves, A., Eds.; Serie Advances in GeoEcology 37; Catena Verlag: Reiskirchen, Germany, 2004; pp. 85–96, ISBN 9783923381500.

48. Boletín Oficial del Estado. *Real Decreto 1492/2011, de 24 de Octubre, Por el Que se Aprueba el Reglamento de Valoraciones de la Ley de Suelo*; Boletín Oficial del Estado: Madrid, Spain, 2011; p. 26. Available online: https://www.boe.es/boe/dias/2011/11/09/pdfs/BOE-A-2011-17629.pdf (accessed on 14 September 2017).
49. Boletín Oficial del Estado. *Real Decreto Legislativo 7/2015, de 30 de Octubre, Por el Que se Aprueba el Texto Refundido de la Ley de Suelo y Rehabilitación Urbana*; Boletín Oficial del Estado: Madrid, Spain, 2015; p. 59. Available online: https://www.boe.es/boe/dias/2015/10/31/pdfs/BOE-A-2015-11723.pdf (accessed on 14 September 2017).
50. Economics for the Environment Consultancy Ltd. (EFTEC). *Developing UK Natural Capital Accounts: Woodland Ecosystem Accounts*; Department for Environment, Food and Rural Affairs (Defra): London, UK, 2015; p. 97. Available online: http://sciencesearch.defra.gov.uk/Default.aspx?Menu=Menu&Module=More&Location=None&Completed=0&ProjectID=18909 (accessed on 14 September 2017).
51. Office for National Statistics; Department for Environment, Food and Rural Affairs (Defra). *Principles of Natural Capital Accounting*; Office for National Statistics: Newport, UK, 2017; p. 52. Available online: https://www.ons.gov.uk/economy/environmentalaccounts/methodologies/principlesofnaturalcapitalaccounting (accessed on 14 September 2017).
52. Ovando, P.; Caparrós, A.; Diaz-Balteiro, L.; Pasalodos, M.; Beguería, S.; Oviedo, J.L.; Montero, G.; Campos, P. Spatial Valuation of Forests' Environmental Assets: An Application to Andalusian Silvopastoral Farms. *Land Econ.* **2017**, *93*, 85–106. [CrossRef]
53. Beguería, S.; Campos, P.; Serrano, R.; Álvarez, A. Producción, usos, renta y capital ambientales del agua en los ecosistemas forestales de Andalucía. In *Biodiversidad, Usos del Agua Forestal y Recolección de Setas Silvestres en los Ecosistemas Forestales de Andalucía*; Campos, P., Díaz, M., Eds.; Memorias científicas de RECAMAN; Editorial CSIC: Madrid, Spain, 2015; Volume 2, memoria 2.2; pp. 102–273, ISBN 9788400100421.

environments

MDPI

Article

What Do Users Really Need? Participatory Development of Decision Support Tools for Environmental Management Based on Outcomes

Richard J. Hewitt [1,2,*] and Christopher J. A. Macleod [1]

[1]   Information and Computational Sciences Group, The James Hutton Institute, Craigiebuckler, Aberdeen AB15 8QH, UK; Kit.Macleod@hutton.ac.uk
[2]   Observatorio para una Cultura del Territorio, 28012 Madrid, Spain
*   Correspondence: richard.hewitt@hutton.ac.uk; Tel.: +44-1224-395436

Received: 5 October 2017; Accepted: 29 November 2017; Published: 6 December 2017

Abstract: There is increasing demand from stakeholders for tools to support outcomes-based approaches in environmental management. For such tools to be useful, understanding user requirements is key. In Scotland, UK, stakeholders were engaged in the development of an Environmental Decision Support System (EDSS) to support the management of land and freshwater resources for multiple policy outcomes. A structured participatory engagement process was employed to determine stakeholder requirements, establish development principles to fulfil these requirements and road-test prototypes. The specification that emerged from this bottom-up process was for an EDSS to be spatially-explicit, free at the point of use, and mobile device compatible. This application, which is under development, does not closely resemble most existing published EDSS. We suggest that there is a mismatch between the way scientists typically conceptualise EDSS and the kinds of applications that are likely to be useful to decision-makers on the ground. Interactive mobile and web-based geospatial information services have become ubiquitous in our daily lives, but their importance is not reflected in the literature on EDSS. The current focus in environmental management on adaptive, stakeholder-centred strategies based on outcomes offers an opportunity to make better use of these new technologies to aid decision-making processes.

Keywords: Environmental Decision Support Systems; applications; outcomes-based approach; adaptive management; user requirements; environmental management; participatory land planning

1. Introduction

Research in environmental science is often undertaken under the premise that scientific information and knowledge is necessary to inform environmental policy and management. This has led to the widespread development of computerized tools to bridge the divide between scientific analysis of the state of the environment (e.g., water quality, biodiversity, land use change) and environmental policy objectives (e.g., European directives on freshwater quality and terrestrial biodiversity conservation measures). These tools fall approximately into two types, though in practice there is a great deal of overlap. Decision Support Systems (DSS) are typically targeted at supporting policy implementation in a specific context, while Policy Support Systems or (PoSS) have broader aims, including policy formulation and strategy [1]. In the context of environmental management, we follow [2], and refer to both of these two types as Environmental Decision Support Systems (EDSS).

In theory, the relationship between science and policy is close and direct [3]. In practice, this is not always the case, especially when science identifies wicked problems [4] to which policy makers or environmental managers are unable to respond in conventional normative ways. Consensus is beginning to emerge around the need for integrative, adaptive approaches to environmental management as a means to tackle these kinds of intractable problems [5,6]. Adaptive environmental management seeks to integrate project design, management, and monitoring, to provide a framework to systematically test assumptions, promote learning, and supply timely information for management decisions (e.g., [7]). Involvement of stakeholders is essential to facilitate the processes of generating and sharing different knowledges to improve understanding of the effectiveness of management actions. Policy makers and land managers require better access to the results of scientific analysis, while scientific stakeholders need to better understand other stakeholders' needs in order to structure and focus their research (e.g., [8]). Private sector stakeholders, like land-based businesses, are increasingly interested in easy to use web and mobile-based dashboard "business intelligence" tools to better manage their holdings, creating new demands and opportunities to bridge the relationship between policy and management, and the supply and provision of scientific analysis. A need is therefore emerging for new software tools and applications to respond to these demands, bringing new ways of working [9]. For example, up-to-date information on the state of the physical environment (moisture, erosion, crop growth), and land-based policies and incentives can be accessed directly by land managers through smartphone and tablet applications. This helps them play a more proactive role in environmental management and may reduce the need for intervention by regulators. At the same time, scientists can move from passive provision of information for policy makers, to on-the-ground facilitation of knowledge exchange between all stakeholders. Making applications web-based facilitates access, which might be expected to lead to faster and more widespread adoption by taking advantage of existing internet infrastructure and appealing to users of modern mobile devices.

In this paper, we discuss our recent progress in responding to the challenge of providing web-based digital tools that meet user needs for their adaptive management of natural resources. To this end, we address the following three key research questions:

1. How can we develop decision-support tools that align better with adaptive management and outcomes-based approaches to environmental management: what are the key requirements of such a system?
2. How can we make better use of well-established web-based and mobile devices and software for supporting environmental decisions?
3. How can we involve key stakeholders like scientists, regulators and land managers to better understand these requirements?

The paper is structured as follows. In the next section, we describe some of the most important limitations of EDSS as currently conceived from an adaptive management perspective. Subsequently, we show how we have tried to address these limitations and answer our research questions by engaging policy stakeholders, regulators and resource managers to co-develop an environmental decision-support application for understanding the effectiveness of environmental policy interventions in river catchments in Scotland. We present the results of this process and finish with a general discussion summarizing key points and lessons learnt. Finally we offer some brief recommendations for future development of EDSS.

2. Background

The application of computerized tools to these kinds of problems has a long history (see e.g., [10–12]). The use of computer modelling to facilitate adaptive environmental management was advocated by [13]. Early DSS (see e.g., [14,15]) were conceived as computerized tools to manage operational decisions where either the decision formulation or the solution, or both, were arguable or directly contested [2]. The literature contains many different types of EDSS, depending on the kinds of decisions they are intended to support, their anticipated or declared end user, and the stage of the environmental policy process to which they are directed. Some, like the SimLucia model ([16,17]), developed on behalf of the United Nations Environment Programme (UNEP), are intended to allow high-level policy makers to formulate appropriate long-term responses to environmental change. Others, like QUICKScan ([18]) are targeted at more local scales and over shorter timeframes e.g., to help multiple stakeholders to negotiate, compare options and understand trade-offs in the implementation of concrete policies. Others, such as the COLLAGE tool, are more specific. COLLAGE is designed to help local stakeholders plan renewable energy installations by balancing generation capacity against local spatial planning concerns ([19]). While the involvement of end-users is clearly a key aspect of many such systems, this does not necessarily imply the democratization of knowledge and decision-making that lies at the heart of stakeholder-centred approaches to environmental management, i.e., in participatory, collaborative or mediated modelling (e.g., [20–22]). An EDSS designed under the conventional paradigm of DSS is a tool to help the competent authorities solve environmental problems, but does not necessarily emphasize sharing and co-construction of knowledge with stakeholders or social learning (*sensu* [23]) as a strategy for managing disagreement resulting from the diverse perspectives and requirements of multiple stakeholders. This means that frequently, such systems are not targeted at stakeholder needs [24]. From an adaptive management point of view, in which it is often desirable or even essential to share knowledge effectively amongst diverse stakeholder groups, this is problematic. Recently, digital catchment observatories have been suggested as a means to improve knowledge sharing and co-construction with stakeholders [25].

At the same time, many EDSS, ostensibly intended for non-scientific stakeholders to pick up and apply to their specific problems or needs, are not used for that purpose [2,26,27]. While the reasons for the lack of uptake by intended end users are diverse [2,28] lack of stakeholder involvement at the design stage is clearly a significant factor. Volk et al [24], in their analysis of four EDSS in landscape and catchment management contexts, concluded that "the appropriate and methodological stakeholder interaction and the definition of 'what end-users really need and want' have been documented as general shortcomings of all four examples of DSS." Understanding and managing the diverse requirements a software application needs to meet—the requirements problem—is recognized as a major challenge in software development generally [29]. Responding to this challenge by improving the integration of the end user in the development process is a key underlying motivation of the User-Centered Systems Design (UCSD) paradigm [30,31]. However, while user-centered approaches have become mainstream in software development circles, they still lack general application in an EDSS context.

Further, since publication of [2], the role of software in supporting decisions of various kinds in everyday life has grown substantially. Mobile and web-based applications have become ubiquitous in all kinds of contexts, e.g., purchasing or contracting goods and services, banking, navigation, social networking etc. Typically these tools are mobile, web-based, built on Free and Open Source Software (FOSS), and touch user interface-optimized. Yet EDSS for the most part, lag behind the innovation curve. Most EDSS are still stand-alone desktop systems, many require expensive proprietary software and frequently are not optimized or not available for touch-enabled devices. In agriculture, a new generation of decision-support tools has begun to emerge, in the form of Farm Management Information Systems [32–34], allowing farm managers to adapt their operations to variables like temperature and precipitation, market prices or policy measures. However, there remains a substantial mismatch between the computationally intensive and conceptually challenging modelling approaches typically used by the scientific community (e.g., [1,35]), and the lightweight, web and mobile-based, touch-enabled applications for smart devices that we all use in our daily lives.

Finally, a key limitation of conventional EDSS from our point of view relates to the difficulty of applying them in the context of outcomes-based approaches i.e., the environmental, social or financial improvements that management actions aim to make. The likely result, in terms of environmental improvement or otherwise of a particular measure, should be a determining factor in the choice of measure and where implemented. This requires us to understand not only the potential spatial distribution of particular environmental variables in the landscape e.g., water body or terrestrial habitat status as determined by environmental policies, in order to suggest locations for management measures and interventions, but also the causal logic that connects these measures to the desired outcomes. Conventional EDSS approaches have emphasized the former, while approaches like logic modelling (also known as results chain logic modelling) are frequently applied to the latter [36,37]. Logic modelling, which has its origins in program theory [38] is widely used for planning and evaluation of environmental and agricultural policy measures in the UK and elsewhere [39]. At present, however, we are unaware of any EDSS software or framework that successfully integrates the aspatial process-based approaches found in logic modelling software like Miradi with the spatial, Geographical Information Systems (GIS)-type approach found in EDSS for rural planning at the level of the management unit (e.g., [10]).

In the following paper, we present our recent work in this area, and argue that a reappraisal of the process of developing software for EDSS is necessary to take into account adaptive management and outcomes-based approaches to environmental decision-making and the wide range of new smart applications.

3. Case Study Background

The work presented here forms part of a long-term Scottish Government Strategic Research Programme (SRP) project which aims to understand and improve the management of Scotland's land and freshwater resources for multiple policy and management outcomes. In Scotland, the Land Use Strategy aims to take a more integrated approach to management of natural capital to deliver multiple benefits [40]. Since 2007, Scottish Government policy objectives have been framed as a set of national level outcomes to guide all policy, management and applied research [41]. This emphasis on outcomes stems from the increasing interest across research, policy and management communities to improve not only how we plan individual and multiple landscape management actions, but also how we evaluate and learn from their successes and failures (see e.g., [42]). In Scotland, as elsewhere in Europe, land managers are eligible for grants in return for implementing environmental improvement measures under Pillar 2 of the Common Agricultural Policy (CAP). A wide range of options are available relating to livestock, croplands and vegetation, among others (see e.g., [43]). Choosing the most appropriate measure for a given land holding is a complex problem involving multiple choices, actors and possible outcomes.

4. Methods

In order to respond to these three research questions, we conducted research and stakeholder engagement activities in two phases where Phase 1 addressed the first and second research questions and Phase 2 addressed the third. Future work will be addressed in Phase 3. The conceptual design of the process is shown in Figure 1, in which the specific actions described in this paper are numbered. These are described in Sections 4.2 and 4.3.

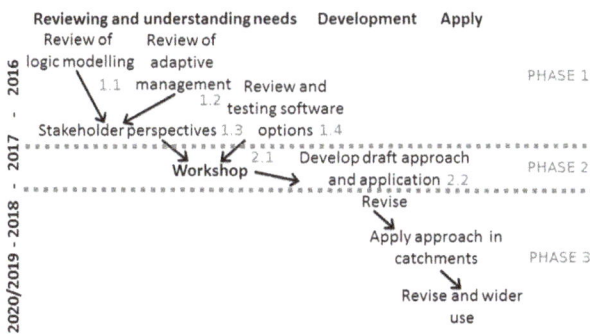

Figure 1. Conceptual process design for stakeholder-driven development of an outcomes-based Environmental Decision Support System (EDSS).

4.1. Ethical Statement

All stakeholders gave their informed consent for inclusion before they participated in the study. The study was conducted in accordance with the Declaration of Helsinki, a widely accepted worldwide standard intended to ensure ethical conduct in scientific research on human subjects and approved by the Scottish government rural and environment science and analytical services division (RESAS) (project code: SRP RD1.4.3d). The ethical code followed was the "Research Ethics Policy for Human Participants" of the James Hutton Institute.

4.2. Phase 1: Stakeholder Engagement and Review of Software Options

Following a review of outcomes-based logic modelling ([37], 1.1 in Figure 1) adaptive management ([44], 1.2 in Figure 1), meetings were held with policy stakeholders in Scottish Government to discuss how logic modelling could support their outcomes-based approach in the National Performance Framework [41]. To start the co-construction of an outcomes-based logic modelling approach to aid decisions about landscape interventions to deliver multiple benefits e.g., water quality, terrestrial biodiversity and land manager income, a series of interviews were carried out with 13 national and regional level stakeholders about adaptive management and outcomes-based approaches ([45], Figure 1: 1.3). The results of this work are extensive and lie beyond the scope of this paper, but provided a useful starting point for understanding the rationales behind the outcomes-based approach.

Using initial insights from stakeholders of their needs to be addressed, experiences from colleagues involved in participatory GIS projects and our expertise in integrative modelling with stakeholders we started to narrow down the wide range of software options (1.4 in Figure 1). Potential software options were selected for their ability to meet seven initial requirements which formed our screening criteria (Table 1). Review of existing software options for supporting outcomes-based environmental management was undertaken from published, unpublished and internet sources. Where possible, we downloaded software and experimented with it in order to better inform our review. A short list of software options was then explored and an interactive prototype was developed to be demonstrated to stakeholders at the workshop.

4.3. Phase 2: Participatory Workshop

In Phase 2, we organized a half-day workshop with stakeholders with expertise in outcomes-based environmental management, in order to deepen understanding of requirements and begin a process of co-development (2.1 in Figure 1). The workshop was attended by six participants representing key domains in land and water management in Scotland, comprising representatives from the Cairngorm National Park Authority, Scottish Natural Heritage and Scottish Environmental Protection Agency, and three researchers with >50 years collective experience of developing practical tools to support land and water management for multiple benefits.

The workshop was based around a series of linked activities. In the first activity the participants were asked to rate a list of 17 "needs" for outcomes-based environmental management from information provided by stakeholders during the earlier interviews [45]. Participants were also asked to add additional needs that they felt to be necessary. For the second activity, working in pairs comprising a researcher and a practitioner, participants were invited to provide suggestions for the key principles that our approach would use to address the identified needs. The idea of "principles" is central to UCSD (see e.g., [31,46]) and is distinct from users' functional requirements or needs in that it specifies how a system should be developed, not what the system aims to do. Following other authors, see e.g., [18,47] we developed a draft list of guiding principles for development based on own understanding of what was needed (Figure 2).

Approach: facilitated, integrative and adaptive.

Digital application: accessible, relevant,
practical, transparent, evidence and
outcomes-based.

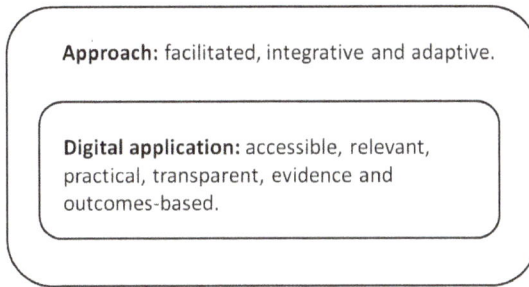

Figure 2. Researchers' own draft principles, based on literature.

A sheet of existing principles from related approaches or applications was provided to the participants as a guide. As we were interested to see what principles the participants would come up with independently, we did not share our draft list. Principles suggested by the participants were discussed by the whole group, and at the end of the activity, the participants were invited to comment on the draft list of principles we had developed ourselves prior to the workshop.

The third activity included an overview of hands-on advanced testing of software options, and an overview of the process of development of the interactive digital prototype. The last activity involved facilitated discussion around the prototype on a large touch table. The workshop ended with a general discussion of the main themes explored. The outcome of the workshop involved production of a report [48], which was circulated to participants for their approval, which they were happy to give. The outcome of Phase 2 was the draft approach and prototype application (2.2 in Figure 1).

Audio and video recordings were made and photographs were taken with the full agreement of all participants. A wide range of material was collected from each activity including *pro-formas* completed by participants for each activity including commentary on post-it notes, a list of principles and detailed documentation of the discussions, recorded on wall-charts by facilitators during the workshop. A report containing all of this information was circulated to workshop participants who approved its content. A selection of this information forms the basis of the results section of this paper.

Table 1. Assessment of software options for tool development.

Options	Criteria							Selected: Yes or No	Links
	Free to Use [1]	Use on Touch Devices [2]	Interact Spatially [3]	Logic Model [4]	Develop/ Extend [5]	Innovation [6]	Status [7]		
Existing adaptive management logic model applications									
Miradi	100-999	No	No	Yes	No	Low	Active	No	https://www.miradi.org/
Mobile version	Maybe [8]	Yes	Maybe	Yes	Maybe	Maybe	Active	No	http://monitoringapp.net/
Existing standalone participatory modelling applications									
Quickscan	100-999	No	Yes	Maybe	No	Low-med	Active	No	http://www.quickscan.pro/products
Metronamica	>10,000	No	Yes	Maybe	No	Low-med	Active	No	http://www.metronamica.nl/
Geodesign	0	Maybe	Yes	Maybe	Maybe	Med-high	Active	No	https://www.geodesignhub.com/
GIS modelling applications									
Community Viz ArcMap plugin	100-999	No	Yes	No	Yes	Med	Active	No	http://placeways.com/communityviz/index.html
MCDA4ArcMap	0 [9]	Maybe	Yes	No	Yes	Med	Active	No	https://mcda4arcmap.codeplex.com/
EMDS (US EPA)	0	Maybe	Yes	Maybe	Maybe	Low-med	Active	No	https://esenvironmental.com/projects_epa_emds.htm
Standard ArcMap	100-999	No	Yes	No	Yes	Low-med	Active	No	http://www.arcgis.com/
Standard QGIS	0	Yes with plugin	Yes	Yes	Yes	Med	Active	No	http://www.qgis.org/en/site/
ROAM (QGIS app)	0	Yes	Yes	Maybe	Yes	Med-high	Active	Yes	https://github.com/DMS-Aus/Roam
Software packages and applications for developing web-based applications									
Python back-end based Leaflet applications e.g., Django or Flask	0	Yes	Yes	Yes	Yes	High	Active	Yes	https://www.djangoproject.com/ http://flask.pocoo.org/
Javascript back-end based Leaflet applications using Node	0	Yes	Yes	Yes	Yes	High	Active	Yes	https://nodejs.org/en/
Kivy	0	No [10]	Yes	Yes	Yes	High	Active	No	https://kivy.org/#home
RShiny, RShinydashboard	0	Yes	Yes	Yes	Yes	High	Active	Yes	https://shiny.rstudio.com/ https://rstudio.github.io/shinydashboard/
Jupyter dashboard	0	Maybe	Yes	Yes	Yes	High	Active	Yes	https://github.com/jupyter/dashboards
Tableau	0	Yes	Yes	Yes	Yes	Med-high	Active	Yes	https://www.tableau.com/
Spotfire	1000-9999	Yes	Yes	Yes	Yes	Med-high	Active	No	http://spotfire.tibco.com/
Plotly (Dash)	0	Yes	Yes	Yes	Yes	Med-high	Active	Yes	https://plotly/products/dash/

[1] Approximate cost (£): 0, 1–99, 100–999, 1000–9999, and > 10,000; [2] Is the user interface designed for non-expert/group touch user interface, i.e., is the navigation easy to use? yes or no; [3] Is there part of the screen that includes a map? yes or no; [4] Can logic models potentially be viewed and edited? yes or no; [5] Can the project team with their Python, R and JavaScript expertise develop the application and extend it? yes or no; [6] What is the potential for scientific innovation? low = old application and widely used, med = established application but not widely used, high = new application; [7] Is it still being supported? inactive = little evidence of activity over the past year, active = evidence of activity over the past year; [8] A "maybe" indicates that information was not available to judge against these criteria; [9] ArcMap not free to use; [10] Not on Windows mobile devices.

5. Results

5.1. Phase 1

The stakeholder interviews confirmed the need for tools that supported an outcomes-based approach with a central role for facilitation in place-based studies. There is a wide range of software tools and applications that are or could be used to aid facilitated discussions of place-based studies. Using the list of screening criteria (Table 1) we reduced a long list of 19 software options to seven for further hands-on testing and development. The long list of software options were classified into four broad groups. These were: existing adaptive management logic modelling applications e.g., Miradi (https://www.miradi.org/), existing desktop participatory modelling applications e.g., QUICKScan (http://www.quickscan.pro/), GIS applications e.g., QGIS (http://www.qgis.org/en/site/), and a range of software packages for developing web based applications e.g., R Shinydashboard (https://rstudio.github.io/shinydashboard/). In order to review these options in a structured way, we developed a set of seven criteria. These criteria were informed by our own previous work in this area, the interviews with stakeholders [45] and conversations with colleagues. These criteria are more a result of the researchers' reflections following early conversations with stakeholders, than the product of detailed analysis. Nonetheless, they comprise a useful starting point for the development of the activities described in this paper. These criteria were as follows:

The tool/software:

1. *Should be free at the point of use.* This is an essential pre-requisite of the outcomes-based approach, which requires transparency and effective exchange of knowledge and data between stakeholders at all points in the process chain. Commercially licensed software is typically not free to use or share and thus cannot meet this basic requirement.
2. *Should work on touch devices like mobile phones, tablets and larger touch tables.* Touch-enabled devices are now very widely used, which is of itself an important reason why their use should be considered for EDSS. Touch user interfaces are popular because they facilitate user interaction, and, on larger devices, allow multiple stakeholders to interact at the same time. To ensure that any application can be widely shared, it should function on a range of devices, but without touch capability, the EDSS is restricted to ordinary PCs. At the same time, touch-enabled applications can easily be used with a keyboard and mouse if necessary, but standard applications frequently do not work on touch user interface devices. At present, many applications that have potential for use as EDSS (e.g., QGIS) are not designed for touch screen use, and we found that these tended to function poorly on touch devices. While a large touchtable is ideal for facilitating knowledge exchange between multiple participants (i.e., as in our workshop), outdoor workers, e.g., in forestry or agriculture, are likely to prefer an application which can run on portable devices like smartphones or tablets. In general, applications that are responsively designed to run on an ordinary smartphone may often be suitable for other devices as well.
3. *Should have map-based functionality for users to interact with spatial information e.g., information on fields and other features related to land and water management.* The focus on spatial information is crucial, since nearly all environmental decision-makers use maps as a means of communicating information about interventions in the landscape. At the same time, to establish a genuine process of knowledge exchange, information needs to flow in both directions, so some form of user interaction is necessary, either in terms of allowing the user to pull data from the system, or in actually offering the possibility to upload and work with users' own data.
4. *Should include functionality for outcomes-based logic models i.e., linking land management to a range of outcomes.* This apparently very specific requirement can be seen to have broader implications for environmental decision-support. The software needs to be able to link the information provided to the user to desirable or required outcomes in a dynamic way. In this way it is possible to understand the implications of different interventions—e.g., for example, by adding a connection between land use type in a holding and sediment or nutrient run-off, a user would potentially

be able to evaluate the likely outcomes of land use changes in terms of water quality and other policy and management objectives.

5. *Should allow developers and end-users to develop and extend the software/existing application.* The ability to modify and extend the software to add or change functionality was thought to be important, since environmental management is a dynamic sphere of action in which new datasets, plans and policies and strategies are frequently emerging. It should be clarified that we did not anticipate that all users should want or need to do this, but that it should in principle be possible for the development team.

6. *Should have potential for scientific innovation.* This is a constraint that may have little interest from the point of view of an end-user, but is important for researchers in today's competitive science environment. A tool or approach that cannot show some potential to advance scientific practice or understanding cannot be expected to hold the interest of the scientific community, and makes it more difficult to obtain funding. It should be emphasized that we are not necessarily talking about technological innovation—finding new ways for stakeholder groups to interact around a problem would also be a worthwhile scientific innovation.

7. *Should be actively maintained, preferably through a large, open user community.* Given the previously mentioned interest in software that is free (at least at the point of use), it is important to distinguish between projects that are no longer actively maintained and those that are, preferable with a large or active user community. This is an important factor in ensuring the ability to modify or extend the software in the future—changes in PC hardware or software over time mean that software needs to be actively maintained. The multi-purpose software environment R and the popular free GIS package QGIS, for example, are actively maintained, with vibrant user communities, whereas the GIS software ILWIS, while still maintained, seems to be less actively developed and the user community, by comparison, is small.

Using these screening criteria we selected seven potential software options for more detailed hands-on testing (see Table 1). Six of these seven options were software packages and applications for developing web-based applications. The other option was ROAM which is QGIS application designed to aid field-based collection of data using touch devices. Software options were classified in four broad groups (Table 1), which are discussed as follows.

5.1.1. Existing Adaptive Management Logic Model Applications

The only software known to the authors in this group is Miradi, a software application intended to help adaptive management of conservation projects. Though there are a range of EDSS designed to operate within adaptive management contexts, e.g., the InVEST nutrient delivery model [49], these do not include explicit treatment of logic modelling approaches. Miradi was thus included as the only software explicitly designed to facilitate the logic modelling outcomes-based approach. Logic chain modelling is explicitly implemented through a flowchart view, and other functionality to assist work planning and budgeting and assessment of viability of targets. At present the software has no spatial component, but stated future plans for the software include development of a spatial mapping capability.

5.1.2. Existing Standalone Participatory Modelling Applications

This group comprises a small selection from the large range of modelling applications currently in existence that include explicit participatory modelling functionality or that have been used in participatory processes and is not intended to be representative or comprehensive (for more detailed reviews of participatory modelling software options and approaches see e.g., [50,51]). QUICKScan is a well-known commercial software application developed by Alterra, at Wageningen Netherlands, which is widely applied, including in Scottish cases [52]. It offers a rule-based approach to help stakeholders negotiate trade-offs in resource management contexts, and is explicitly spatial.

This seemed quite appropriate for our case, but as a non-modifiable and non-free software, could not be included in our selection. Metronamica is another software application with 20 years + development behind it, very popular in the land-use modelling community. Though it does not offer specific participatory functionality, it is frequently used in decision-support contexts (e.g., [1]), and for working with stakeholders [35,53], even if its complexity means that stakeholders rarely use the model themselves.

5.1.3. GIS Modelling Applications

This group includes very well-known commercial applications like ArcMap as well as the popular open source application QGIS. These have been included because there are many spatial decision support applications built around these platforms, see e.g., for ArcMap [19,54] and for QGIS [55,56], but unfortunately, we found both of these to be poorly adapted to touchscreen environments, both in terms of operation (sluggish and buggy) and in terms of the user interface (tiny buttons, missing right mouse button functionality). The 3D Community Viz Public Participation GIS plugin for ArcMap was unusable on our touch table. The ROAM application, which provided a touch-screen friendly interface for QGIS was much more promising, and was selected for further investigation.

5.1.4. Software Packages and Applications for Developing Web-Based Applications

This last group stood apart from the other three software groups in that all of the tools reviewed in this group were web-based, and many were designed for mobile devices as well. Though these tools have the potential disadvantage of requiring adequate internet coverage, they can be designed to work off-line (e.g., progressive web applications). They represent a significant innovation in terms of user interaction and inter-platform flexibility. For all of the other groups, non-Windows operating system users are disadvantaged (with the notable exception of QGIS), and mobile device support is nearly non-existent. This group of applications stood out as addressing much more closely stakeholder's needs, e.g., mobile, touch-enabled and Free-and-Open source (FOSS). With these applications a clear trade-off could be identified between being able to use "as-is" and customizability—in many cases there seemed to be good potential to flexibly incorporate and link maps, tables and charts. One application in particular, Rshiny, and its dashboard extension, Rshinydashboard, was very promising in this regard, since these tools are part of the well-known FOSS environment "R", which is very popular with scientific users and has a large and active user community. For this reason, the advanced prototype was developed in Rshiny.

5.1.5. Summary of Findings from Software Survey

Overall, we found that by applying general criteria arising out of consultation with stakeholders we quickly reduced a long list of options to a handful of choices. This is not a reflection of a general scarcity of EDSS *per se*, rather an indication that the criteria we applied are not common features of EDSS.

5.2. Phase 2

5.2.1. Stakeholders' Principles for EDSS

Workshop participants identified 12 principles for development of a software application for supporting environmental decision-making to improve the management of land and freshwater resources in Scotland. These are shown in Table 2. The key points that emerged from this activity and the subsequent discussion are described as follows.

Table 2. Workshop participants' 12 principles related to stakeholder requirements (centre) and our own list of 7 proposed criteria (right).

No.	Principle	What Stakeholder Requirement Does This Relate to?	Which of the Seven Criteria Does This Relate to?
1	Updateability	None	2. Work on touch-enabled devices 5. Allow users to develop and extend 6. Potential for scientific innovation 7. Actively maintained
2	Agile development. Multiple versions	• Dashboard format • Support/ incentives and regulation	1. Free at the point of use. 5. Allow users to develop and extend 7. Actively maintained
3	Provides cost-benefits of options	• Helps indicate areas to "invest" in interventions	3. Map-based functionality for users to interact with spatial information 4. Include functionality for outcomes-based logic models i.e., linking land management to a range of outcomes
4	Clarity of objective: public interest, land manager, community	• Policy and on the ground connections	3. Map-based functionality for users to interact with spatial information. 4. Include functionality for outcomes-based logic models i.e., linking land management to a range of outcomes
5	Iterative: exploration of multiple scenarios and consequences	• Iterative evaluation of options	4. Should include functionality for outcomes-based logic models i.e., linking land management to a range of outcomes
6	Credible/transparent	• Provide evidence of benefits • Information needs to be provided in a digestible format	1. Free at the point of use. 2. Work on touch-enabled devices 3. Map-based functionality for users to interact with spatial information.
7	Credible and trustworthy	• Information needs to be provided in a digestible format [2]	4. Should include functionality for outcomes-based logic models i.e., linking land management to a range of outcomes
8	Practical—relates to users experience and needs	• Provide evidence of benefits [2] • Spatial location of interventions	3. Map-based functionality for users to interact with spatial information. 4. Include functionality for outcomes-based logic models i.e., linking land management to a range of outcomes
9	Accessible/easy to use	• Information needs to be provided in a digestible format [3] • Place-based • Information needs to be provided in a digestible format [4] • Spatial location of interventions [2]	1. Free at the point of use. 2. Work on touch-enabled devices 3. Map-based functionality for users to interact with spatial information.
10	Clear on operational scale	• Integrated approach for incentives, regulation and voluntary measures/ balance is key	3. Map-based functionality for users to interact with spatial information.
11	Dealing honestly with uncertainty: avoid large-scale uniformity of outcome, diversity manage via incentives/regulations	• Tools should promote diverse outcomes, even to similar places and objectives	4. Should include functionality for outcomes-based logic models i.e., linking land management to a range of outcomes 5. Allow users to develop and extend
12	Limited need for time-consuming tool. Must not take ages to develop	• Spatial location of interventions [3] • Information needs to be provided in a digestible format [5] • Tools should promote diverse outcomes, even to similar places and objectives [2] • Information needs to be provided in a digestible format [6]	2. Work on touch-enabled devices 3. Map-based functionality for users to interact with spatial information. 5. Allow users to develop and extend

Note: The number in superscript in the middle column denotes the number of times this requirement was listed. In some cases the same requirement appears more than once for a single principle, because this requirement was selected by more than one of the three groups.

1. Updateability: This was regarded as very important due to the rapidly evolving nature of environmental policy and the need to be adaptable and responsive. Participants emphasized the highly changeable nature of both the policy context and the data (Principle 1 in Table 2).
2. Practicability and usefulness: It was agreed that the tool should be both practical and useful. Since usefulness is subjective and therefore dependent on the end-user, involving end-users in the development process seemed like a good way to achieve this, e.g., through AGILE development principles [57] (Principle 2 in Table 2).
3. Clarity of purpose and objectives: Participants highlighted the need for clarity about the intended purpose and objective of the application. This is likely to depend on specific end users, further highlighting the need for a flexible, AGILE-type approach (see above).
4. Uncertainty, credibility and trustworthiness: Participants emphasized the need to deal honestly with uncertainty, both of data and of model/application outputs. This was regarded as having important implications for the credibility and trustworthiness of the information provided by the decision-support application (Principle 6 in Table 2). It was suggested that data should be filtered before providing it, since users may lack the knowledge to appreciate uncertainty. However, it was also appreciated that absolute certainty was not attainable, since scientists themselves frequently disagree about the trustworthiness of data or the reliability of the results. However, this was regarded as an important development principle even if it is not achievable in practice.
5. Openness about limitations/transparency: It was regarded as important to be transparent about what a tool can and cannot do. Transparency of the tool and of the decision-making process was a key principle (Principle 6 in Table 2). This is somewhat related to the issue of trust and credibility (above).
6. Specify intended scale of data and operation: In an issue related to the need for openness about uncertainty, participants raised the need to be clear about the intended cartographic scale of operation for the application. This can be partially addressed in practice by ensuring that datasets do not display at zoom levels that exceed their nominal scale.
7. Cost-benefit estimates for policy interventions: The need to provide cost-benefit information was highlighted by one of the participants, but others disagreed. One argument that was made against this idea was that it would be difficult to determine which actors would be best placed to make the cost-benefit assessment, since each actor would require a different cost-benefit analysis to suit their land/circumstances.
8. Diversity vs commonality: Although there was a high level of agreement about some aspects, for example, the importance of credibility (Principles 6 and 7 in Table 2) and accessibility (Principle 9 in Table 2), there was disagreement about others, e.g., appropriateness of cost-benefit analysis. This reminds us that end users are likely to have widely varying objectives in mind, and that it will be difficult to please all of them with a single application.
9. Decision support, not decision automation: Participants were keen to emphasise the need to be clear that the tool supports decisions, but does not actually make them.

In discussion, our own, more generalized principles (Figure 2) found general agreement. Several of our own suggested principles were also included by the participants, e.g., transparent, accessible and practical. However, participants observed that more clarity could be provided by setting them out as sentences (rather than as single words). Following the workshop, we drew up a revised set of general principles incorporating the principles suggested by stakeholders and following their advice (Table 3). This is intended as a synthesis, not a replacement, of the detailed principles set down in Table 2.

Table 3. Integrated general principles for application development.

Approach
The approach will help facilitate decisions about land and water resources.
It will aim to be integrative through considering a range of environmental and financial outcomes.
It will aim to support adaptive management though clarity of objectives/outcomes, and linking with evidence that supports exploration of those options to achieve those objectives/outcomes.
The approach will be designed to be easy to use and efficient.

Digital application
It will be accessible for anyone to use.
It will be relevant and practical for land managers.
It will aim to be credible, with transparency in the information and methods used.
It will be designed to be updateable with new information as it comes available.

5.2.2. Facilitated Discussion around the Prototype Application

The prototype application (Figure 3), as we had hoped, since we had developed it in response to stakeholder criteria, was well-received. Though the application meets many of these requirements, being FOSS, optimized for touchscreen use and multi-platform, it does depend on the R software environment, which is not currently supported as standard on mobile phones, though it does appear to be possible to run it [58]. Stakeholders particularly appreciated the smooth map scrolling provided by the leaflet application, a GIS map server package available under R. The users appreciated the dashboard format, with a large map window occupying most of the screen and user controls and a linked chart output in a box to the right. The application was tested with land use data (Land Cover Map 2007 [59], which it allowed users to interact with (enquiry of land use type, clipping out a polygon and querying land use area in the polygon area). The land use data generated an interesting discussion, and stakeholders highlighted the importance of providing scale information, clearly naming the source of the land use data, and adequately informing about its known limitations. Participants were not widely agreed about the reliability of the national land use dataset, which led to a detailed discussion around reliability and trustworthiness of information presented.

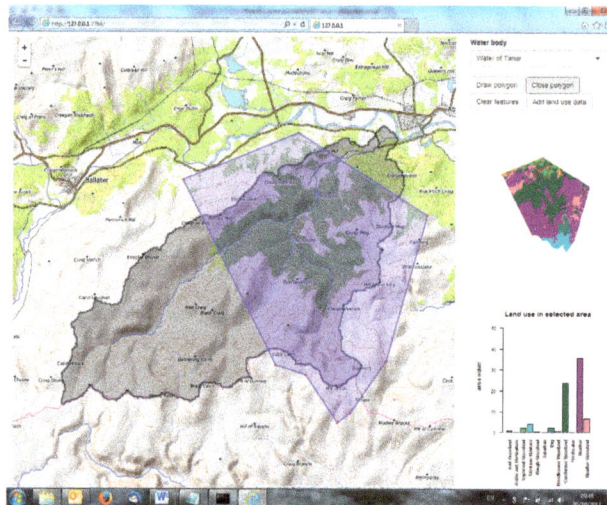

Figure 3. Prototype application running on Windows. The screen capture shows the Water of Tanar sub-catchment of the river Dee in Aberdeenshire, Scotland. The user has drawn a polygon, which has been clipped out of the land use base map (LCM2007) and is displayed in the box to the right of the map window. Below it, the area of each land use in the polygon is displayed as a histogram.

6. Discussion

Outcomes-Based Environmental Management: What Do Users Really Want?

Our process is ongoing, and an exhaustive review of EDSS is beyond the scope of this paper. For this reason, our findings should be considered as propositions for discussion, not statements of certainty. Nonetheless, our work does suggest that some important aspects of EDSS design may often not receive adequate attention. The new breed of mobile and web-based tools offers new opportunities for application development, but at the same time, to develop software systems effectively it is necessary to understand user's functional and non-functional requirements. This lies at the heart of producing useful software and is an established principle in software development that developers of EDSS could learn from. In order to understand user requirements, structured participatory processes like the one we have presented in this paper are necessary. The process of co-construction of an application typically passes through a series of key steps relating to facilitated participatory activities. Stakeholders can analyze their needs as environmental decision-makers, identify key development principles for a software application designed to address them and road-test an advanced prototype application.

We discovered that the tool our stakeholders wanted does not much resemble most published examples of EDSS used by the scientific community (see e.g., [60]). Our application should be free to use, touch and mobile-device enabled, part of an existing active development chain or community and be clearly oriented to understanding and evaluating environmental policy options in terms of outcomes, e.g., showing how a riparian buffer strip can improve water quality in a given location, or showing how land use change may influence diffuse pollution. It will be intended for use, probably in the field, by land managers and regulators, e.g., farmers, protected area managers or utilities operators.

One possible reason why the EDSS that has been conceptualized by our stakeholder group is so different to existing EDSS found in the literature is because of the specific requirements of our study and its context, e.g., catchment management in Scotland, adaptive management framework, and emphasis on understanding effectiveness of interventions through outcomes. But these aspects are also recognized as important in environmental management further afield. Though adaptive management has some well-recognized limitations (see e.g., [61]) it is still widely regarded, at least in modified guise of adaptive co-management (ACM) as playing a valuable role in good environmental governance [62]. At the same time, outcomes-based approaches are increasingly seen as best practice in environmental assessment, decision-making and stakeholder engagement [63]. Therefore the apparent absence of appropriate tools probably does indicate a real need.

In fact, there seems to be a genuine mismatch between what environmental decision-makers like our stakeholders want, and what EDSS developers, who are typically scientists, tend to produce (Table 4). This impression is supported by other researchers in EDSS (e.g., [18]), and is relevant to the science-policy divide in general (see e.g., [64]).

Table 4. The mismatch between EDSS that scientists have tended to develop, and EDSS that environmental decision-makers may want.

EDSS that Scientists Have Tended to Develop	EDSS that Decision-Makers May Want
Platform	
Micro-computer oriented, Often Windows operating system only	Mobile device oriented (touchscreen), multi-platform
Cost model	
Proprietary	Free and Open-Source
Application type	
Stand-alone, out of the box solution that the user cannot easily modify	Flexible, integrated (web-based), easily customizable
Purpose	
Understanding a complex problem	Deciding from a range of options
Exploring future scenarios	Evaluating cost-benefit
Predicting tendencies	Linking science and policy to show clear outcomes
	Synthesizing complex information into a digestible message
Implicit role of scientist	
Selling knowledge systems	Facilitator and provider of free and open advice
External expert	Participant and stakeholder

There are four key reasons for this mismatch in general terms.

1. *Problem-solving stage.* The kinds of EDSS that help stakeholders understand complex interactions, explore future scenarios, or predict existing tendencies are generally not suitable, or really intended, for use directly by practitioners, but are rather intended to supply information to them in the form of reports and briefings based on model results. Generally, practitioners do not need or want to run land use simulations for 2050 or experiment with the effect of different types of forcing mechanisms on climate. An EDSS that a farmer might hold in their hand to allow them to explore a range of options for managing a river catchment belongs to a different part of the process at a later stage in the policy process. This kind of EDSS operates under the assumption that first type described is already embedded in the policy-making process. This type of EDSS belongs to a different problem-solving stage, where, in the words of [65], "we agree to stop worrying about generalities and focus instead on impacts, mitigation, compensation, and accommodation". There is a fundamental need to better understand the requirements of users at the policy implementation stage.

2. *Type of decision-maker and level of action.* A related question is the type of stakeholder to whom the EDSS is directed and the level of action at which they are typically operating. Water regulators with responsibility for large land areas, for example, may have advanced scientific training, and want to make use of integrated hydrological models like SWAT [66]. Local level actors, like municipal planners, conservation volunteers, or farmers, are unlikely to have such specific training and will have radically different questions they need to answer. The hypothetical farmer referred to above is making decisions at the scale of the land parcel, unlike, say, an agriculture minister, who is interested to know, in broader terms, about land use in general to inform land use policy more broadly. Many EDSS found in the scientific literature are concerned with these higher level or strategic decisions, while EDSS directed at the scale of the land parcel tend to be under-represented in the scientific literature (though see e.g., [10]). One of the most important priorities for our stakeholder group was the provision of reliable scientific information in an accessible and digestible format, rather than solution of complex analytical problems, or estimations of long-term tendencies.

3. *Technological development time-lag.* The significant time lag between submitting research proposals and the award of funding, and between development of a model or system and publishing it in a journal means that the most recently published EDSS are based on conceptualizations of technology (if not actual availability) from three or four years ago. Given the pace of change in information systems development, there is a significant likelihood that EDSS of the kind that emerged from our participatory process are currently being developed but are poorly represented in literature. An interesting review of this topic in the context of rural development can be found in [67].

4. *Change in the role of the scientist.* As can be appreciated from Table 4, the type of EDSS conceptualized on the right requires the scientist to play a different, perhaps less glamorous role, than that of aloof, objective expert of popular imagination. To help achieve sustainability objectives in land and water management scientific stakeholders should focus on facilitation and knowledge transfer at the level of implementation of concrete actions, as well as knowledge production activities higher up the chain.

Designers of EDSS need to move away from developing tools for standard desktop software and systems and make full use of mobile and web-based technologies with which people are becoming increasingly familiar. Web dashboards, already popular as "business intelligence" tools, provide the user with a range of integrated graphical outputs, e.g., a map, a histogram and a data table and offer an idea of how such applications might look. FOSS tools like Rshiny and Rshiny dashboard which facilitate web-development from inside widely supported scientific software environments with active user communities (in this case, R), seems like a promising direction of travel.

Providing complex information under high uncertainty in an easily digestible format is a major challenge. Our stakeholders were clear that the limitations and level of uncertainty in complex scientific information needed to be clearly explained in any application, and that great care should be taken not to provide information out of context or at inappropriate spatial scales.

Finally, our results highlight the importance of salience, credibility and legitimacy (e.g., [68]) in the development of EDSS. The group had major concerns around the trustworthiness of the data presented in the application (credibility). A process of co-development can help to increase credibility, especially if the data behind the application are well-known to, or even provided by, the stakeholders themselves. This also enhances legitimacy, by helping stakeholders to take ownership of the process and the tool. We suggest, as the title of our article indicates, that finding out "what users really need" is essential to ensure salience (relevance) of EDSS to its potential users, and that co-construction of any such tool, through a process similar to that described here, is a useful way forward.

7. Conclusions

Developing software applications to support land managers decisions under outcomes-based approaches to adaptive environmental management requires better stakeholder engagement practices to understand their needs and development requirements, and a stronger focus on new digital tools and open software and data. We have presented an example of a stakeholder-centered process to develop one such tool. By focusing on understanding and managing outcomes of environmental improvement measures, a preliminary review based on initial requirements led to the rejection of a large range of currently available tools and software as unsuitable. Many do not run effectively on the mobile and web-based devices that land managers typically use, and most are proprietary systems that require an upfront financial outlay and do not allow users to modify the source code or customize the application to suit their own needs. These aspects restrict their use to mostly scientific stakeholders and do little to facilitate environmental decision-making at the scale of the land parcel. What stakeholders really need are mobile and web-based devices that are free at the point of use and flexible enough to be easily adapted to meet changing requirements and different user needs.

Under the late nineties and early noughties "out-of-the-box toolkit" paradigm, scientists sought to package their knowledge into marketable products and systems under the assumption that clients would materialize with fully-formed problems requiring solutions. Since then, the world has moved on. Scientists now understand that this kind of hands-off back-office knowledge delivery is ineffective; these kinds of tools are rarely used. Not only do land-managers not want to use a complex computational model, they need software that answers specific questions like "how much will it cost to build a riparian buffer strip along the river that bounds my land, and to what extent will this reduce surface run-off from my arable fields?", or "what specific interventions would be most appropriate in this part of the catchment, and what incentives and regulations are likely to be relevant?" To provide better answers to these questions, scientific stakeholders need to play strong facilitation roles at all points in the development process, including problem-framing, stakeholder engagement, application development and use.

Acknowledgments: The research described was funded by the Rural & Environment Science & Analytical Services Division of the Scottish Government. We are very grateful to all of the stakeholders who participated as part of this process.

Author Contributions: Richard J. Hewitt and Christopher J. A. Macleod carried out the research described and wrote the paper.

Conflicts of Interest: The authors declare no conflict of interest.

References

1. Kok, K.; van Delden, H. Combining two approaches of integrated scenario development to combat desertification in the Guadalentin watershed, Spain. *Environ. Plan. B Plan. Des.* **2009**, *36*, 49–66. [CrossRef]
2. McIntosh, B.S.; Ascough, J.C.; Twery, M.; Chew, J.; Elmahdi, A.; Haase, D.; Harou, J.J.; Hepting, D.; Cuddy, S.; Jakeman, A.J.; et al. Environmental decision support systems (EDSS) development—Challenges and best practices. *Environ. Model. Softw.* **2011**, *26*, 1389–1402. [CrossRef]
3. Holmes, J.; Clark, R. Enhancing the use of science in environmental policy-making and regulation. *Environ. Sci. Policy* **2008**, *11*, 702–711. [CrossRef]
4. Winder, N. Successes and problems when conducting interdisciplinary or transdisciplinary (= integrative) research. In *Interdisciplinarity and Transdisciplinarity in Landscape Studies: Potential and Limitations*; Delta Program: Wageningen, The Netherlands, 2003; pp. 74–90.
5. Macleod, C.K.; Blackstock, K.; Haygarth, P. Mechanisms to improve integrative research at the science-policy interface for sustainable catchment management. *Ecol. Soc.* **2008**, *13*, 48. [CrossRef]
6. Allen, C.R.; Fontaine, J.J.; Pope, K.L.; Garmestani, A.S. Adaptive management for a turbulent future. *J. Environ. Manag.* **2011**, *92*, 1339–1345. [CrossRef] [PubMed]
7. Allan, C.; Stankey, G.H. *Adaptive Environmental Management*; Springer the Netherlands: Dordrecht, The Netherlands, 2009.
8. Rosenhead, J.; Mingers, J. *Rational Analysis for a Problematic World Revisited*; John Wiley & Sons: Hoboken, NJ, USA, 2001.
9. Shim, J.P.; Warkentin, M.; Courtney, J.F.; Power, D.J.; Sharda, R.; Carlsson, C. Past, present, and future of decision support technology. *Decis. Support Syst.* **2002**, *33*, 111–126. [CrossRef]
10. Sharifi, M.A.; van Keulen, H. A decision support system for land use planning at farm enterprise level. *Agric. Syst.* **1994**, *45*, 239–257. [CrossRef]
11. Matthews, K.B.; Sibbald, A.R.; Craw, S. Implementation of a spatial decision support system for rural land use planning: Integrating geographic information system and environmental models with search and optimisation algorithms. *Comput. Electron. Agric.* **1999**, *23*, 9–26. [CrossRef]
12. Theobald, D.M.; Hobbs, N.T.; Bearly, T.; Zack, J.A.; Shenk, T.; Riebsame, W.E. Incorporating biological information in local land-use decision making: Designing a system for conservation planning. *Landsc. Ecol.* **2000**, *15*, 35–45. [CrossRef]
13. Holling, C.S. *Adaptive Environmental Assessment and Management*; John Wiley & Sons: Hoboken, NJ, USA, 1978.
14. Simon, H. *The New Science of Management Decision*; Harper Brothers: New York, NY, USA, 1960.
15. Gorry, G.A.; Scott Morton, M.S. A Framework for Management Information Systems. 1971. Available online: https://dspace.mit.edu/bitstream/handle/1721.1/47936/frameworkformana00gorr.pdf (accessed on 5 October 2017).
16. Engelen, G.; White, R.; Uljee, I.; Drazan, P. Using cellular automata for integrated modelling of socio-environmental systems. *Environ. Monit. Assess.* **1995**, *34*, 203–214. [CrossRef] [PubMed]
17. Engelen, G.; Uljee, I.; White, R. Vulnerability Assessment of Low-lying Coastal Areas and Small Islands to Climate Change and Seal Level Rise—Phase 2: Case Study St. Lucia. 1997. Available online: http://www.dpi.inpe.br/cursos/tutoriais/modelagem/software_demos/simlucia/SimLuciaManual.pdf (accessed on 5 October 2017).
18. Winograd, M.; Pérez-Soba, M.; Verweij, P. QUICKScan: A Pragmatic Approach for Decision Support in Ecosystem Services Assessment and Management. In *Handbook on the Economics of Ecosystem Services and Biodiversity*; Edward Elgar Cheltenham: Cheltenham, UK, 2013.
19. Flacke, J.; De Boer, C. An Interactive Planning Support Tool for Addressing Social Acceptance of Renewable Energy Projects in The Netherlands. *ISPRS Int. J. Geoinf.* **2017**, *6*, 313. [CrossRef]
20. Voinov, A.; Bousquet, F. Modelling with stakeholders. *Environ. Model. Softw.* **2010**, *25*, 1268–1281. [CrossRef]
21. Van den Belt, M. *Mediated Modeling: A System Dynamics Approach to Environmental Consensus Building*; Island Press: Washington, DC, USA, 2004.
22. Barreteau, O.; Antona, M.; D'Aquino, P.; Aubert, S.; Boissau, S.; Bousquet, F.; Daré, W.; Etienne, M.; Le Page, C.; Mathevet, R.; et al. Our Companion Modelling Approach. 2003. Available online: http://jasss.soc.surrey.ac.uk/6/2/1.html (accessed on 31 March 2003).

23. Reed, M.S.; Evely, A.C.; Cundill, G.; Fazey, I.; Glass, J.; Laing, A.; Newig, J.; Parrish, B.; Prell, C.; Raymond, C.; et al. What is social learning? *Ecol. Soc.* **2010**, *15*, r1. [CrossRef]
24. Volk, M.; Lautenbach, S.; van Delden, H.; Newham, L.T.; Seppelt, R. How can we make progress with decision support systems in landscape and river basin management? Lessons learned from a comparative analysis of four different decision support systems. *Environ. Manag.* **2010**, *46*, 834–849. [CrossRef] [PubMed]
25. Mackay, E.B.; Wilkinson, M.E.; Macleod, C.J.; Beven, K.; Percy, B.J.; Macklin, M.G.; Quinn, P.F.; Stutter, M.; Haygarth, P.M. Digital catchment observatories: A platform for engagement and knowledge exchange between catchment scientists, policy makers, and local communities. *Water Resour. Res.* **2015**, *51*, 4815–4822. [CrossRef]
26. Borowski, I.; Hare, M.P. Exploring the gap between water managers and researchers: Difficulties of model-based tools to support practical water management. *Water Resour. Manag.* **2007**, *21*, 1049–1074. [CrossRef]
27. Uran, O.; Janssen, R. Why are spatial decision support systems not used? Some experiences from the Netherlands. *Comput. Environ. Urban Syst.* **2003**, *27*, 511–526. [CrossRef]
28. Bagstad, K.J.; Semmens, D.J.; Waage, S.; Winthrop, R. A comparative assessment of decision-support tools for ecosystem services quantification and valuation. *Ecosyst. Serv.* **2013**, *5*, 27–39. [CrossRef]
29. Leffingwell, D.; Widrig, D. *Managing Software Requirements: A Unified Approach*; Addison-Wesley Professional: Boston, MA, USA, 2000.
30. Norman, D.A.; Draper, S.W. *User Centered System Design*; Lawrence Erlbaum Associates: Hillsdale, NJ, USA, 1986.
31. Gulliksen, J.; Göransson, B.; Boivie, I.; Blomkvist, S.; Persson, J.; Cajander, Å. Key principles for user-centred systems design. *Behav. Inf. Technol.* **2003**, *22*, 397–409. [CrossRef]
32. Sørensen, C.G.; Fountas, S.; Nash, E.; Pesonen, L.; Bochtis, D.; Pedersen, S.M.; Basso, B.; Blackmore, S.B. Conceptual model of a future farm management information system. *Comput. Electron. Agric.* **2010**, *72*, 37–47. [CrossRef]
33. Sørensen, C.G.; Pesonen, L.; Bochtis, D.D.; Vougioukas, S.G.; Suomi, P. Functional requirements for a future farm management information system. *Comput. Electron. Agric.* **2011**, *76*, 266–276. [CrossRef]
34. Kaloxylos, A.; Groumas, A.; Sarris, V.; Katsikas, L.; Magdalinos, P.; Antoniou, E.; Politopoulou, Z.; Wolfert, S.; Brewster, C.; Eigenmann, R.; et al. A cloud-based Farm Management System: Architecture and implementation. *Comput. Electron. Agric.* **2014**, *100*, 168–179. [CrossRef]
35. Hewitt, R.; Van Delden, H.; Escobar, F. Participatory land use modelling, pathways to an integrated approach. *Environ. Model. Softw.* **2014**, *52*, 149–165. [CrossRef]
36. Margoluis, R.; Stem, C.; Swaminathan, V.; Brown, M.; Johnson, A.; Placci, G.; Salafsky, N.; Tilders, I. Results chains: A tool for conservation action design, management, and evaluation. *Ecol. Soc.* **2013**, *18*, 22. [CrossRef]
37. Macleod, C.J.A. How Can Logic Modelling Improve the Planning, Monitoring and Evaluation of Policy Measures and Wider Interventions for Multiple Benefits? 2016. Available online: http://www.hutton.ac.uk/sites/default/files/files/RESAS124_O3_D3_1_v0_5Final.pdf (accessed on 5 October 2017).
38. Funnell, S.C.; Rogers, P.J. *Purposeful Program Theory: Effective Use of Theories of Change and Logic Models*; John Wiley & Sons: Hoboken, NJ, USA, 2011.
39. Treasury, H.M. The Magenta Book: Guidance for Evaluation. 2011. Available online: https://www.gov.uk/government/uploads/system/uploads/attachment_data/file/220542/magenta_book_combined.pdf (accessed on 5 October 2017).
40. Scottish Government. Getting the Best from Our Land. A Land Use Strategy for Scotland 2016–2021. 2016. Available online: http://www.gov.scot/Resource/0050/00505253.pdf (accessed on 22 March 2016).
41. Scottish Government. Scottish Budget Spending Review 2007. Chapter 8: A National Performance Framework. 2007. Available online: http://www.gov.scot/Publications/2007/11/13092240/9 (accessed on 5 October 2017).
42. Pahl-Wostl, C. A conceptual framework for analysing adaptive capacity and multi-level learning processes in resource governance regimes. *Glob. Environ. Chang.* **2009**, *19*, 354–365. [CrossRef]
43. Scottish Government. Rural Payments and Services: Management Options and Capital Items. 2017. Available online: https://www.ruralpayments.org/publicsite/futures/topics/all-schemes/agri-environment-climate-scheme/management-options-and-capital-items/#32412 (accessed on 27 January 2017).

44. Macleod, C.J.A.; Blackstock, K.; Brown, K.; Eastwood, A.; Gimona, A.; Prager, K.; Irvine, J. Adaptive Management: An Overview of the Concept and Its Practical Application in the Scottish Context. 2016. Available online: http://www.hutton.ac.uk/sites/default/files/files/research/srp2016-21/RESAS_srp143_aD1_ReportOnRelevantAdaptiveManagementApproachesForScotland_v0.8Final.pdf (accessed on 5 October 2017).

45. Macleod, C.J.A.; Hewitt, R.J. Summary of research on developing a more integrated approach to land and water management using incentives and regulations for the delivery of multiple benefits: Exploring national and regional level stakeholder views and needs. Unpublished work, 2017.

46. Gould, J.D.; Lewis, C. Designing for usability: Key principles and what designers think. *Commun. ACM* **1985**, *28*, 300–311. [CrossRef]

47. Ball, J. Towards a methodology for mapping 'regions for sustainability' using PPGIS. *Prog. Plan.* **2002**, *58*, 81–140. [CrossRef]

48. Macleod, C.J.A.; Hewitt, R. Workshop Summary: Developing an Outcome-Based Approach for Understanding the Effectiveness of Interventions in Catchments for Multiple Benefits. Unpublished work, 2017.

49. InVEST. Nutrient Delivery Ratio Model. Available online: http://data.naturalcapitalproject.org/nightly-build/invest-users-guide/html/ndr.html (accessed on 5 October 2017).

50. Voinov, A.; Kolagani, N.; McCall, M.K.; Glynn, P.D.; Kragt, M.E.; Ostermann, F.O.; Pierce, S.A.; Ramu, P. Modelling with stakeholders—Next generation. *Environ. Model. Softw.* **2016**, *77*, 196–220. [CrossRef]

51. Basco-Carrera, L.; Warren, A.; van Beek, E.; Jonoski, A.; Giardino, A. Collaborative modelling or participatory modelling? A framework for water resources management. *Environ. Model. Softw.* **2017**, *91*, 95–110. [CrossRef]

52. Dick, J.; Verweij, P.; Carmen, E.; Rodela, R.; Andrews, C. Testing the ecosystem service cascade framework and QUICKScan software tool in the context of land use planning in Glenlivet Estate Scotland. *Int. J. Biodivers. Sci. Ecosyst. Serv. Manag.* **2017**, *13*, 12–25. [CrossRef]

53. Volkery, A.; Ribeiro, T.; Henrichs, T.; Hoogeveen, Y. Your vision or my model? Lessons from participatory land use scenario development on a European scale. *Syst. Pract. Action Res.* **2008**, *21*, 459–477. [CrossRef]

54. Graymore, M.L.; Wallis, A.M.; Richards, A.J. An Index of Regional Sustainability: A GIS-based multiple criteria analysis decision support system for progressing sustainability. *Ecol. Complex.* **2009**, *6*, 453–462. [CrossRef]

55. Novaczek, I.; MacFadyen, J.; Bardati, D.; MacEachern, K. *Social and Cultural Values Mapping as a Decision-Support Tool for Climate Change Adaptation*; The Institute of Island Studies, University of Prince Edward Island: Charlottetown, PE, Canada, 2011.

56. Wangdi, K.; Banwell, C.; Gatton, M.L.; Kelly, G.C.; Namgay, R.; Clements, A.C. Development and evaluation of a spatial decision support system for malaria elimination in Bhutan. *Malar. J.* **2016**, *15*, 180. [CrossRef] [PubMed]

57. Cockburn, A. *Agile Software Development*; Addison-Wesley: Boston, MA, USA, 2002.

58. Teoh, T. Running R and RStudio from an Android Mobile Phone or Tablet. Available online: http://tteoh.com/technology/2016/10/20/r-rstudio-mobile-android/ (accessed on 20 October 2016).

59. Land Cover Map 2007. Available online: https://www.ceh.ac.uk/services/land-cover-map-2007 (accessed on 5 October 2017).

60. Grêt-Regamey, A.; Sirén, E.; Brunner, S.H.; Weibel, B. Review of decision support tools to operationalize the ecosystem services concept. *Ecosyst. Serv.* **2017**, *26*, 306–315. [CrossRef]

61. Allan, C.; Curtis, A. Nipped in the bud: Why regional scale adaptive management is not blooming. *Environ. Manag.* **2005**, *36*, 414–425. [CrossRef] [PubMed]

62. Plummer, R.; Baird, J.; Dzyundzyak, A.; Armitage, D.; Bodin, Ö.; Schultz, L. Is Adaptive Co-management Delivering? Examining Relationships between Collaboration, Learning and Outcomes in UNESCO Biosphere Reserves. *Ecol. Econ.* **2017**, *140*, 79–88. [CrossRef]

63. Baldwin, C.; Hamstead, M. *Integrated Water Resource Planning: Achieving Sustainable Outcomes*; Routledge: Abingdon, UK, 2014.

64. Bradshaw, G.A.; Borchers, J.G. Uncertainty as Information: Narrowing the Science-Policy Gap. *Conserv. Ecol.* **2000**, *4*, 7. [CrossRef]

65. Cartledge, K.; Dürrwächter, C.; Hernandez Jimenez, V.; Winder, N. Making sure you solve the right problem. *Ecol. Soc.* **2009**, *14*, r3. [CrossRef]

66. Arnold, J.G.; Srinivasan, R.; Muttiah, R.S.; Williams, J.R. Large area hydrologic modeling and assessment part I: Model development. *J. Am. Water Resour. Assoc.* **1998**, *34*, 73–89. [CrossRef]
67. Salemink, K.; Strijker, D.; Bosworth, G. Rural development in the digital age: A systematic literature review on unequal ICT availability, adoption, and use in rural areas. *J. Rural Stud.* **2017**, *54*, 360–371. [CrossRef]
68. Cash, D.; Clark, W.C.; Alcock, F.; Dickson, N.M.; Eckley, N.; Jäger, J. Salience, Credibility, Legitimacy and Boundaries: Linking Research, Assessment and Decision Making. 2002. Available online: https://papers.ssrn.com/sol3/papers.cfm?abstract_id=372280 (accessed on 5 October 2017).

environments

MDPI

Article

Rapid Urban Growth in the Kathmandu Valley, Nepal: Monitoring Land Use Land Cover Dynamics of a Himalayan City with Landsat Imageries

Asif Ishtiaque [1,*], Milan Shrestha [2] and Netra Chhetri [3]

1   School of Geographical Sciences and Urban Planning, Arizona State University, Tempe, AZ 85287, USA
2   School of Sustainability, Arizona State University, Tempe, AZ 85287, USA; Milan.Shrestha@asu.edu
3   School for the Future of Innovation in Society, Arizona State University, Tempe, AZ 85287, USA;
    Netra.Chhetri@asu.edu
*   Correspondence: Asif.Ishtiaque@asu.edu; Tel.: +1-480-358-5962

Received: 11 September 2017; Accepted: 7 October 2017; Published: 8 October 2017

Abstract: The Kathmandu Valley of Nepal epitomizes the growing urbanization trend spreading across the Himalayan foothills. This metropolitan valley has experienced a significant transformation of its landscapes in the last four decades resulting in substantial land use and land cover (LULC) change; however, no major systematic analysis of the urbanization trend and LULC has been conducted on this valley since 2000. When considering the importance of using LULC change as a window to study the broader changes in socio-ecological systems of this valley, our study first detected LULC change trajectories of this valley using four Landsat images of the year 1989, 1999, 2009, and 2016, and then analyzed the detected change in the light of a set of proximate causes and factors driving those changes. A pixel-based hybrid classification (unsupervised followed by supervised) approach was employed to classify these images into five LULC categories and analyze the LULC trajectories detected from them. Our results show that urban area expanded up to 412% in last three decades and the most of this expansion occurred with the conversions of 31% agricultural land. The majority of the urban expansion happened during 1989–2009, and it is still growing along the major roads in a concentric pattern, significantly altering the cityscape of the valley. The centrality feature of Kathmandu valley and the massive surge in rural-to-urban migration are identified as the primary proximate causes of the fast expansion of built-up areas and rapid conversions of agricultural areas.

Keywords: urbanization; Kathmandu; Nepal; land use; land cover; sustainability; remote sensing

---

1. Introduction

The entire Himalayan region has been undergoing significant socio-economic changes in the last five decades [1,2]. However, the pace and ways some major cities in the foothills of Himalayas have transformed in the recent decades are unprecedented, raising sustainability concerns [3,4]. The Kathmandu Valley (KV) epitomizes this extraordinary urban growth occurring in the Himalayas. Located in the central hills of Nepal Himalayas, Kathmandu is the capital city of Nepal, and it combines with the Lalitpur and Bhaktapur metropolitan areas, along with several other smaller cities and towns to form the KV as a cosmopolitan and sprawling valley. This valley has been experiencing several new environmental challenges in the recent decades, such as traffic congestion, air pollution, declining water table, and loss of open space [5,6]. The impacts of rapid urbanization in the Himalayan cities, particularly sprawl and other types of pell-mell urban growth, go beyond the urban footprints [7]. Monitoring of land use and land cover (LULC) in these cities, therefore, is not only a pragmatic way to detect and quantify landscape-level transformation, but also a window to understanding the complex social-ecological relationships in the region characterized by its fragility and sensitivity to hazards and disaster, such as earthquakes, landslides.

Nepal is recorded as one of the top ten fastest urbanizing countries in the world [8]. The 2011 census recorded the population of the KV alone at one million [9] and is projected to double by 2030 [8]. The population growth in newly developed peri-urban areas is significantly higher when compared to the historic urban core of the valley. According to Muzzini & Aparicio [10], in 2011, annual population growth was high in the peripheral municipalities of Kirtipur (5%) and Madhyapur Thimi (5.7%). The growth of population and the rapid expansion of built-up area in recent decades have caused a substantial LULC change in KV. With 3.94% urban growth rate between 2010 and 2014, the KV is going through significant transformation of its landscapes in recent years [8] making it important to understand the dynamics of LULC change processes, including their interactions with local and regional environmental change.

Despite such rapid growth in population and urban area (interchangeably used as *built-up area*), only a few LULC change studies have been conducted on KV to date. For instance, Haack and Rafter [11] analyzed the land use changes that occurred between 1978 and 2000 using GIS tools and found around 450% growth of urban areas in these years. In another study, Haack [1] showed the similar trend of urban expansion by comparing maps from different years in the period of 1955 and 2000. More recently, Thapa and Murayama [6] analyzed LULC change patterns of the KV between the period of 1967 and 2000. They found that the urban growth increased particularly after the 1980s. Rimal [12] reported similar findings analyzing the urban growth pattern between 1976 and 2009. Thapa & Murayama [13] projected that the urban area will continue to increase along the major roads. While these studies highlighted the land change trajectories of the KV from the 1980s to the 2000s, much of the new conversions of agricultural land to housing development is left for speculation. When considering the massive expansion of urban areas and simultaneous diminution of agricultural lands that occurred especially during the first decade of 21st century, a systematic assessment of the LULC change patterns is crucial, as it would also help to interpret and contextualize much of the new urbanization patterns taking place in the Himalayan region. Our objective in this paper, hence, is to examine the LULC change dynamics in the period of 1989–2016 using Landsat imageries and pixel-based analysis methods and highlight the value of monitoring the urban growth of Himalayan cities to foster dialogue in the management of urban growth.

2. Study Area

Located between 27°36' and 27°48' N, between 85°12' and 85°31' E, the KV is a rapidly urbanizing mountain basin in the Himalayas. Surrounded by the Himalayan mountain range, the valley of Kathmandu is comprised of three districts: Kathmandu, Lalitpur, and Bhaktapur, together with expanding an area of 899 km^2 [14]. The generally flat floor of the valley is at the average elevation of 1300 m surrounded by mountains that are 1900–2800 m tall [11], except for a narrow winding outlet of the Bagmati River towards the south, three mountain passes of about 1500 m altitude on the east and west of the valley. This valley is a tertiary structural basin that is covered by fluvial and lacustrine sediments and encircled by mountains on all sides [14]. The KV, therefore, can be pictured as a bowl-shaped depression with an elevated basin and a plateau with a rim. More than 20% of the KV has slopes >20°, and half of the area has >5° slopes [6]. The valley floor has two types of landforms- alluvial plains and elevated river terraces [15] and is known for growing cereals and vegetables. The climate of the KV is influenced by several factors, including the south Asian monsoon affecting precipitation and wind direction. The valley receives more than 80% of its annual rainfall during the four summer months—June through September. This is followed by a clear, sunny fall, cold winter, dry spring, and humid summer. The annual average rainfall and temperature in the valley are 1407 mm and 18.1 °C, respectively.

Administratively, the KV is divided into five municipal areas: Kathmandu, Lalitpur, Kirtipur, Madhyapur Thimi, and Bhaktapur, and several small village development committees (VDCs). With the history and culture dating back 2000 years, the cities within the KV rank among the oldest human settlements in the central Himalayas. The KV shares the characteristics with many other rapidly urbanizing cities in the region. These include, unregulated urban development, inadequate enforcement of land use policies, poorly maintained city infrastructures, the massive influx of people from surrounding rural areas and hinterlands, land speculation, excessive pressure of commercial activities, and gaps in supply and demand for basic services. The previous studies i.e., [1,6,11,12,14] indicated that the urban growth was happening in the central KV area and the mountainous areas were sparsely populated patches of agricultural settlements and mixed forests. As a result, disregarding the forested steep slopes (>20°) we focused our study on the central KV comprising an area of 422.84 km^2 (Figure 1).

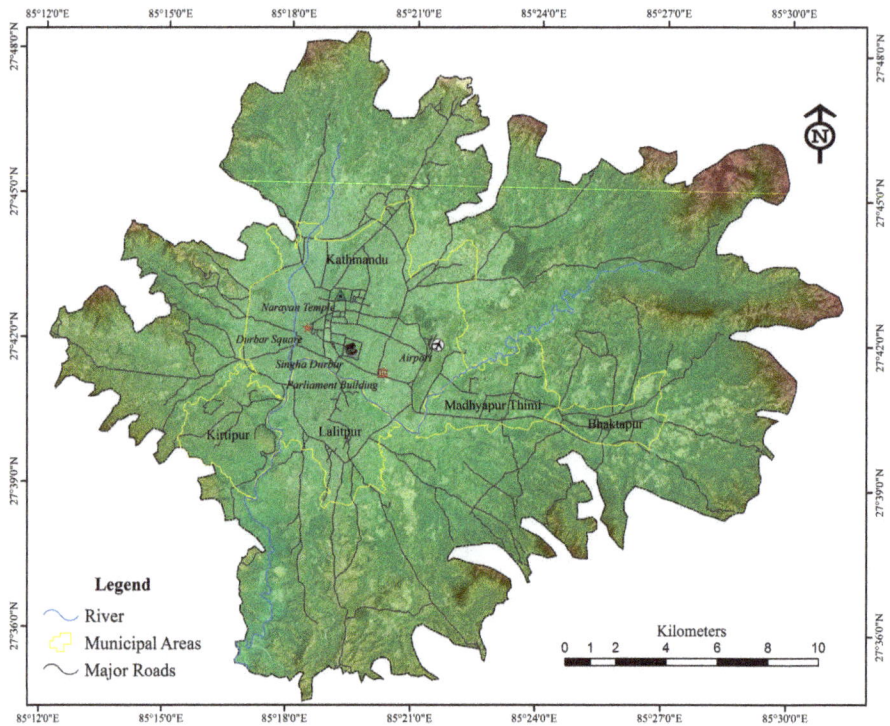

Figure 1. The Kathmandu valley study area.

3. Methods

LULC change detection becomes a key research priority with multi-directional impacts on both natural and human systems [16]. Among many techniques of change detection, remote sensing is the most commonly used technique because of its cost-effectiveness and timesaving characteristics [17]. In this study, we used pixel-based approach to detect LULC change pattern. Lu et al. [18] and Tewkesbury et al. [19] provided an exhaustive list of pixel-based change detection methods. Of all these methods pre- and post-classification comparisons have widely been used owing to their comparative advantages, see [19–24]. In the pre-classification techniques, LULC changes are detected through the differences in the pixel reflectance values between dates of interest; however, these techniques are not efficient in showing the nature of change [18,25]. On the other hand, post-classification comparison provides a complete matrix for change detection, which makes it the widely most used change detection techniques [19,23,26]. For this reason, we selected post-classification comparison method to understand the nature and pattern of change. The accuracy of the LULC change in this technique largely depends on the individual accuracy of each classification [18].

3.1. Satellite Data

We used four Landsat images of the year 1989, 1999, 2009, and 2016 in this study (Table 1). These scenes were acquired from the freely available Landsat archive of United States Geological Survey (USGS) (http://earthexplorer.usgs.gov/). The KV is located at path 141, row 041 in worldwide reference system type 2 (WRS2). All of these scenes were already georeferenced to the Universal Transverse Mercator (UTM) map projection (Zone 45N) with WGS84 datum and ellipsoid. The acquisition qualities of these images were the highest (09, meaning no quality issues/errors detected). The cloud cover was insignificant in 2009 and 2016 images. The 1989 and 1999 images have considerable cloud cover, however. The cloud cover was on the western edge of the Landsat scene while our study area was located at the center of the scene. Therefore, there was no visible cloud cover over our study area. Yet, we did atmospheric correction for haze removal (see Section 3.2).

Table 1. Details of the Landsat images used in this study.

Acquisition Year	Acquisition Date	Sensor	Acquisition Quality	Cloud Cover
1989	2 December	Landsat TM	09	8.14%
1999	28 January	Landsat TM	09	24.66%
2009	16 February	Landsat ETM+	09	0.20%
2016	4 February	Landsat ETM+	09	0.26%

3.2. Image Pre-Processing

Satellite image preprocessing is essential before analyzing it to avoid data distortion or manipulation. Preprocessing is also needed to establish direct linkage between data and biophysical phenomena [21]. Our image preprocessing steps include scan-line corrector (SLC) gap filling, radiometric correction, and image enhancement.

The SLC, which compensates the forward motion of Landsat 7, permanently failed in May 2003. As a result of that, Landsat ETM+ scenes have zigzagged pattern gaps on both sides, which cause around 20% loss of data [27]. These unscanned data could be replaced by data in the overlap portion of adjacent scenes (i.e., lateral overlapping) or by subsequent passes over the same scene (i.e., images of the same area from other dates), with the result that every location would be observed eventually [28]. In this study, the gap filling for 2009 and 2016 Landsat ETM+ images were obtained overlapping the adjacent scenes of the same time.

Atmospheric correction and topographic normalization are required to improve the classification results [29,30]. Atmospheric correction primarily includes the removal of haze, which originates from fractions of water vapor, fog, dust, smoke, or other minute atmospheric particles [31]. The topographic normalization is important for mountainous areas such as KV, because the presence of slopes can cause variations in illumination of identical features [32,33]. We used Atmospheric and Topographic CORrection (ATCOR) feature in ERDAS Imagine for haze removal and topographic normalization [34,35]. ATCOR is used to eliminate atmospheric noise and illumination effects to retrieve physical parameters of the earth's surface, such as surface reflectance. As our study area is a valley land with slight variation in topography, we used ATCOR3, which requires average elevation of the study area for the analysis. We took the average elevation of KV as 1377 m, obtained from ASTER Global DEM with 30 m spatial resolution. The de-hazing algorithm of ATCOR is useful to turn a hazy image into neat one, and the topographic correction results in the output of true reflectance values [36]. Image enhancement is the modification of pixel values to improve visual interpretation by increasing the distinction between features [37]. One of the most commonly used methods- histogram equalization (HE) was used in this study. The main idea of HE-based methods is to re-assign the intensity values of pixels to make the intensity distribution uniform to the utmost extent [38].

3.3. Classification Procedure

In pixel-based image classification, a combination of both supervised and unsupervised classification provides a more accurate land use classification scheme [39]. In this study, we used hybrid classification approach to derive five major land use classes, to be consistent with the classes commonly identified for Nepal (Table 2). Our hybrid approach comprises of unsupervised classification of images followed by supervised classification.

Table 2. Land use/land cover classes.

Land Use/Cover Class	Land Uses and Land Covers Included in Class
Urban/Built-up area	Structures of all types: residential, industrial, commercial, airports, and roads/highways
Agricultural area	Croplands and temporary grasslands used for agriculture
Forest/Tree covered area	Forest, parks and permanent tree covered area
Bare ground (BG)	Vacant lands, open area, and fallow lands
River	River

The hybrid classification of land uses in this study started with an unsupervised Iterative Self-Organizing Data Analysis (ISODATA) clustering into 75 clusters. Hyperclustering, which uses a much higher number of clusters than the desired classes [40] was chosen because the exact number of spectral classes in the data set was unknown yet [41]. These clusters were labeled as urban area, agricultural area, forest, bare ground, and river based on the Google Earth observation and other land use maps (see [6,11]) of the study site. Spectrally similar classes of the identical land use type were merged. The comprehensive set of the spectral class signature was used in the second stage as training data for supervised classification [42]. Next, we selected at least 100 training samples for each class. These spectral signatures were considered satisfactory only when the confusion among the land use was minimal [43]. We then conducted supervised classification using the maximum likelihood classifier (MLC) algorithm. The classification provided thematic raster layers, which were later used for post-classification change detection.

4. Results and Discussion

4.1. Classification Accuracy Assessment

Atmospheric and topographic disturbances are the two of many factors that could affect the accuracy of LULC change detection in mountain regions. The validity of the classification can be determined by accuracy assessment [44]. In this study, the accuracy of each classified image was assessed by a set of 450 points selected through stratified random sampling with at least 50 points for each class. We took at least 90 points for agricultural and built-up category, as they comprise most of our study area. Error matrix for cross-tabulation of the mapped class vs. reference class was used to assess classification accuracy. Overall accuracy, kappa coefficient, tau coefficient, producer's, and user's accuracy were derived from the error matrix. The summary of accuracy assessment is provided in Table 3.

Table 3. Summary of classification accuracies (in %).

LULC Class	1989		1999		2009		2016	
	Producer's	User's	Producer's	User's	Producer's	User's	Producer's	User's
Built-up Area	100	88.24	100.0	91.89	98.08	85.0	99.03	91.89
Agriculture	80.49	90.83	81.37	86.46	81.48	83.81	89.38	86.52
Forest	94.74	92.31	97.53	95.18	95.89	97.22	96.30	95.12
BG	86.54	81.82	75.47	83.33	86.21	85.71	85.25	89.66
River	98.23	95.00	100.00	100.00	97.50	90.89	98.60	95.00
Overall Accuracy	88.12		88.22		86.27		88.69	
Kappa Coefficient	0.85		0.85		0.82		0.85	
Tau Coefficient (*Equal probability*)	0.85		0.85		0.83		0.86	

LULC: land use and land cover.

4.2. LULC Change Trajectories in the KV

Until the 1980s, the urban areas (interchangeably used as built-up areas) of KV were limited within the confines of the historic settlements of the five municipalities. The outward expansion of the urban area began in the early 1990s and accelerated at the turn of the 20th century. In the 2000s, the built-up areas continued to expand further along the major roads that link the outskirts of the five municipalities and there is no sign that it is going to stop anytime in the near future, as more complex road networks are being planned for the future. Most of the newly expended built-up areas are replacing agricultural lands that once were considered to be the most fertile and productive in the country.

Table 4 summarizes the LULC change trends from 1989 through 2016 obtained from the classified images. From 1989 to 2016, built-up area increases from 2153 to 11,019 hectares (ha)—an increase of 412% is a very significant change, because much of the new developments occurred as an expansion of the existing city cores (Figure 2). At present, built-up area comprises about 26% of the study site, whereas agricultural area diminished at a rate of 1.8% per year resulting in a total 32% loss during the period of 1989–2016. Currently, the extent of the agricultural area is around 55% of the central KV compared to 82% in 1989 (Table 4). The rate of conversions of agricultural land to other types of land uses remains high throughout the study period (see details in Section 4.3). The loss of agricultural land in the KV resembles the worldwide trend of the urban conversions of agricultural lands reported elsewhere [45–47], but much is unknown about how these conversions will affect fragile ecosystems of this mountainous valley, including with the loss of green space, sealing of soil, disturbance to stream corridors, and alteration of agro-ecological services (e.g., water retention, vegetation, air circulation).

Table 4. Summary of land use land cover change in the period of 1989–2016 (areas are presented in hectares).

LULC Class	1989		1999		2009		2016	
	Area	%	Area	%	Area	%	Area	%
Built-up Area	2153.79	5.10	4712.88	11.15	10,216.20	24.16	11,020.62	26.06
Agriculture	34,057.40	80.54	31,069.20	73.48	27,007.37	63.87	23,387.06	55.30
Forest	4138.56	9.79	4172.76	9.89	3627.99	8.58	6227.37	14.73
BG	1854.54	4.39	2252.7	5.34	1355.13	3.21	1576.73	3.73
River	80.00	0.19	76.80	0.18	74.50	0.18	73.00	0.17
Total	42,284.30	100.00	42,284.30	100.00	42,284.30	100.00	42,284.30	100.00

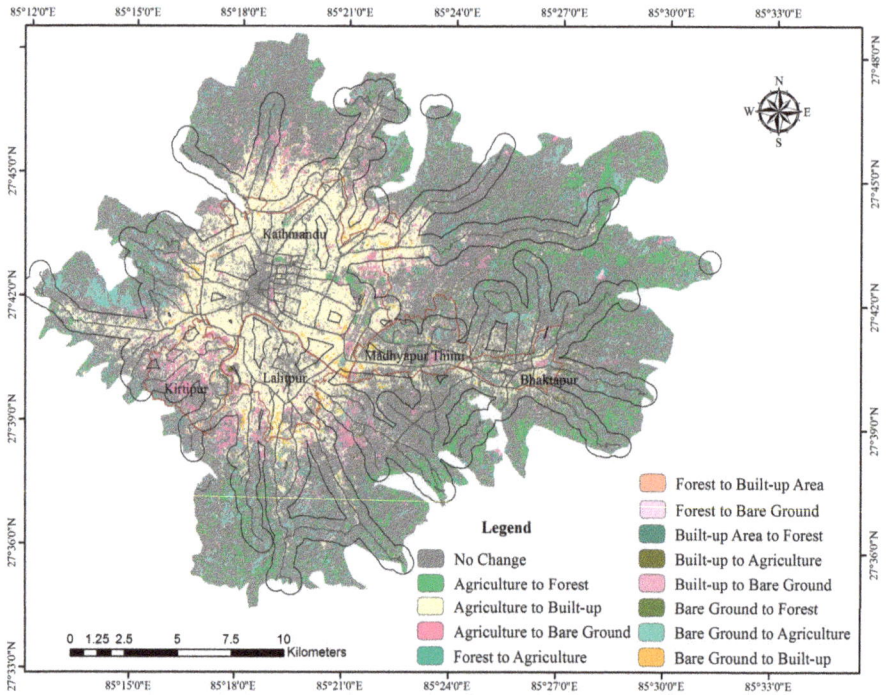

Figure 2. Land use land cover change analysis of Kathmandu valley (1989–2016). The roads are buffered to 500 m to show as evidence that conversion of agricultural lands to built-up areas is particularly happening along the major roads.

The forested or tree covered area comprises a small portion of the central KV, which include urban parks (e.g., Bhandarkhal, Boudha), forest resorts (e.g., Mrigasthali), forest reserves (e.g., Gokarna), and some forest patches on the steep slopes. Forest patches in the central KV are well maintained and only observed a marginal loss in last four decades. Covering around 4% area of the central KV, bare grounds are scattered largely at the outskirts of the Kathmandu city. The use of these privately-owned lands depends on the location of these lands. The bare grounds proximate to hilly areas are used for agriculture, whereas, the bare grounds located at the outskirts of the city are mostly cleared up to expand the built-up area.

4.3. LULC Change Pattern

To further explore the LULC change pattern, we created confusion matrices for 1989–2016, 1989–1999, 1999–2009, and 2009–2016 changes (see Tables 5–8). In all of the tables, unchanged pixels are located along the major diagonal of the matrix. Conversion values were sorted by area and listed in descending order [48].

The results of Table 5 indicate that the increase in the built-up area mainly came from the conversion of agricultural land to urban area in this 27-year study period, 1989–2016. Figures 2 and 3 show the transition among land use classes in the period of 1989–2016. The built-up area expanded radially towards all direction at the rate of 14.70% per annum. The conversion to built-up area mostly happened within 500 m of major roads in KV (see Figure 2). The conversion of active agricultural land to bare ground or perennial fallows at the urban fringes shows a booming housing market during this period. This type of conversion is the initial step of urban expansion, which subsequently opens up the areas for housing developments. Of the 34,000 ha of agricultural land in 1989, around 26%

(8880 ha) has been converted into built-up area within 2016. This result supports the fact that there was a huge surge in the rural-to-urban migration in the KV between the mid-1990s to the late 2000s—the proximate causes are discussed in Section 4.4.

Table 5. Land use land cover change matrix (1989–2016). (Areas are presented in hectares.)

LULC Class		2016					
		Built-Up Area	Agriculture	Forest	BG	Water	Total
	Built-up Area	1541.79	378.18	210.33	23.49	0.0	2153.79
	Agriculture	8880.39	20,973.10	2907	1296.90	0.0	34,057.39
1989	Forest	262.26	742.32	3100.23	33.75	0.0	4138.56
	BG	334.98	1289.16	9.81	220.59	0.0	1854.54
	River	1.20	4.30	0.00	1.50	73.00	80.00
	Total	11,020.62	23,387.06	6227.37	1576.73	73.00	42,284.30

Table 6. Land use land cover change matrix (1989–1999). (Areas are presented in hectares.)

LULC Class		1999					
		Built-Up Area	Agriculture	Forest	BG	Water	Total
	Built-up Area	1249.38	675.54	216.96	8.91	0.00	2153.79
	Agriculture	3240.18	27,823.60	1166.85	1819.08	0.00	34,057.4
1989	Forest	158.58	1184.22	2787.33	11.43	0.00	4138.56
	Fallow land	63.54	1376.10	1.62	413.28	0.00	1854.54
	River	1.20	2.00	0.00	0.00	76.80	80.00
	Total	4712.88	31,069.20	4172.76	2252.70	76.80	42204.3

Table 7. Land use land cover change matrix (1999–2009). (Areas are presented in hectares.)

LULC Class		2009					
		Built-Up Area	Agriculture	Forest	BG	Water	Total
	Built-up Area	3978.27	615.96	100.08	17.37	0.00	4712.88
	Agriculture	5627.88	23,590.29	959.76	889.29	0.00	31069.22
1999	Forest	92.88	1507.95	2564.6	7.29	0.00	4172.76
	Fallow land	517.14	1290.87	3.51	441.18	0.00	2252.70
	River	0.00	2.30	0.00	0.00	74.50	76.80
	Total	10,216.17	27,007.37	3627.99	1355.13	74.5	42,204.36

Table 8. Land use land cover change matrix (2009–2016). (Areas are presented in hectares.)

LULC Class		2016					
		Built-Up Area	Agriculture	Forest	BG	Water	Total
	Built-up Area	8192.70	1598.4	66.69	358.38	0.00	10,216.17
	Agriculture	2397.78	20,345.24	3386.8	875.16	0.00	27,007.37
2009	Forest	145.53	693.54	2770.70	18.18	0.00	3627.99
	Fallow land	283.41	745.65	3.06	323.01	0.00	1355.13
	River	0.00	0.00	0.00	1.50	73.00	74.50
	Total	11,019.42	23,382.83	6227.29	1574.73	73.00	42,204.27

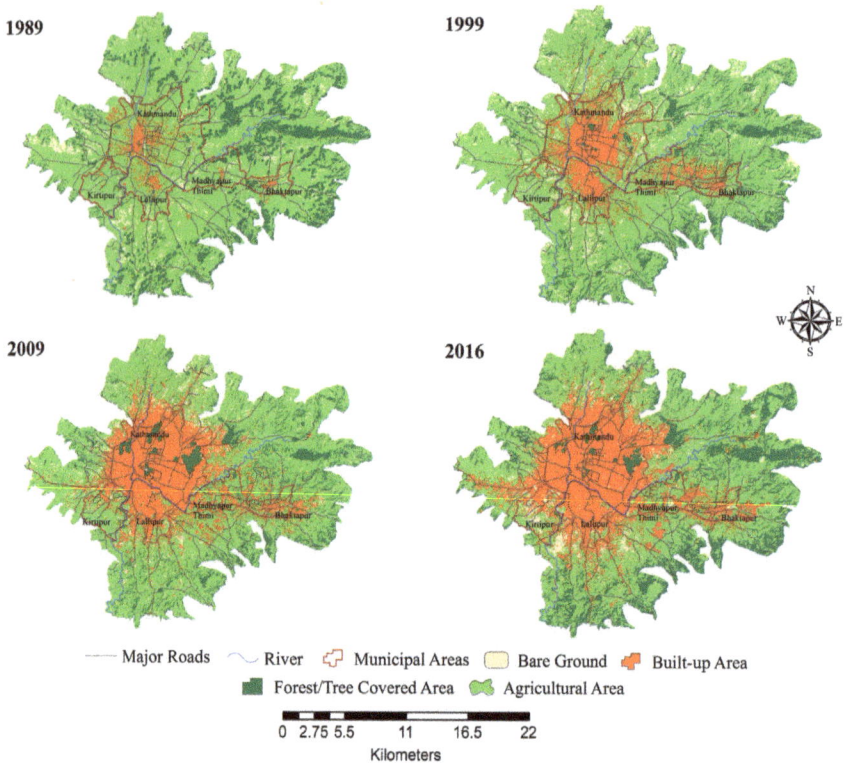

Figure 3. Land use land cover maps of Kathmandu Valley (for the analysis, the roads are merged into "built-up area" category. However, in this figure the major roads are shown as a separate layer to understand the pattern of land use land cover change).

Table 5 and Figure 2 also show some conversions of agricultural lands to forest areas occurring at the outer margins of our study area that are adjacent to the densely forested areas. Although more precise ground validation is needed, this trend indicates the discontinuation of agricultural activities on marginal, steep-sloped lands could pave the way for shrublands and secondary forests (see Figure 2). Similarly, approximately 18% (742 ha) tree covered area has been converted into agricultural lands, and most of these changes occurred in the peri-urban areas with lower elevation and slopes. It is not clear exactly what contributed to this change. It is important to note that this conversion took place in the areas far from the city area. It is also important to note that there has been a minimal change in tree-covered areas within the city limit (see Figure 3, for instance). The shifting from fallow to agricultural is seasonal, largely driven by monsoon rainfall patterns, local economy, and the availability of agricultural labor. The water area occupies very little space of the valley and has not been changed much in last three decades.

During the ten years between 1989 and 1999, the built-up area increased by about 120%, whereby significant expansion took place along the major roads such as local roads, service roads, and access roads that link the outlying towns with the five municipalities (see Figures 2 and 3). The Kathmandu and Lalitpur municipalities experienced substantial growth in the built-up areas. The built-up areas were further expanded into the Madhyapur Thimi and Bhaktapur municipalities, and also in the southern parts of Duwakot and Jhaukhel VDCs. This growth of built-up area was primarily obtained by converting agricultural lands. Around 10% of the agricultural lands were converted into built-up areas in this period (see Table 6). Our analysis also reveals that as distance increases away from these major roads, the built-up areas tend to be less dense with the presence of open space and agriculture land. In other words, households select the locations that reduce their travel time and hence the concentration of settlement along the 500 m of major road networks.

The most aggressive form of urban growth in the KV happened between the period 1999 and 2009, which also coincides with the booming period of the real estate market largely fueled by the influx of migrants from the countryside displaced by political turmoil and/or by stagnant growth in agricultural sector. During this period (1999–2009), the KV saw 117% growth in built-up areas. This expansion came at the expense of 18% agricultural lands (see Table 7). The majority of this expansion occurred in Kathmandu and Lalitpur municipal areas, and for the first time in history, entire areas of these municipalities became urbanized (see Figure 3). The built-up area further stretched to eastward VDCs (i.e., Tathali, Sudal) expanding further beyond the Bhaktapur municipal boundary. In this decade, new built-up areas were established in the southeastern VDCs (e.g., Balkot, Tikathali, Sirutar, Lubhu), and along the major roads connecting the valley to the rest of the country. During this period, the valley also observed a substantial loss of forest cover whereby about 36% of the tree-covered area were cleared up for agriculture purpose. The rapid growth of built-up areas pushed the farmers to clear up the forest and expand agriculture on the foothills (see Figure 3). Overall, this fast growth of built-up areas during 1989–2009 can be attributed to the spike in real estate market, massive urban in-migration compounded by political instability in the countryside.

For the period between 2009 and 2016, two major LULC changes are worth noting: (1) forest area has been in a relatively stable condition in the central KV and is slowly beginning to expand in the outer margins, and (2) the aggressive urban growth of the 1999–2009 period has somewhat slowed down in the last ten years. The further expansion of forest areas in the KV outskirts, particularly in the northwestern part is notable; however, it is unclear which proximate causes are driving this change. One potential cause is that this area covers the Shivpuri Nagarjun National Park, which was formerly a watershed and wildlife reserve, but it was upgraded to a national park status in 2002 to more aggressively protect the forest areas and watershed vital to the water supply of the KV. Although the ground verification is needed, the areas in higher elevation and slopes that are adjacent to these forest areas are likely to have gained forest coverage. There are also some community forests located near those forest patches, where communities are taking more active roles in managing forest resources. One could argue that except for the protected forests, most forest patches are heavily fragmented, which is often consistent with reported cases of land fragmentation in the peri-urban area or urban fringes (see [49,50]).

Our analysis also reveals that urban growth has seen a slight slowdown in certain parts of the valley in recent years. Between 2009 and 2016, the built-up area increased only about 8% (see Table 8). The most noticeable growth was in Kirtipur municipality and the built-up areas in the KV were mostly expanding only along the major roads, radially growing outward from Kathmandu metropolitan area. Thapa & Murayama [13] predicted such outward urban growth and in-filling of existing urban areas. The built-up areas of some of the fastest growing areas, such as Madhyapur Thimi and Bhaktapur municipal areas ceased to expand in this period (see Figure 3). Overall, our results show that the KV cityscape has changed dramatically between 1989 and 2016. Because of the aggressive urban growth experienced since the mid-1990s to the late 2000s, the KV is showing a concentric pattern of urbanization. It is worthwhile to note that there were five distinct municipalities with distinct urban boundaries and plenty of open or green space between them until the late 1980s; however, those municipalities have since coalesced into a large metropolitan area where agricultural lands and open spaces were aggressively converted to residential areas to accommodate growing demands for housing and other urban infrastructures.

4.4. Key Drivers of LULC Change in KV

The results of the LULC change detection clearly establish that this valley has experienced an unprecedented level of urban growth in the last three decades. The final quarter of the 20th century witnessed a rapid expansion of the KV, reflecting the trend of urban growth dominant in the Himalayan region and elsewhere in South Asia [4]. This trend transformed the KV composed of the network of small towns—each with their own place-based identities and sophisticated architectural heritage—into a metropolis of 'concrete jungle', struggling to preserve its historical identity and ecosystem services [5]. Based on a careful review of the existing literature and expert knowledge, two of the co-authors have several years of working experience in Nepal, we identified that several proximate causes that have directly contributed to this transformation, including (a) rural-urban migration, (b) economic centrality, (c) socio-political factors, and (d) booming real estate market. All of these are arguably related to government policies (or the lack thereof).

4.4.1. Rural-Urban Migration

The KV has been experiencing rapid population growth particularly since the 1980s. Being the home of 22.3% Nepal's urban population, KV is the fastest growing urban agglomeration in South Asia [8]. The highest contribution of this growth comes from rural to urban migration, which in turn is driven by the economic opportunities available in the capital relative to the rural areas. For instance, during the 1990s as high as 40% population growth happened due to urban in-migration [51]. Currently, the net inflow of migrants accounts for 36% of KV populations [52]. Hailed from the remote rural areas they mostly migrate because of economic reasons (i.e., better livelihood opportunities) and educational purpose. However, rural push factors play a dominant role in urban in-migration too [10]. Extreme poverty, lack of economic opportunities, low living standard, and an absence of basic amenities in the rural areas are some of the many push factors. Moreover, the civil conflict escalated the migration in recent decades [53]. While farmlands in conflict-affected areas were facing labor shortage, the displaced people also became the driver of LULC change in the KV.

4.4.2. Economic Centrality

The KV is also the administrative and economic hub of the country with a growing middle class [54]. The relatively flat topography, transportation accessibility, economic opportunities, and political and policy factors have consolidated the centrality feature of KV [1,10,11,14]. Among all of these factors, economic centrality is considered as the prime factor of rapid land use change. With the concentration of social services—primarily the growth of higher education and healthcare industry–and growing economic opportunities in tourism and other service sectors, the capital city has remained the most preferred destination for seeking jobs, income generation opportunities, and residence. This was further compounded by limited investment opportunities elsewhere in the country and/or other economic sectors in the cities. Similarly, the entire largest manufacturing cluster is concentrated in the KV, which provides as much as 40% manufacturing employment and 41% nonfarm and service employment [52]. Manufacturing employment per square kilometer is above 600 in the Kathmandu city area [55], the highest in the country, which in turn has attracted people to change their occupations from farming to manufacturing. In addition, centralization of government offices, the growth of foreign aid and tourism, and construction of access roads connecting the KV with the rest of Nepal have further propagated the economic centrality. This capital-centric development model (or urban primacy) that is typical of several low to middle-income countries has been one of the main drivers of LULC change in KV. The KV is Nepal's gateway to tourists, whereby 90 percent of tourists enter the country [56], as the valley also has the rich cultural heritage including the seven designated world heritage sites. Tourism is also a key component of the valley's economy [57], putting pressure on agriculture land to build facilities for continued flow of tourists and a growing middle class further away from the city core.

4.4.3. Socio-Political Factors

There has also been a huge influx of internally displaced people to the valley due to the decade-long civil unrest that began in the mid-1990s [58]. While triggering the socio-political crisis, the conflict disrupted local economic activities by frequent strikes, closures of businesses, extortion, and threats. More than 500,000 people believed to have been displaced during the insurgency period [56]. Nepal's urban centers, especially KV, had to absorb the influx of these migrants. As a capital city, Kathmandu is naturally the political and administrative center of the country, and it also became a safe refuge for those internally displaced people during the political turmoil period of 1996 to 2008. Overall, the KV is the hub for all important socio-economic sectors in the country: tourism, finance, industry, education, transportation, and healthcare.

4.4.4. Real Estate Boom

Nepal in general and KV more specifically, experienced a real estate boom in the recent decades, especially between the mid-1990s to the late 2000s. According to Nepal Land and Housing Association, the land price in the KV risen by 300% since 2003, one of the key drivers of LULC change. Land ownership in the KV can be divided into private, *Guthi* (religious trust), government, and public. With more than 90 percent of cultivated lands and 61 percent of registered lands, private land ownership is a dominant form of tenure arrangement in the valley [59]. This means that there is little government control over land and housing in the KV, and the absence of real land-acquisition laws in practice, the state has not effectively regulated in the booming real estate market. During the fiscal year of 2008–2009 close to 185,000 people and firms bought new land and housing in the valley [60]. While there is no accurate data of how real estate agencies are currently involved in the land market, our interactions with local government officials reveal that there are as much as 150 real estate agencies and about a dozen of housing companies involved in land acquisitions, pooling, and housing in the KV. The increasing number of middle-class families in the KV is demanding new modern facilities such as, housing sub-divisions and colonies with modern amenities (e.g., private parking, modern grocery stores, restaurants) in the suburbs further contributing the LULC change in the fringe areas of the valley.

These key proximate causes of LULCC detected in KV are obviously interlinked and quite complex, but it is safe to argue that the economic centrality and urban population growth swelled primarily by the rural-to-urban migration played the major role in the rapid urban growth of the KV. It is essentially the large differences in economic opportunities between the KV and the rest of the country that resulted in a growing influx of people from rural areas to the KV.

5. Conclusions

Land change trajectories of the KV detected in this study represent a quintessential urbanization trend that is sweeping across the Himalaya region and beyond; this trend is a form of the "urban primacy model" in which a city—typically the capital—controls the flow of all economic and financial transactions, industrial production, and most importantly the governance of a country [4]. The most striking change in the KV is that agriculturally productive peri-urban areas are now being encroached upon by rapid housing development that is expanding outward in a typical concentric zone fashion. The built-up area is expanding rapidly mostly at the cost of agricultural lands. In last three decades, built-up areas increased by 412%, while agricultural land encountered a 31% loss. This change has transformed not only the physical landscapes of the valley, but it also has altered the ecosystem services provided by agricultural lands and open space. Our results on the urban growth rate support the main conclusion of Haack & Rafter [11], who found a 450% urban growth between 1978 and 2000—the urban growth in KV continued to be rapid and largely uncontrolled. Also, our findings of the outward expansion of city area along the major roads confirm the result of Thapa & Murayama [13].

The growth of settlements in the KV is generally spontaneous, with little intervention on the part of government authorities. The current existing land use policy (or constitutional provision) does not allow the government to impose any kind of restriction on the use of private property. Rapid urban expansion coupled with unmanaged settlement development has led to various socio-environmental challenges. The principal reason for such unmanaged developments in the KV is due to ineffective land use, zoning, and land sub-division policy. Additionally, the uncontrolled urban growth of KV during the last three decades due to the reasons discussed above has resulted in severe infrastructure deficits—the KV simply has inadequate infrastructure to support the massive surge in population growth seen in the last four decades. Unplanned urban growth can lead to a loss of open spaces that adversely impacts the urban environment. Given that the KV is projected to grow bigger in the future, failure to formulate sustainable urban development strategies and implement effectively could create severe socio-environmental consequences, including stagnant economic productivity, poor infrastructures, low quality of life, and rise in urban divide. From the perspective of holistic urban management, this may be a major hindrance in the future that needs urgent attention from government and other stakeholders.

This trend clearly shows the need to study the sustainability implications of urban sprawl in this fragile, mountainous landscape. How long can a mountainous valley like Kathmandu sustain the urban growth rate it has experienced the last four decades? It is particularly urgent to examine the impacts of the conversion of agricultural land to the built environment, socio-ecological significance of disappearing open space, fragmentation of habitats and important biological corridors, changes in urban food and diet system, rising urban divide, increasing pollution levels, and most importantly, the governance of urban growth (or the lack thereof).

Acknowledgments: The authors are thankful to Sanjoy Roy, GIS expert at the International Union for Conservation of Nature (IUCN) Bangladesh, and Tasnuba Jerin, PhD student at the Department of Geography, University of Kentucky for their suggestions and assistances. Finally, we thank the three anonymous reviewers for their useful suggestions.

Author Contributions: Milan Shrestha (MS) conceived the plan; MS and Asif Ishtiaque (AI) designed the experiments; AI performed the experiments and analyzed the data; AI, MS, and Netra Chhetri wrote the paper.

Conflicts of Interest: The authors declare no conflict of interest.

References

1.   Haack, B. A history and analysis of mapping urban expansion in the Kathmandu valley, Nepal. *Cartogr. J.* **2009**, *46*, 233–241. [CrossRef]
2.   Tiwari, P.C.; Joshi, B. Urban Growth in Himalaya: Environmental Impacts and Developmental Opportunities. *Mt. Res. Initiat. Newsl.* **2012**, *7*, 29–32.
3.   Goodall, S.K. Rural-to-urban migration and urbanization in Leh, Ladakh: A case study of three nomadic pastoral communities. *Mt. Res. Dev.* **2004**, *24*, 220–227. [CrossRef]
4.   UN-HABITAT. *State of the World's Cities 2012/2013: Prosperity of Cities*; Routledge: Abingdon, UK, 2013.
5.   Haack, B.N.; Khatiwada, G. Rice and bricks: Environmental issues and mapping of the unusual crop rotation pattern in the Kathmandu Valley, Nepal. *Environ. Manag.* **2007**, *39*, 774–782. [CrossRef] [PubMed]
6.   Thapa, R.B.; Murayama, Y. Examining spatiotemporal urbanization patterns in Kathmandu Valley, Nepal: Remote sensing and spatial metrics approaches. *Remote Sens.* **2009**, *1*, 534–556. [CrossRef]
7.   Joshi, B.; Tiwari, P.C. Land-use changes and their impact on water resources in Himalaya. In *Environmental Deterioration and Human Health*; Springer: Dordrecht, The Netherlands, 2014; pp. 389–399.
8.   UNDESA (United Nations, Department of Economic and Social Affairs). *World Urbanization Prospects. The 2014 Revision*; UNDESA: New York, NY, USA, 2015.
9.   CBS (Central Bureau of Statistics). *National Population and Housing Census 2001*; Government of Nepal: Kathmandu, Nepal, 2012.
10.   Muzzini, E.; Aparicio, G. *Urban Growth and Spatial Transition in Nepal: An Initial Assessment*; The World Bank: Washington, DC, USA, 2013.

11. Haack, B.N.; Rafter, A. Urban growth analysis and modeling in the Kathmandu Valley, Nepal. *Habitat Int.* **2006**, *30*, 1056–1065. [CrossRef]
12. Rimal, B. Application of Remote Sensing and GIS, Land use/land cover change in Kathmandu. *J. Theor. Appl. Inf. Technol.* **2011**, *23*, 80–86.
13. Thapa, R.B.; Murayama, Y. Urban growth modeling of Kathmandu metropolitan region, Nepal. *Comput. Environ. Urban Syst.* **2011**, *35*, 25–34. [CrossRef]
14. Thapa, R.B.; Murayama, Y. Drivers of urban growth in the Kathmandu valley, Nepal: Examining the efficacy of the analytic hierarchy process. *Appl. Geogr.* **2010**, *30*, 70–83. [CrossRef]
15. ICIMOD (International Centre for International Mountain Development). *Economic and Environmental Development Planning for the Bagmati Zone*; ICIMOD: Kathmandu, Nepal, 1993; Volume 3, p. 85.
16. Turner, B.L.; Lambin, E.F.; Reenberg, A. The emergence of land change science for global environmental change and sustainability. *Proc. Natl. Acad. Sci. USA* **2007**, *104*, 20666–20671. [CrossRef] [PubMed]
17. Lambin, E.F.; Geist, H.J.; Lepers, E. Dynamics of land-use and land-cover change in tropical regions. *Annu. Rev. Environ. Resources* **2003**, *28*, 205–241. [CrossRef]
18. Lu, D.; Mausel, P.; Brondizio, E.; Moran, E. Change detection techniques. *Int. J. Remote Sens.* **2004**, *25*, 2365–2401. [CrossRef]
19. Tewkesbury, A.P.; Comber, A.J.; Tate, N.J.; Lamb, A.; Fisher, P.F. A critical synthesis of remotely sensed optical image change detection techniques. *Remote Sens. Environ.* **2015**, *160*, 1–14. [CrossRef]
20. Singh, A. Review article digital change detection techniques using remotely-sensed data. *Int. J. Remote Sens.* **1989**, *10*, 989–1003. [CrossRef]
21. Coppin, P.; Jonckheere, I.; Nackaerts, K.; Muys, B.; Lambin, E. Digital change detection methods in ecosystem monitoring: A review. *Int. J. Remote Sens.* **2004**, *25*, 1565–1596. [CrossRef]
22. Dewan, A.M.; Yamaguchi, Y. Land use and land cover change in Greater Dhaka, Bangladesh: Using remote sensing to promote sustainable urbanization. *Appl. Geogr.* **2009**, *29*, 390–401. [CrossRef]
23. Hussain, M.; Chen, D.; Cheng, A.; Wei, H.; Stanley, D. Change detection from remotely sensed images: From pixel-based to object-based approaches. *ISPRS J. Photogramm. Remote Sens.* **2013**, *80*, 91–106. [CrossRef]
24. Mahmud, M.S.; Haider, F.; Ishtiaque, A.; Masrur, A. Remote sensing & GIS based spatio-temporal change analysis of Wetland in Dhaka City, Bangladesh. *J. Water Resour. Prot.* **2011**, *3*, 781–787.
25. Ridd, M.K.; Liu, J. A comparison of four algorithms for change detection in an urban environment. *Remote Sens. Environ.* **1998**, *63*, 95–100. [CrossRef]
26. Gómez, C.; White, J.C.; Wulder, M.A. Optical remotely sensed time series data for land cover classification: A review. *ISPRS J. Photogramm. Remote Sens.* **2016**, *116*, 55–72. [CrossRef]
27. Pringle, M.J.; Schmidt, M.; Muir, J.S. Geostatistical interpolation of SLC-off Landsat ETM+ images. *ISPRS J. Photogramm. Remote Sens.* **2009**, *64*, 654–664.
28. USGS (United States Geological Survey). Preliminary Assessment of the Value of Landsat 7 ETM+ Data following Scan Line Corrector Malfunction. Available online: https://landsat.usgs.gov/documents/SLC_off_Scientific_Usability.pdf (accessed on 16 July 2013).
29. Song, C.; Woodcock, C.E.; Seto, K.C.; Lenney, M.P.; Macomber, S.A. Classification and change detection using Landsat TM data: When and how to correct atmospheric effects? *Remote Sens. Environ.* **2001**, *75*, 230–244. [CrossRef]
30. Hale, S.R.; Rock, B.N. Impact of topographic normalization on land-cover classification accuracy. *Photogramm. Eng. Remote Sens.* **2003**, *69*, 785–791. [CrossRef]
31. Makarau, A.; Richter, R.; Müller, R.; Reinartz, P. Haze detection and removal in remotely sensed multispectral imagery. *IEEE Trans. Geosci. Remote Sens.* **2014**, *52*, 5895–5905. [CrossRef]
32. Tan, B.; Masek, J.G.; Wolfe, R.; Gao, F.; Huang, C.; Vermote, E.F.; Ederer, G. Improved forest change detection with terrain illumination corrected Landsat images. *Remote Sens. Environ.* **2013**, *136*, 469–483. [CrossRef]
33. Vanonckelen, S.; Lhermitte, S.; Van Rompaey, A. The effect of atmospheric and topographic correction methods on land cover classification accuracy. *Int. J. Appl. Earth Observ. Geoinf.* **2013**, *24*, 9–21. [CrossRef]
34. Richter, R.; Schläpfer, D. Geo-atmospheric processing of airborne imaging spectrometry data. Part 2: atmospheric/topographic correction. *Int. J. Remote Sens.* **2002**, *23*, 2631–2649. [CrossRef]
35. Richter, R.; Schläpfer, D. Atmospheric/Topographic Correction for Satellite Imagery. DLR-IB. 2005. Available online: http://www.rese.ch/pdf/atcor3_manual.pdf (accessed on 11 September 2017).

36. Richter, R. *Atmospheric/Topographic Correction from Satellite Imagery (ATCOR 2/3 User Guide, Version 8)*; DLR-German Aerospace Center: Wessling, Germany, 2011.

37. Lillesand, T.; Kiefer, R.W.; Chipman, J. *Remote Sensing and Image Interpretation*; John Wiley & Sons: New York, NY, USA, 2014.

38. Cheng, H.D.; Shi, X.J. A simple and effective histogram equalization approach to image enhancement. *Digit. Signal Process.* **2004**, *14*, 158–170. [CrossRef]

39. Rozenstein, O.; Karnieli, A. Comparison of methods for land-use classification incorporating remote sensing and GIS inputs. *Appl. Geogr.* **2011**, *31*, 533–544. [CrossRef]

40. Bauer, M.E.; Burk, T.E.; Ek, A.R.; Coppin, P.R.; Lime, S.D.; Walsh, T.A. Satellite inventory of Minnesota forest resources. *Photogramm. Eng. Remote Sens.* **1994**, *60*, 287–298.

41. Cihlar, J. Land cover mapping of large areas from satellites: Status and research priorities. *Int. J. Remote Sens.* **2000**, *21*, 1093–1114. [CrossRef]

42. Kuemmerle, T.; Radeloff, V.C.; Perzanowski, K.; Hostert, P. Cross-border comparison of land cover and landscape pattern in Eastern Europe using a hybrid classification technique. *Remote Sens. Environ.* **2006**, *103*, 449–464. [CrossRef]

43. Gao, J.; Liu, Y. Determination of land degradation causes in Tongyu County, Northeast China via land cover change detection. *Int. J. Appl. Earth Observ. Geoinf.* **2010**, *12*, 9–16. [CrossRef]

44. Congalton, R.G.; Green, K. A Practical Look at the Sources of Confusion in Error Matrix Generation. *Photogramm. Eng. Remote Sens.* **1993**, *59*, 641–644.

45. Sun, C.; Wu, Z.F.; Lv, Z.Q.; Yao, N.; Wei, J.B. Quantifying different types of urban growth and the change dynamic in Guangzhou using multi-temporal remote sensing data. *Int. J. Appl. Earth Observ. Geoinf.* **2013**, *21*, 409–417. [CrossRef]

46. Pandey, B.; Seto, K.C. Urbanization and agricultural land loss in India: Comparing satellite estimates with census data. *J. Environ. Manag.* **2015**, *148*, 53–66. [CrossRef] [PubMed]

47. Hegazy, I.R.; Kaloop, M.R. Monitoring urban growth and land use change detection with GIS and remote sensing techniques in Daqahlia governorate Egypt. *Int. J. Sustain. Built Environ.* **2015**, *4*, 117–124. [CrossRef]

48. Yuan, F.; Sawaya, K.E.; Loeffelholz, B.C.; Bauer, M.E. Land cover classification and change analysis of the Twin Cities (Minnesota) Metropolitan Area by multitemporal Landsat remote sensing. *Remote Sens. Environ.* **2005**, *98*, 317–328. [CrossRef]

49. Irwin, E.G.; Bockstael, N.E. The evolution of urban sprawl: Evidence of spatial heterogeneity and increasing land fragmentation. *Proc. Natl. Acad. Sci. USA* **2007**, *104*, 20672–20677. [CrossRef] [PubMed]

50. Shrestha, M.K.; York, A.M.; Boone, C.G.; Zhang, S. Land fragmentation due to rapid urbanization in the Phoenix Metropolitan Area: Analyzing the spatiotemporal patterns and drivers. *Appl. Geogr.* **2012**, *32*, 522–531. [CrossRef]

51. ADB (Asian Development Bank). *Unleashing Economic Growth: Region-Based Urban Strategy for Nepal*; ADB: Manila, Philippines, 2010.

52. CBS (Central Bureau of Statistics). *Report on the National Labor Force Survey 2008*; Government of Nepal: Kathmandu, Nepal, 2009.

53. Nathalie, W.; Pradhan, M.S. Political Conflict and Migration: How Has Violence and Political Stability Affected Migration Patterns in Nepal? Presented at the Third Annual Himalayan Research Policy Conference, Madison, WI, USA, 16 October 2008.

54. Liechty, M. *Suitably Modern: Making Middle-Class Culture in Kathmandu*; Martin Chautari: Kathmandu, Nepal, 2008.

55. CBS (Central Bureau of Statistics). *Nepal—Census of Manufacturing Establishments 2006–2007*; Government of Nepal: Kathmandu, Nepal, 2007.

56. Shrestha, B. The Land Development Boom in Kathamndu Valley, CDS/CIRAD/ILC. 2011. Available online: http://www.landcoalition.org/sites/default/files/documents/resources/CDS_Nepal_web_11.03. 11.pdf (accessed on 18 January 2017).

57. Pant, P.R.; Dangol, D. "Kathmandu Valley Profile", Briefing Paper. Prepared for Governance and Infrastructure Development Challenges in the Kathmandu Valley. 2009. Available online: https://www.eastwestcenter.org/fileadmin/resources/seminars/Urbanization_Seminar/Kathmandu_ Valley_Brief_for_EWC___KMC_Workshop__Feb_2009_.pdf (accessed on 11–13 February 2009).

58. Bohra-Mishra, P.; Massey, D.S. Individual decisions to migrate during civil conflict. *Demography* **2011**, *48*, 401–424. [CrossRef] [PubMed]
59. USAID (United States Agency for International Development). *Kathmandu Valley: Urban Land Policy Study*; PADCO (Planning & Development Collaborative International): Washington, DC, USA, 1986.
60. Sharma, M.M. Realty Business Grows Across Major Cities Migration by Choice, Force Spurs Transactions. Republica. 2009. Available online: http://www.myrepublica.com/portal/index.php?action=news_details& news_id=10666 (accessed on 18 January 2017).

environments

MDPI

Article

Urban Land Allocation Model of Territorial Expansion by Urban Planners and Housing Developers

Carolina Cantergiani * and Montserrat Gómez Delgado

Department of Geology, Geography, and Environment, University of Alcala; 28801 Alcalá de Henares, Spain; montserrat.gomez@uah.es
* Correspondence: carolina.carvalho@uah.es; Tel.: +34-91-885-5263

Received: 21 September 2017; Accepted: 27 December 2017; Published: 29 December 2017

Abstract: Agent-based models have recently been proposed as potential tools to support urban planning due to their capacity to simulate complex behaviors. The complexity of the urban development process arises from strong interactions between various components driven by different agents. AMEBA (agent-based model for the evolution of urban areas) is a prototype of an exploratory, spatial, agent-based model that considers the main agents involved in the urban development process (urban planners, developers, and the population). The prototype consists of three submodels (one for each agent) that have been developed independently and present the same structure. However, the first two are based on a land use allocation technique, and the last one, as well as their integration, on an agent-based model approach. This paper describes the conceptualization and performance of the submodels that represent urban planners and developers, who are the agents responsible for officially launching expansion and defining the spatial allocation of urban land. The prototype was tested in the *Corredor del Henares* (an urban–industrial area in the Region of Madrid, Spain), but is sufficiently flexible to be adapted to other study areas and generate different future urban growth contexts. The results demonstrate that this combination of agents can be used to explore various policy-relevant research questions, including urban system interactions in adverse political and socioeconomic scenarios.

Keywords: urban land allocation model; agent-based model; spatial simulation of urban growth; urban planners; developers; urban modeling

---

1. Introduction

Urban growth is considered a complex phenomenon due to the strong interactions between different economic, social, environmental, cultural, and institutional components. A deeper understanding of how, why, and where these interactions occur is required in order to be able to plan the territory and create better futures for a constantly growing population. Thus, urban planning is fundamental, as is the need to comprehend and anticipate territorial changes. Modeling offers an alternative means to achieve this due to its capacity to create different possible future scenarios and depict the consequences of urban growth [1–3].

Recent decades have witnessed exponential growth in the use of exploratory models based on different techniques (e.g., stochastic, artificial intelligence, and fuzzy tolerance) in the field of simulation. Although cellular automata (CA) predominate [2], agent-based models (ABM) open new perspectives. Both were originally developed in the field of artificial intelligence with the aim of reproducing the knowledge and reasoning of several heterogeneous agents with the capacity for autonomous action in a specific environment, who at the same time need to coordinate to jointly solve planning problems [4].

However, the use of modeling as a tool to support urban planning has remained limited due to historical skepticism about modeling designed for this purpose [5–7]. Nevertheless, although few ABMs have been employed to address urban issues, other cellular models such as CA have gained more widespread acceptance [8–11]. This is confirmed in a review of urban simulation modeling techniques by Triantakonstantis and Mountrakis [2], who found that more than 80% of the publications analyzed had used a CA, whereas ABM only accounted for less than 10%. These statistics must be placed in context, since CAs have a long history of being used for spatial modeling, whereas ABM are comparatively new in this field. Nevertheless, CAs pose some problems, for example when considering long distance interactions [3], thus paving the way for the use of ABM in the urban modeling arena.

ABM techniques are particularly suitable in this field to gain a better understanding of the urban development process because they allow individual simulation of each agents' behavior and show how, in conjunction, their behavior produces changes in the territory in the form of urban growth. Hence, ABM provides a good laboratory for developing new models of cities, since they elucidate how different city elements interact and thus enable planners to better understand what might happen in the future [3,12,13].

Another important issue to consider is the geographical scope of urban ABM, which varies according to purpose and encompasses multiple spatial scales, although always considering individual behavior. Thus, the literature contains models aimed at exploring local issues such as residential mobility [14] or emergency situations [3,15], and others with a broader, regional scope, for example those aimed at determining urban segregation or informal settlement formation [16–19]. Most have been designed to simulate housing and land market dynamics [20–27]. Consequently, the application of ABM has proved very versatile at different scales, including the sub-regional scale, here understood as the geographical level between the municipal and regional scale.

The sub-regional scale is particularly suitable for urban planning, since urban growth and its impacts can be better controlled at this level. This is especially useful in Spain, where planning is legally proposed at municipal level, with few examples at regional scale [28]; however, it is at this latter level where not only urban growth phenomena but also their territorial, environmental, and other associated impacts could be managed more efficiently. The lack of legal instruments at sub-regional scale has created serious problems, especially during the recent real estate boom [29–31].

Besides scale, a further advantage of ABM when simulating urban growth processes is their capacity to reproduce the behavior, actions, and interactions of the multiple agents involved, who may present diverse profiles (e.g., social, political, and economic). Urban planners and developers are unquestionably among the most important agents in urban growth, since they decide the amount of land to be converted to urban uses, as well as where development should first take place. The population represents another important agent in this process, since the population's behavior is responsible for the demand for new housing, in this case occupying the available dwellings generated by the two abovementioned agents.

Implementing an ABM that simultaneously considers both policy-making and the residential development process poses an enormous challenge in modeling due to the large number of variables to consider (accounting for both internal status and externalities), the spatial (sub-regional) and temporal scale (behavior may occur at different time intervals), and above all, the combination of top-down and bottom-up approaches. With respect to the spatial and temporal scale, it is assumed that these are strictly linked to the kind of data used. Many models developed to study changes observed at the urban level consider stylized data in a hypothetical world. Artificial cities and rules are constantly being created in order to test these models and their underlying theories [21,22,26,32].

Although some independent allocation models have been used to simultaneously simulate both levels of the planning process (policy-making by urban planners and residential expansion by developers), only a few ABMs have been aimed at considering both when simulating urban growth at the sub-regional scale [33–35]. Some of these have considered the environment or urban planners as agents, albeit not with the same goal [18,20,36,37]. In contrast, many urban ABMs consider developers as agents, mostly focusing on land market dynamics [21–23,32]. Similar models have accounted for both agents [36–39], although neither coincided with the purpose of modeling, which is to test and explore the urban process or the system itself. The closest example would be the study by Ligmann-Zielinska et al. [27], who developed a model using these three agents, although an independent deterministic model rather than an ABM was developed separately in order to simulate the typical top-down process of planning.

In line with these precedents, the objective of the present study was to create a model (AMEBA) in order to address the abovementioned gaps by simulating the urban development process at the sub-regional scale considering urban planners, developers, and the population, while simultaneously addressing the challenge of acting at heterogeneous intervals. This paper discusses the first and second submodels of the prototype, which adopt a more deterministic rather than stochastic approach, and were thus based on an urban land allocation model. The population submodel represents a parallel stage of this research and takes a stochastic approach to the occupation of newly developed areas, rather than the territorial expansion presented here. Since this latter forms part of the final integrated model, it is briefly introduced in the explanation of the model structure, but is not discussed further.

AMEBA has been developed as a prototype with three submodels developed independently on the same ABM platform, endowing it with sufficient flexibility to be adapted to other study areas and reproduce different future urban growth contexts. The submodels and the final prototype were created, fed, and tested using empirical data on a specific area in Spain, the *Corredor del Henares*. The sub-regional profile of the study area presents substantial socioeconomic and urban development diversity, and thus represents a suitable area to test the model.

Following this introduction, Section 2 gives a conceptual overview of the prototype (with three integrated submodels), and describes the study area, model structure, and other technical issues (platform and database). Section 3 reports on the implementation of the submodels that address territorial expansion (individually and in conjunction), the simulation experiments, and the validation process. Section 4 concludes the paper with a discussion of the submodels' capabilities and limitations, and directions for future research.

2. Testing Area, Conceptual Model, and Data

As previously mentioned, ABM are suitable for studying complex issues, and therefore could serve as a valuable tool to support urban planning processes. This section presents the conceptual model of the prototype (AMEBA), describing its structure, platform, and database, as well as the variables used to test the model in the study area.

2.1. Testing Area

The metropolitan area of Madrid has become one of the most dynamic areas of urban growth in the Iberian Peninsula and Europe [40]. It has three main urban growth corridors, each with their own defining characteristics (Figure 1). The study area encompassed 18 municipalities in the Region of Madrid selected from the northeast corridor, known as the *Corredor del Henares*, which runs from Madrid to Guadalajara (the capital of the neighboring region). This area is home to almost 600,000 inhabitants [41], covers 624 km², and is strongly influenced by its proximity to the city of Madrid. In addition, it is a particularly dynamic area in urban and demographic terms, with substantial internal sociodemographic and urban development differences. This complexity is further increased by the lack of regulation aimed at solving territorial problems at the regional level, rendering it an ideal laboratory to test the prototype reported here.

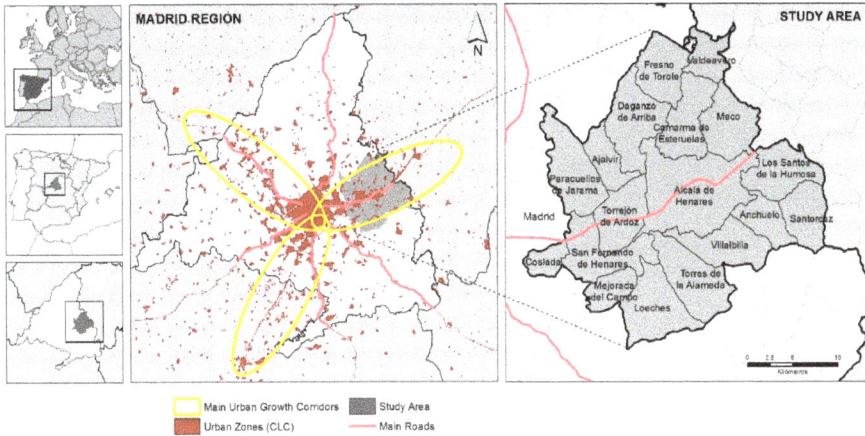

Main Urban Growth Corridors Study Area

Urban Zones (CLC) Main Roads

Figure 1. Urban corridors in the region of Madrid and study area.

The general urban growth trends identified in the main Spanish cities in the period of economic wealth (1997–2007) indicate rapid and chaotic urban development. This coincided with the housing bubble, subsequently followed by a stable period of consolidation [42–44]. Such trends have been witnessed in many other European urban areas in countries, such as Italy, Croatia, and France, mainly associated with their principal cities and coastal zones [45,46].

2.2. Prototype Structure, Platform and Database

The urban growth process at the regional scale could be theoretically and logically described as a three-stage process involving three types of agent: urban planners, developers, and the population [47]. The first two are responsible for the territorial expansion itself, while the latter represents the occupation of the territory, and all three converge to shape a complex urban system. The urban growth process starts with the identification of new potential urban areas by urban planners in line with national, regional, and local policies. In the second stage, developers choose where to promote new residential developments. Finally, the population selects where to live based on their individual preferences and according to their income possibilities. Although other elements may also be involved in the urban development process (such as infrastructure expansion or political guidelines), these would have been difficult to include in our model, so we decided to limit our agents to these three. Each of them is considered in an independent submodel with different characteristics. The first represents the typical deterministic, top-down process, while the second may be considered a transition between deterministic and stochastic approaches, and the third could be considered the closest representation of an ABM as we understand it (using a bottom-up approach), where complex interactions between agents are represented stochastically. The three of them are coordinated separately before subsequent integration in a final model, where outputs and inputs feed each other continuously. It is worth mentioning that the model is explicitly spatial, so all variables are spatially distributed.

The prototype is intended to be as useful as possible, and therefore it was constructed using an adaptable structure of submodels, each covering a critical variable in the urban growth process. This approach was adopted in response to the low impact that urban growth simulations have had on real-life planning processes, due to their closed architecture and the high amount of data needed to run them, not commonly available in many regions or countries [2].

As presented in Cantergiani et al. [48] the submodels are described below:

The first submodel (urban planner) simulates the urban planner´s decision-making process, which consists of selecting new areas for urban development according to physical restrictions (e.g., protected areas, high slopes, or proximity to water bodies), distance to elements of interest (e.g., roads or consolidated urban areas), and the amount of growth required to meet existing demand. These criteria are set as parameters and can be modified at initialization to generate different scenarios.

The second submodel (developer) focuses on the developer's decision-making process regarding new residential developments. As part of their behavior, developers must decide where to build new housing, how many new developments must be built, their capacity, and their target economic group. This process takes into account the legal status of the territory (defined by urban planners in the first submodel), as well as the areas that will optimize their profits.

The third submodel (population) simulates the process of residential location choice and occupation by the population. In this case, different agents look for the best place to move according to their purchasing power and location preferences, which may include distance to the public transport network, education facilities, and other factors.

The flowchart in Figure 2 shows how the submodels are integrated and where the result of each submodel feeds into the next in a chain-like process.

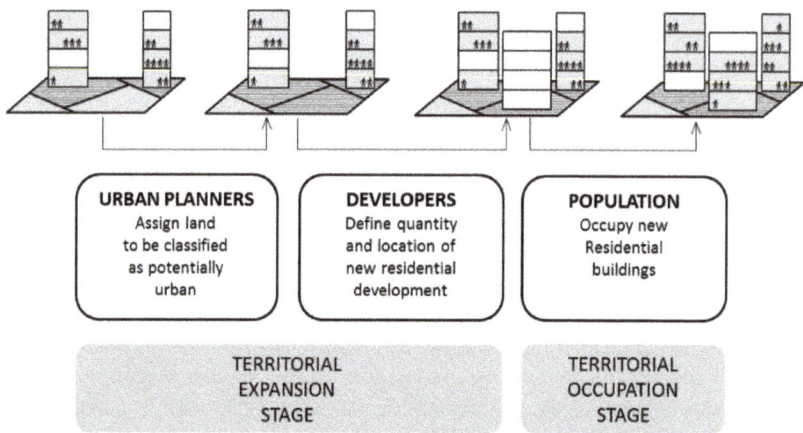

Figure 2. Integrated decision model to simulate urban growth.

The conceptual model considers two components for the three agents: territorial availability (TA) and spatial interest criteria (SIC). The first (TA) indicates potential areas taking into account spatial restrictions. For urban planners, it considers the limits imposed by legal regulations (e.g., master plans, law on protected areas) and physical urban development restrictions (e.g., along rivers, steep slopes) to spatially delineate potential areas for development. For developers, TA restricts the area to those classified as potentially urban. The second component (SIC) represents attractiveness, which indicates the preferences of each agent regarding the distance to given spatial elements. For urban planners, these include roads, public facilities, hazardous areas, agricultural productivity, and existing urban fabric, while for developers they include roads, public transport stops, existing urban fabric, and status of the existing residential stock. Due to the flexibility of each submodel of the prototype, several variables can be selected to represent these components and run the model [47], although here we only report those tested in the study area.

The platform used to develop the prototype was NetLogo v.5.3.1 (Center for Connected Learning and Computer-Based Modeling, Northwestern University, Evanston-IL, IL, USA) [49], which met the technical requirements and presents many advantages. To name a few, it has an intuitive and simple interface, it is widely used for similar applications, it is user-friendly, and it is open source. Moreover, it enables simple connection with geographical databases through a Geographic Information System (GIS) extension, fundamental for spatial analysis. For instance, this connection facilitates the flexibility required for adaptation to other study areas, with different input data in terms of extension or resolution.

The submodel interface consists of three sections: (i) an input box where geographical information is called up and spatially represented in a visualization panel; (ii) an area where the user sets the initial parameters that will define the scenarios to simulate; and finally (iii) a third area where the results are represented in the form of graphs, spatial distribution, and an external matrix containing statistical references (Figure 3).

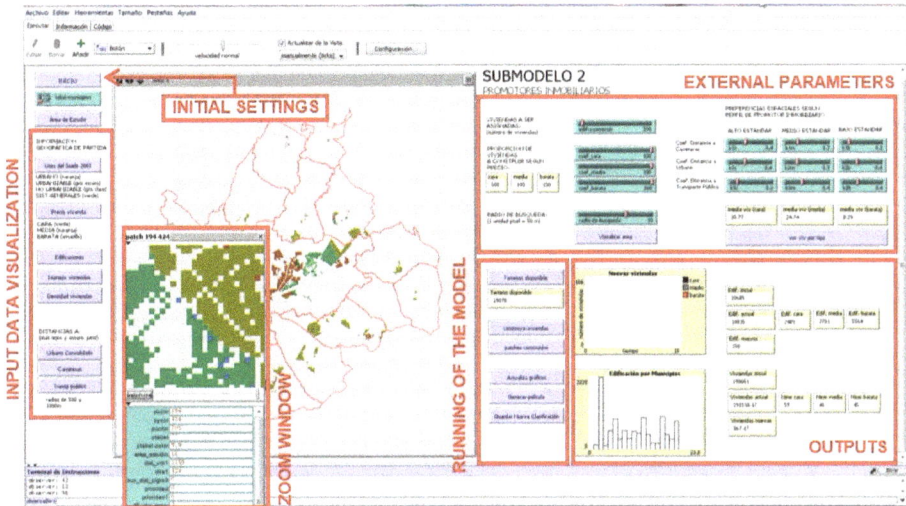

Figure 3. Distribution of the functional blocks in the interface (common for all submodels).

All submodels have a similar interface, differing only in the number and kind of external parameters to be set, and obviously in the expected outputs. Since a set of parameters can be freely modeled, users can adapt the model to their specific context and generate different future urban growth scenarios.

Bearing in mind the regional scale and main drivers of urban growth in the context cited and other similar contexts, a considerable effort was made to compile the most appropriate alphanumerical and geographical information, always considering the territorial component. Although they might form part of the process, statistical components were only incorporated into the model when they could be spatially represented (such as price and income distribution). Other variables, such as traffic flows, global climate change, mobility, and time variation, were not included at this stage for the abovementioned reasons, although we recognize they must play a partial role in the evolution of the phenomenon studied.

The input data used to implement the submodels were selected to reflect the drivers of urban growth, empirical knowledge of urban growth trends in the study area [47,48,50,51], and the most commonly used variables in other studies [8,33]. These latter included the municipal boundaries (surface), zoning status (surface), and housing distribution (pixel) used by agents to assess specific environments and conditions, as well as the other spatial information indicated in Table 1 below. These were represented by pixels measuring 50 × 50 m. Once data had been collected, they were transformed into determinants for subsequent use in the model.

Table 1. Description of determinants, based on input data, incorporated into the prototype.

Data Layer	Scale	Origin	Determinant and Description
Census zones and associated statistical data	1:1000	INE (2001)	Distance to consolidated urban zones Contributes to determination of agents' preferences
Zoning status	1:50,000	Department of Environment, Housing and Land Use Planning, Region of Madrid (2006)	Land classification (reclassified as urban, non-building, and potentially urban) Functions as the legal constraint on agents' actions
Natural Protected Zones (Spanish initials: ENP)	1:50,000	Environment Database of Spanish Ministry of the Environment (2012)	Restricted areas (sum of individual restrictions spatially distributed) These act as a mask under which agents cannot intercede
Buffer along main water bodies	1:25,000	BCN 25/CNIG, IGN (2008)	
Slope (origin: Digital Terrain Model)	1:50,000	BCN 25/CNIG, IGN (2012)	
Land use and land cover	1:100,000	CORINE Land Cover Project of IGN (2000–2005)	Agricultural productivity generated from reclassification of the land use database Agents state their preferences according to the high, medium, or low level of productivity
Urban facilities (health and education, public transport, locally unwanted land uses)	1:25,000	NomeCalles Database of Statistical Institute of Region of Madrid (2005)	Distance to each type of facility This may indicate a higher or lower level of interest
Cadastral data containing information at parcel level (code, municipality, use, year of construction, area, centroid, etc.)	1:1000	General Directorate for Land Register, Ministry of Finance and Public Administration (2006)	Aggregated data by pixel represent the initial residential building distribution Agents behave according to the existing buildings
Network of national and regional roads	1:25,000	BCN 25/CNIG, IGN (2008)	Accessibility calculated as distance to the road network This may indicate a higher or lower level of interest

INE: National Statistics Institute; BCN: National Cartographic Database; CNIG: National Center for Geographical Information; IGN: National Geographic Institute; ENP: Protected Natural Spaces.

3. Implementation of the Territorial Expansion Simulation

3.1. Urban Planner Submodel

The urban planner submodel simulates regional urban growth and represents the most general scale of the three submodels, indicating areas where it is legally permissible to construct new residential buildings. This submodel presents a simple and deterministic structure, since the behavior of urban planners generally reflects a top-down approach whereby decisions are made within a general plan by an institutional agent rather than by individuals (or by an individual representing an institution). Nonetheless, the allocation model was developed in NetLogo in order to facilitate integration into future submodels that might simulate agents' behavior using the bottom-up approach.

The expected result is useful for any urban planner, and represents a laboratory that allows the selection of new areas for reclassification as potentially urban (Figure 4), generated from their decision regarding the most suitable areas according to criteria that may differ depending on the agent's profile. Rather than testing the model with different types of urban planner, we only considered one type presenting different profiles that in turn could generate numerous scenarios.

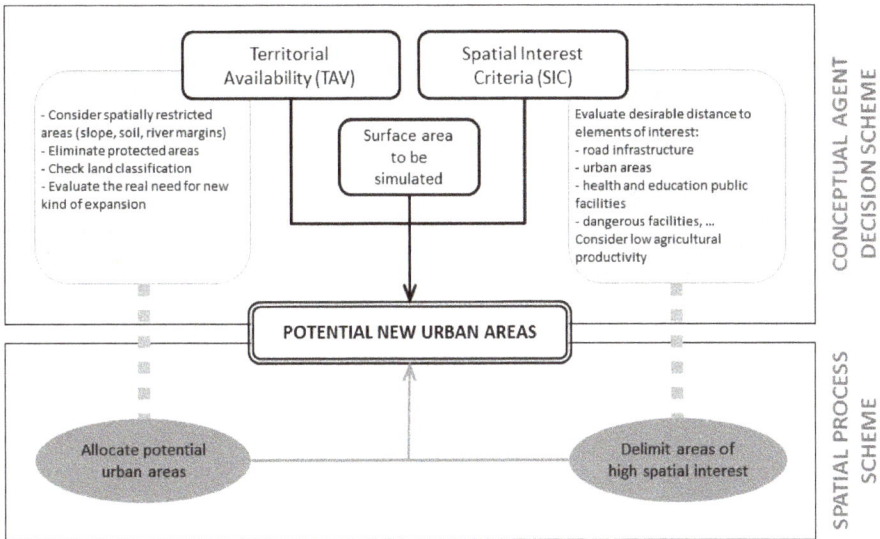

Figure 4. Conceptual decision structure of the urban planner submodel.

Besides defining the amount of land for potential reclassification as urban, the interface allows the user to set initial conditions that characterize each profile (Figure 5). Hence, the interface settings allow the user to establish weights for variables such as distance to urban areas, locally unwanted land uses (e.g., hazardous waste dumps, prisons, and trash disposal plants), road infrastructures, and health/education services, and to define priority areas for potential reclassification as urban according to their agricultural productivity.

Figure 5. Variables that can be set when initializing the model to define different urban planner profiles.

The agent must convert the amount of land specified by the user in the interface, and the simulation outcomes should show where these changes will take place. The model mechanism works as follows: from among the non-restricted surface areas—limited according to user selection from among the three available options—urban planners first analyze those areas with the highest values of interest. These values are graduated using the average of the weights indicated on the sliders at the interface for each pixel (Figure 5). Then, the last coefficient (agricultural productivity) is set, and the model assumes that the proposed percentage is homogeneously distributed spatially and randomly throughout each area according to its agricultural productivity, defined as high (forest and natural areas), medium (agricultural areas), or low (all other categories) based on a reclassification of the

CORINE Land Cover land use categories considered artificial [52]. The final selection is then taken as the best option.

Having described and implemented the conceptual model in NetLogo, the simulation experiments conducted to test the prototype are reported below. Several authors have suggested that the validation process should consist of different components: verification, validation, and a sensitivity analysis [12,53–58]. However, since this paper presents a prototype of an ABM, it focuses less on predicting the future and more on understanding and exploring the behavior of the urban system and reproducing specific scenarios. Consequently, only verification of correct model construction and validation by comparing the results with real data will be discussed further. Verification (also known as internal validation) refers to the correctness of the internal structure, ensuring that the model has been developed in a formally correct manner (e.g., system diagrams, units of measurement, and equations) in accordance with a specified methodology [58]. This implies corroborating that the implemented model conforms to the specifications by running it after changing some initial parameters, as done by other modelers [33]. Statistical validation, or just validation, refers to determining if the model is correct by checking whether it achieves the expected level of accuracy in its predictions. This involves analyzing whether the structure of the model is appropriate for its intended purpose from a conceptual and operational point of view by comparing results to real data.

The goal of these procedures was twofold: on the one hand, to confirm that the results of specific scenarios could help determine whether the model had been correctly constructed, and on the other, to confirm that the model has the capacity to generate the expected simulation, given the scale, available data, and initial set of conditions.

For verification, three different contexts were set for each model by varying some of the initial parameters (e.g., adopting a sustainable, economically conservative, or speculative approach), so that they were sufficiently diverse to generate different outputs. It is important to note that throughout this paper, the term "scenarios" is used to refer to different future contexts, rather than according to the definition of scenario in the strict sense of the term. Moreover, all of the prototype submodels represent tools in the form of open possibilities that can reproduce different urban development scenarios (as reported here) and generate a sequence of products that could be used for a validation process such as a sensitivity analysis, employed by many scientists for this purpose [16,18,36]. We propose to conduct such an analysis on the final model once all three submodels have been integrated.

Hence, although very different results could arise from minor changes in these initial conditions, the usefulness of the simulations resides in the fact that they were not aimed at identifying or analyzing specific urban growth shapes or statistics for the *Corredor del Henares*, but at determining whether the model was correctly constructed. If this latter were to be demonstrated, it would support the notion that ABM constitute a powerful tool for use in real situations and would confirm that the prototype achieves the proposed goal. The flexibility and high number of setting options increase the possibility of employing this tool in different contexts and using real data on other areas.

For the validation process, we compared the results of the different simulations (of the two submodels once integrated) with the corresponding urban land data available for the study area, and obtained the percentage of coincidence.

With these goals in mind, we simulated the expansion of potential development areas using three different urban planner profiles defined by the tendency to employ: (i) a sustainable approach, (ii) an economically conservative approach (typical behavior in a crisis situation), or (iii) a speculative approach (corresponding to a business-as-usual scenario), in accordance with the storyline and scenarios described in Plata Rocha et al. [50]. In order to run the three simulations, the same initial conditions were used, but different parameters were set to characterize each profile (Table 2), based on the work cited. As expected, the simulations yielded some notable differences for the three urban planner profiles, all of them with growth concentrated close to the most dynamic existing urban areas, but distributed differently throughout the territory. Other future contexts could be easily defined by changing values for each parameter.

Table 2. Initial conditions for three different urban planner profiles.

Variable/Urban Planner Profile	Reference	Sustainable	Conservative	Speculative
Distance to locally unwanted land uses *	0.0 to 2.0	1.0	0.0	0.0
Distance to urban areas *	0.0 to 2.0	1.0	2.0	0.5
Distance to roads *	0.0 to 2.0	2.0	1.5	0.5
Distance to health and education facilities *	0.0 to 2.0	2.0	1.0	0.0
Agricultural productivity:				
Conversion to potential building land in zones with high agricultural productivity (%)	0 to 100	0	5%	20%
Conversion to potential building land in zones with medium agriculture productivity level (%)	0 to 100	10%	20%	40%
Conversion to potential building land in zones with low agricultural productivity (%)	0 to 100	90%	75%	40%

* Zero indicates that this factor is not relevant, while two indicates maximum relevance. Intermediate values show proximity to minimum or maximum influence.

Therefore, the simulation results are acceptable since they are spatially and conceptually coherent. An aggregated visualization of real data on the evolution of past and recent urban plans (the municipalities in the study area have relatively old plans and many updates are under discussion) and simulation with a five-year horizon for the three planner profiles (Figure 6) indicate alternative distributions that could be used as a reference for urban planning decision-making.

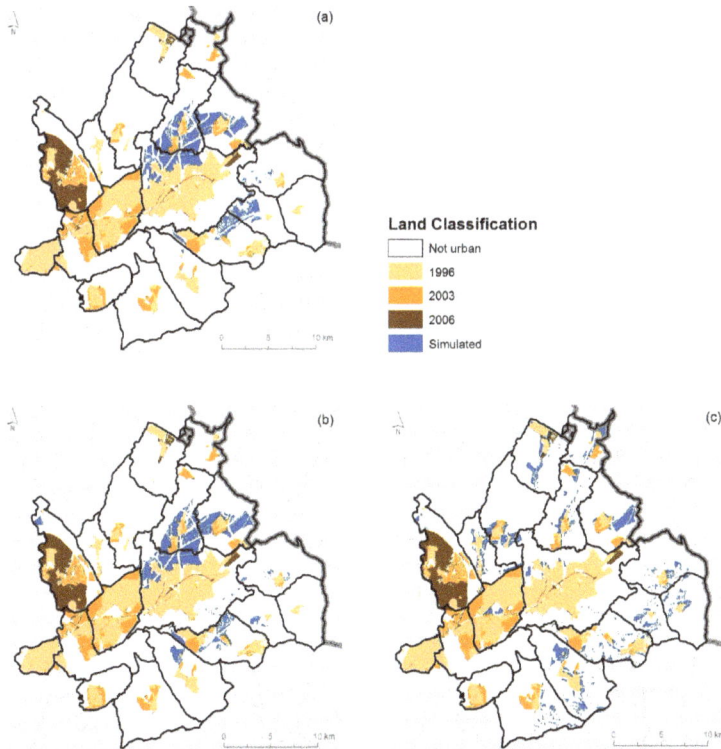

Land Classification
- Not urban
- 1996
- 2003
- 2006
- Simulated

Figure 6. Comparison of simulation results (five years) and real data on past urban plans for the three planner profiles: (**a**) sustainable, (**b**) conservative, (**c**) speculative.

3.2. Developer Submodel

The developer submodel considers a bottom-up rather than a top-down approach [27], assuming a more spatially-restricted knowledge of the territory (in the urban planner submodel, agents are assumed to be familiar with the entire territory in order to make a decision). Simulation of their behavior is expected to yield a new distribution of low, medium, and high standard residential buildings.

In the first instance, the only restrictions (TA) considered referred to regulations; thus, only already-designated areas were candidates for new residential development. As regards attractiveness (SIC), the preferred distance to elements of spatial interest was defined according to the standard of the building for which expansion was to be simulated. In this model, land value and neighborhood characteristics were the two main elements that strongly affected the decision to assign a new area for development (Figure 7).

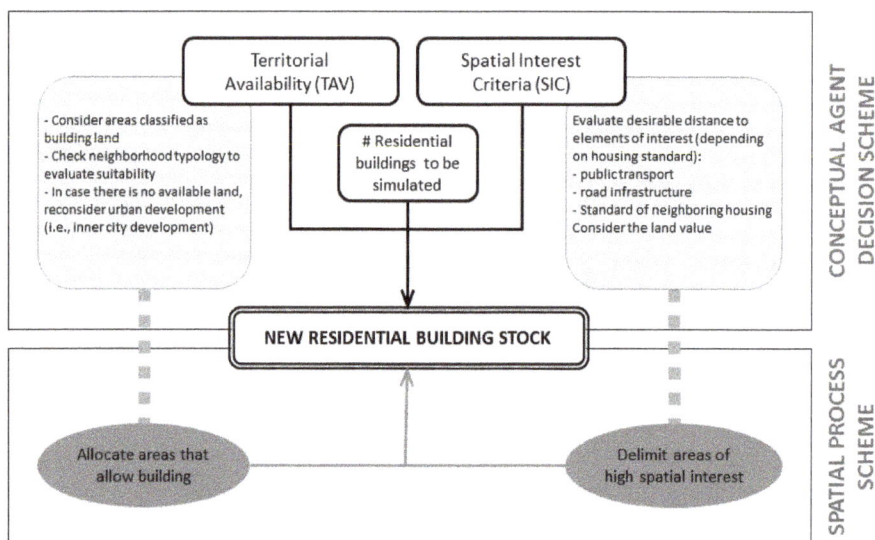

Figure 7. Conceptual decision structure of the developer submodel. #: number.

As in the previous case, agents in this submodel present a unique type of behavior regarding the variables considered, although users may change some of the parameters depending on whether they aim to simulate the construction of high-, medium-, or low-standard housing. In addition, the interface again allows the user to establish the initial conditions by changing parameters such as the quantification of new buildings, specification of respective percentage of each type, and calibration of available coefficients for each standard (distance to roads, urban areas, and public transport). The user can also define the maximum search area that developers should take into account, considering the elements necessary to identify a suitable location (where no ideal location is identified within the area defined, the model allows the user to redefine the search area). The combination of these parameters defines the future scenario.

The decision flow chart (Figure 8) leads to the final location where new housing should be built.

Unlike the urban planner submodel, the developer submodel starts with the selection of the typology of building proposed, i.e., whether the agent will seek free land to build high, medium, or low standard housing. This information is vital to confirm the area of interest, which is different for each case, and later it will also be useful to define the number of dwellings assigned to each new building typology. This choice must respect the maximum allowed for each typology. In the next step, the model focuses on a random point that must obey the restrictions and analyzes the neighborhood, considering an extension which is also indicated in the interface by the user. The preference criteria include the degree of similarity with nearby buildings (attraction for similar, or in some cases, rejection for different) and the spatial interest defined as in the previous submodel. Whenever there is an option that complies with these criteria, and the maximum number of runs has not yet been reached, the new building is assigned to the corresponding pixel. The number of dwellings assigned will depend on building typology, going from high density (low standard) to low density (high standard).

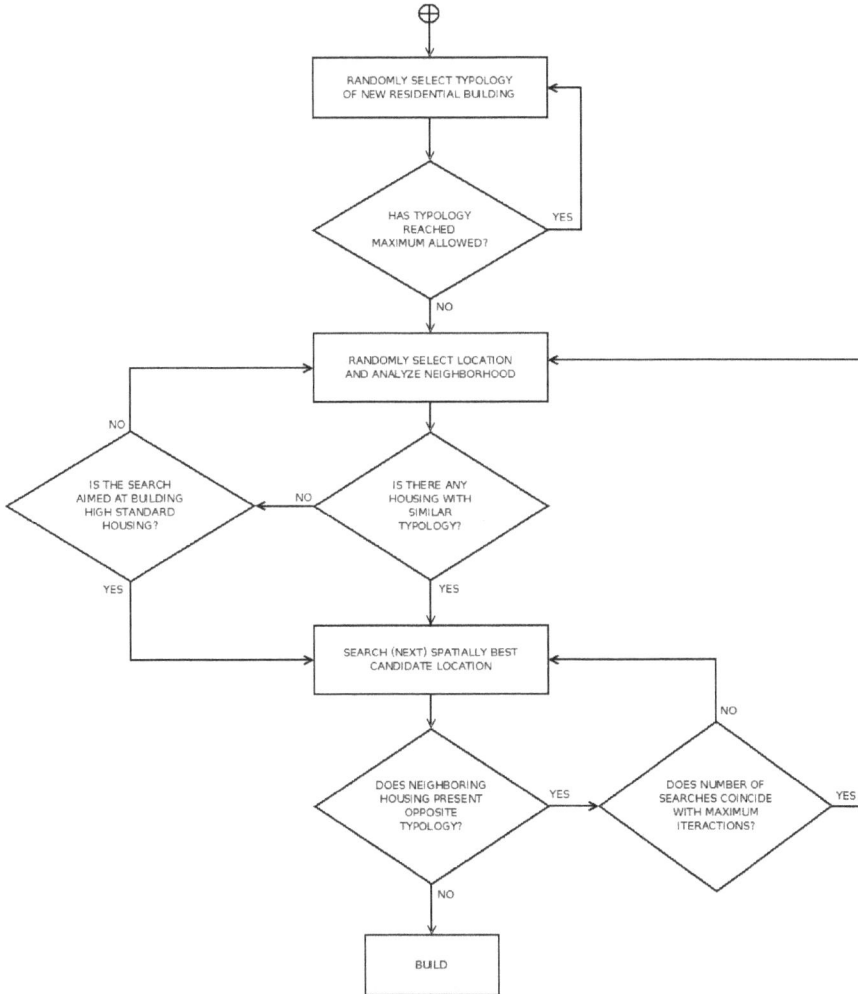

Figure 8. Developer's decision flow chart.

As in the previous case, in order to confirm prototype flexibility and correct construction, scenarios were defined according to the external environment rather than the agent's profile (as in submodel 1). In this case, two parameters could be modified: the residential growth predicted for each housing standard and the maximum search area. In order to test the model, different settings were defined for these parameters in order to simulate the real estate market under three different situations: (i) sustainable (environmental protection approach), (ii) crisis (conservative approach) and (iii) speculative (business as usual). Each scenario reflects different building dynamics, housing typology distribution balance, and urban growth shape, all defined by adjusting the values for the initial submodel conditions (Table 3).

Table 3. Initial settings and description of submodel spatial results from simulation of the three developer scenarios.

Scenario	Initial Settings			Description of Results
Characteristics that represent the scenario	Building dynamics	Housing typology distribution balance	Urban growth shape	
Equivalent parameter in the submodel interface	Number of residential buildings (100–3000)	% of growth for high, medium, or low standard buildings (0–100%)	Compact or disseminated represented by the search area size (50–1500 m)	
Scenario 1: Sustainable	High (3500 new buildings)	Balanced (high for all, mean: 70%)	Narrow (500 m)	Group of buildings in a compact shape, usually connected to existing buildings
Scenario 2: Crisis	Low (500 new buildings)	Unbalanced (low for all, mean: 30%)	Narrow (500 m)	Few buildings, distributed evenly, most of them low standard
Scenario 3: Speculative	High (3500 new buildings)	Balanced (high for all, mean: 70%)	Wide (1500 m)	New buildings dispersed throughout the territory

The last column in Table 3 presents a short description of the results, indicating satisfactory agreement between the proposal and the expected and simulated results, mainly considering the spatial response for the three situations (Figure 9). Furthermore, the final distribution shows the actual and projected housing together, indicating the lack of development in restricted areas (non-building areas or roads, for example). The spatial results for the developed area were generated from pixel distribution over the surface designated for that purpose (urban and developable land). We did not consider illegal settlement dynamics at this stage of the model.

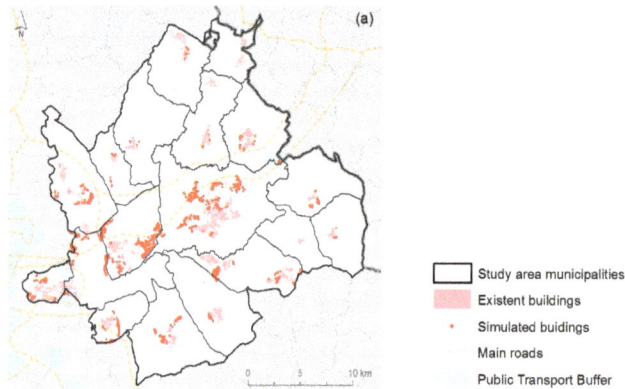

Figure 9. *Cont.*

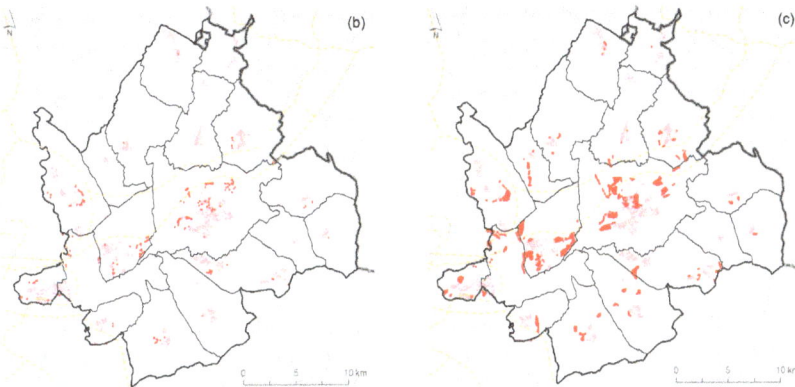

Figure 9. Spatial results for the three simulation scenarios: (**a**) sustainable, (**b**) crisis, and (**c**) speculative.

3.3. Integration of the Urban Planner and Developer Submodels

Although we only integrated two of three prototype submodels, this still presented a challenge due to the problem of how to combine two different submodels in one, which must also be clear and robust. Below, we describe how this difficulty was solved and the two submodels combined to represent the territorial expansion process:

- The submodels were combined through continuous feedback of inputs and outputs. Thus, the developer submodel uses the output of the urban planner submodel (land classification) as one of the inputs; hence, depending on the established criteria, an updated layer is periodically obtained showing areas classified as potential building land. As developers behave according to these legal restrictions, the results of the independent submodels and integrated model should differ.

- Temporal resolution also presents a challenge because the two submodels use different time intervals to reflect the fact that developers build housing in a shorter period of time than urban planners take to develop new planning proposals. By law in Spain, urban plans must be revised at least every ten years with some periodical, partial reviews; thus, five years was the interval considered for the urban planner submodel, although it could easily be modified according to where the prototype is being employed. For the developer submodel, we considered an interval of one year to reflect the time taken to construct new residential housing, although we are aware that this is often a continuous process.

- Successful integration of the two submodels depended not only on which inputs/outputs were considered, but also on when and how they were interchanged, since the submodels employ different time intervals. The rate is determined by compliance with criteria that interrupt the running of one submodel to start that of the other. In order to obtain the first results, to skip from the developer to the urban planner submodel, either (i) the free area suitable for development should not exceed a defined percentage of the total, or (ii) one of the submodels should reach a given number of runs (Figure 10).

In summary, the model starts with a territory classified into urban, potentially urban, and other, and also with a fixed distribution of buildings. The prototype internally calculates the percentage of potentially urban areas occupied by residential buildings and the number of runs for each (starting from zero). If the result exceeds the amount defined by the user (meaning that new areas are in demand), then the urban planner submodel is launched. Otherwise, the developer submodel is run continuously until that percentage is reached. Both submodels run in accordance with the independent procedures described previously. In this case, the NetLogo platform was selected in order to better integrate both of these with each other and with the third population submodel in an ABM context, since other techniques such as CA or multi-criteria models might not be able to solve integration.

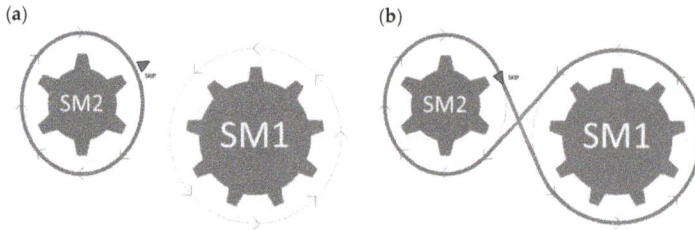

Figure 10. Sequential flow considering compliance or not with the criteria in order to skip from one model to the other: (**a**) Continuous cycle, (**b**) Cycle when "skip" conditions are met. SM1: submodel 1—urban planners; SM2: submodel 2—developers.

In order to conduct initial integration of the submodels, we considered a fixed sustainable scenario for both. In the first instance, this should represent the ideal approach to urban development, taking into consideration legal and strategic factors. Although this variable would depend entirely on the user's interests, for this simulation experiment we set the urban planner submodel to run once for each three runs of the developer submodel, with the latter running at one-year intervals. As expected, the results of this integrated simulation indicated urban growth along roads and around existing developments: potential urban land expansion occurred close to residential buildings in the smaller municipalities, and mainly in the central ones (Figure 11).

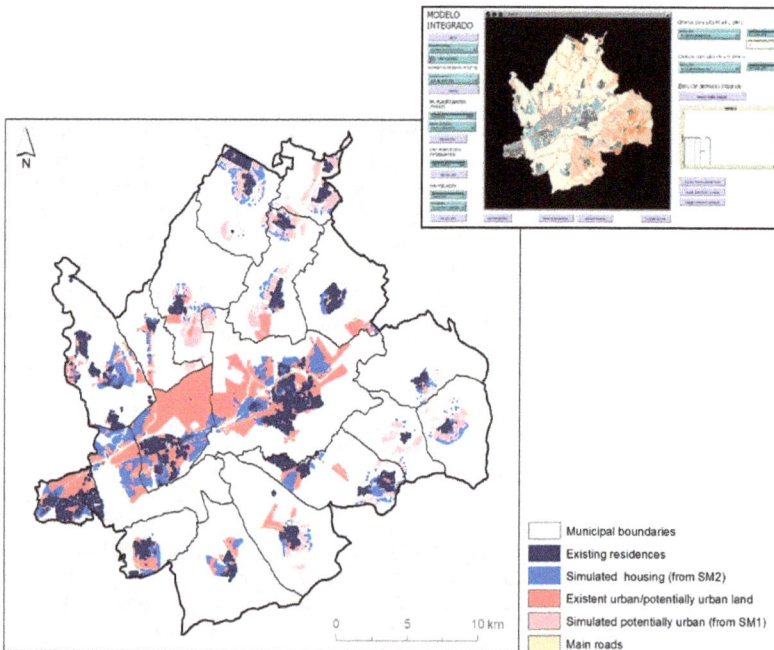

Figure 11. Interface of the model that integrates the urban planner and developer submodels and results shown in a Geographic Information System (GIS).

The tests performed on these models represent the only verification that it was possible to conduct at this time, according to the existing data and structure described for these two submodels. Verification therefore consisted of testing model reliability by running it with different future scenarios in order to confirm that the results complied with the established rules and expected operation. This would demonstrate that the model generates acceptable simulations according to the available established parameters, and was thus considered sufficient verification for a prototype. Many partial modifications were performed throughout the construction of both submodels and the integrated model; for each procedure, additional queries were hosted in the programming to verify correct operation at different stages. An iterations routine was also introduced, and the test results achieved a mean of 90% coincidence after 10 sequential runs. This high level of coincidence is explained by the fact that the structure of both models was based on a deterministic approach.

Before reporting the validation process, it is important to note that the combined outputs of the two submodels comprise a partial result, indicating the proposed developable surface, with which there is no available real data for comparison (a comparison between different scenario outputs is given in Figure 6), and the new distribution of individual built-up pixels in those areas. With this in mind, a validation step was performed comparing the new simulated buildings with real land use data. To this end, we used the only updated data available: the Spanish Land Cover and Use Information System (Spanish initials: SIOSE) produced by the National Geographic Institute of Spain (Spanish initials: IGN), available for 2011 with a reference scale of 1:25,000. This includes the basic land use categories, such as artificial zones, crops, grassland, woodland, scrub, and land without vegetation [59]. Although the model result represents built-up areas in pixels, while the only available data on observed spatial urban growth for the same year refer to surfaces, the intersection of these shows the level of correspondence between model products and reality.

This comparison (Figure 12) revealed an acceptable level of coincidence, bearing in mind that the difference in data shape (pixel and surface) limited coincidence. In terms of percentages of correspondence, our results show that the model closely reflected the real situation in the study area tested, obtaining the highest percentage for the crisis scenario (62.3%), which mirrored the profound real estate crisis that hit Spain during the simulation period. Furthermore, the results for the speculation scenario (39.9%) and the sustainable scenario (48.8%) were coherent, because the former represents the scenario furthest from reality in the study area, while the sustainable scenario indicates more conservative growth, closer to the crisis scenario (see the parameter established for each scenario in Tables 2 and 3).

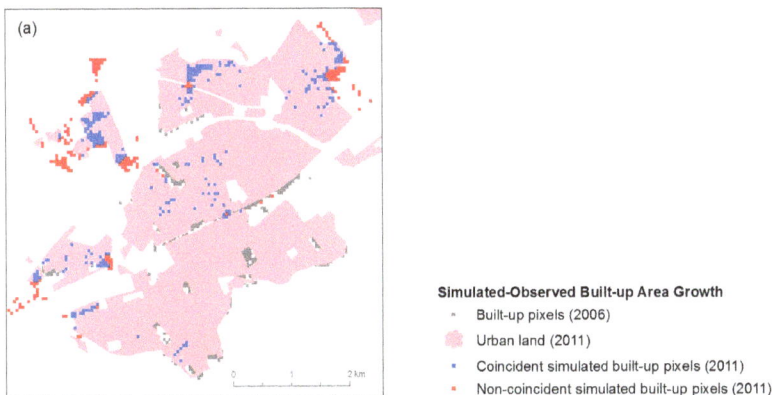

(a)

Simulated-Observed Built-up Area Growth

- Built-up pixels (2006)
- Urban land (2011)
- Coincident simulated built-up pixels (2011)
- Non-coincident simulated built-up pixels (2011)

Figure 12. *Cont.*

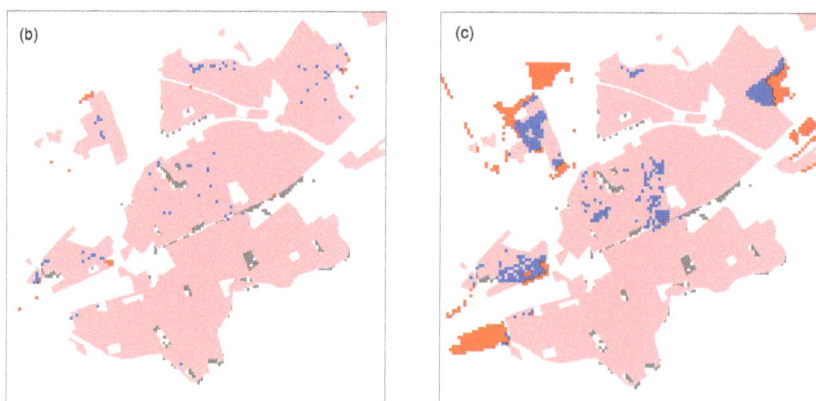

Figure 12. Simulated and observed built-up area growth in a zoomed area (municipality: Alcalá de Henares): Comparison between the simulated residential buildings for the three scenarios (pixels)—(**a**) sustainable, (**b**) crisis, and (**c**) speculative—and the urban land cartography (surface) from the SIOSE database (2011) [59].

4. Discussion

It was a challenge to design a model that simulated the urban system considering its territorial expansion process, since it was necessary to integrate two agents that act at different scales and with different goals, although their products and behaviors are closely related. A larger number of examples of application at similar scales would be required to discuss its possibilities further; nonetheless, the existing ones serve as references, to which we have added empirical knowledge about the region and the urban process in Spain, enabling us to create, run, and test both submodels and the integrated model.

The structure of the agent-based model described here consists of three agents, constructed independently for subsequent integration. When analyzed separately, the territorial expansion reported here is more deterministic, while the stochastic element is given by the third submodel (population) and the integration of the first two (as presented here) or all three submodels. The novelty of this model resides in its capacity to simulate the urban growth decision-making process in an ABM context, and to integrate the submodels, which would be problematic if using other urban growth simulation techniques such as CA or multicriteria evaluation.

It is worth mentioning that the decision to work with separate submodels, running them individually first, endowed us with more control at each step in the entire process, and in some cases it also enabled us to forestall errors. Unlike other authors, who have considered similar agents and goals [27], we used three submodels—urban planner, developer, and population—to represent the main agents when modeling the urban development process. In this paper, we have reported the first two, which together show the spatial distribution of urban growth. These agents' decisions are based on a set of criteria and weighted preferences that together with other interaction elements would be difficult to simulate using urban growth models based on multicriteria evaluation, cellular automata, or other techniques. A further innovation is the interaction of the submodels using input and output feedback in the same framework, generating a continuous process of land reclassification and residential building expansion focusing on the influence of one behavior on another. The ABM architecture facilitates the representation of agent interactions, which will be completed by integrating the population submodel, designed to represent occupation of the available residences generated by the developers.

We believe that this model, which is sufficiently flexible to combine the top-down approach of urban planners and the bottom-up approach of developers, could serve as a useful tool to stimulate lively discussion about planning support modeling at the intermediate scale. In addition, it could also serve to further analyze the impact of land use changes or their relationship to transport system evolution using this bottom-up perspective.

As indicated throughout this paper, this structure enabled us to combine agents' behavior and transform it into mathematical rules that could be entered independently into the urban planner and developer submodels and then transferred to the integrated model. NetLogo was the best platform for this, since its simple interface and GIS extension made it possible to obtain a user-friendly model. This capacity to connect an ABM with GIS represents an important advance for urban simulation, since there are usually clear limitations on cartographic representation in an ABM. Furthermore, the processing power of NetLogo, which may sometimes be considered a limitation for more demanding models [60,61], was sufficient in this case.

In all stages, our model works with territorial information, and thus is explicitly spatial as regards not only input but also output data. Even if the best mechanisms were available to develop an ABM, the data would represent a significant constraint if they were poor quality or did not comply with geographical and scale needs. One of the submodels' outstanding properties is their flexibility in terms of the amount and kind of data that can be used, since this allows the user to easily adapt the prototype to other similar study areas.

An analysis of the physical distribution of housing and potential urban land or a thematic analysis of the results were not among the goals of this study, although some such analyses were performed in order to confirm that the model was correctly constructed and generated accurate products. ABM validation is a controversial issue [12,53,54,56,62–64], and will continue to be so until more research is conducted in this field. Nevertheless, we performed a verification procedure by means of programming tests, empirical reviews, and comparison of results, simulating growth under different contexts (setting different initial parameter values and constituting future scenarios). Furthermore, we performed a validation by comparing the simulated results to real world data, which confirmed our expectations. This step was considered sufficient since, our goal at this stage was simply to test the submodels, considering that further validation will be required in the future when the model (AMEBA) is complete, i.e., once the population submodel has been integrated.

5. Conclusions

In sum, the conceptual prototype, the prototype structure, and implementation of the first two submodels together represent a first attempt to create a model capable of simulating urban growth that could actually be used to support territorial planning, rather than a purely theoretical ABM. Its application at sub-regional scale using real data was a novel experiment. Therefore, considering these submodels in a wider context, the model could be useful for similar studies in different regions, provided that they presented equivalent characteristics or introduced the necessary modifications.

Acknowledgments: The findings presented here form part of the SIMURBAN2 Project (Urban Simulation Project-Part 2): Geosimulation and environmental planning tools for territorial management in metropolitan areas. Application at intermediate level (Ref.: CSO2012-38158-C02-01). One researcher was partially funded by Santander-IELAT 2015/2016. We would also like to thank the anonymous reviewers for their comments and suggestions.

Author Contributions: Carolina de Carvalho Cantergiani designed and developed the model, processed and analyzed the results under the direction of Montserrat Gómez Delgado. Both authors discussed the results and contributed on the conclusions and revisions of the document.

Conflicts of Interest: The authors declare no conflict of interest.

References

1. Berling-Wolff, S.; Wu, J. Modeling urban landscape dynamics: A review. *Ecol. Res.* **2004**, *19*, 119–129. [CrossRef]
2. Triantakonstantis, D.; Mountrakis, G. Urban growth prediction: A review of computational models and human perceptions. *J. Geogr. Inf. Syst.* **2012**, *4*, 555–587. [CrossRef]

3. Crooks, A.T.; Patel, A.; Wise, S. Multi-agent systems for urban planning. In *Technologies for Urban and Spatial Planning: Virtual Cities and Territories*; Pinto, N.N., Tenedório, J.A., Antunes, A.P., Roca, J., Eds.; IGI Global: Hershey, PA, USA, 2014; pp. 29–56.

4. Bousquet, F.; Le Page, C. Multi-agent simulations and ecosystem management: A review. *Ecol. Model.* **2004**, *176*, 313–332. [CrossRef]

5. Couclelis, H. "Where has the future gone?" Rethinking the role of integrated land-use models in spatial planning. *Environ. Plan. A* **2005**, *37*, 1353–1371. [CrossRef]

6. Fernández Güell, J.M. Recuperación de los estudios del futuro a través de la prospectiva territorial. *Ciudad y Territorio* **2011**, *167*, 11–32.

7. Ligtenberg, A.; Wachowicz, M.; Bregt, A.K.; Beulens, A.J.M.; Kettenis, D.L. A design and application of a multi-agent system for simulation of multi-actor spatial planning. *J. Environ. Manag.* **2004**, *72*, 43–55. [CrossRef] [PubMed]

8. Barredo Cano, J.I.; Gomez Delgado, M. Towards a set of ipcc sres urban land-use scenarios: Modelling urban land-use in the madrid region. In *Modelling Environmental Dynamics: Advances in Geomatic Solutions*; Paegelow, M., Camacho Olmedo, M.T., Eds.; Springer: Berlin, Germany, 2008; pp. 365–385.

9. Díaz Pacheco, J. Ciudades, Autómatas Celulares y Sistemas Complejos. Evaluación de un Modelo Dinámico de Cambio de usos de Suelo Urbano en Madrid. Ph.D. Thesis, Universidad Complutense de Madrid, Madrid, Spain, 2015.

10. Norte Pinto, N.E.; Dourado, J.; Natálio, A. Cellular automata modeling for urban and spatial systems. In Proceedings of the International Symposium on Cellular Automata Modeling for Urban and Spatial Systems—CAMUSS, Oporto, Portugal, 8–10 November 2012.

11. Hewitt, R.; van Delden, H.; Escobar, F. Participatory land use modelling, pathways to an integrated approach. *Environ. Model. Softw.* **2014**, *52*, 149–165. [CrossRef]

12. Heppenstall, A.; Crooks, A.T.; See, L.M.; Batty, M. *Agent-Based Models of Geographical Systems*; Springer: Dordrecht, The Netherlands, 2012.

13. Batty, M. *Cities and Complexity: Understanding Cities with Cellular Automata, Agent-Based Models, and Fractals*; Massachusetts Institute of Technology: Cambridge, MA, USA, 2005.

14. Jordan, R.; Birkin, M.; Evans, A. Agent-based modeling of residential mobility, housing choice and regeneration. In *Agent-Based Models of Geographical Systems*; Heppenstall, A., Crooks, A.T., See, L.M., Batty, M., Eds.; Springer: Berlin, Germany, 2012; pp. 511–524.

15. Moussaïd, M.; Kapadia, M.; Thrash, T.; Sumner, R.W.; Gross, M.; Helbing, D.; Hölscher, C. Crowd behaviour during high-stress evacuations in an immersive virtual environment. *J. R. Soc. Interface* **2016**, *13*. [CrossRef] [PubMed]

16. Barros, J. Exploring urban dynamics in latin american cities using an agent-based simulation approach. In *Agent-Based Models of Geographical Systems*; Heppenstall, A., Crooks, A.T., See, L.M., Batty, M., Eds.; Springer: Berlin, Germany, 2012; pp. 571–589.

17. Feitosa, F.F.; Le, Q.B.; Vlek, P.L.G. Multi-agent simulator for urban segregation (masus): A tool to explore alternatives for promoting inclusive cities. *Comput. Environ. Urban Syst.* **2011**, *35*, 104–115. [CrossRef]

18. Malik, A.; Crooks, A.T.; Root, H.; Swartz, M. Exploring creativity and urban development with agent-based modeling. *J. Artif. Soc. Soc. Simul.* **2015**, *18*. [CrossRef]

19. Patel, A.; Crooks, A.; Koizumi, N. Slumulation: An agent-based modeling approach to slum formations. *J. Artif. Soc. Soc. Simul.* **2012**, *15*. [CrossRef]

20. Filatova, T. Empirical agent-based land market: Integrating adaptive economic behavior in urban land-use models. *Comput. Environ. Urban Syst.* **2015**, *54*, 397–413. [CrossRef]

21. Sun, S.; Parker, D.C.; Huang, T.; Filatova, T.; Robinson, D.T.; Riolo, R.L.; Hutchins, M.; Brown, D.G. Market impacts on land-use change: An agent-based experiment. *Ann. Assoc. Am. Geogr.* **2014**, *104*, 460–484. [CrossRef]

22. Ettema, D. A multi-agent model of urban processes: Modelling relocation processes and price setting in housing markets. *Comput. Environ. Urban Syst.* **2011**, *35*, 1–11. [CrossRef]

23. Magliocca, N.; Safirova, E.; McConnell, V.; Walls, M. An economic agent-based model of coupled housing and land markets (chalms). *Comput. Environ. Urban Syst.* **2011**, *35*, 183–191. [CrossRef]

24. Fontaine, C.M.; Rounsevell, M.D.A. An agent-based approach to model futures residential pressure on a regional landscape. *Landsc. Ecol.* **2009**, *24*, 1237–1254. [CrossRef]

25. Parker, D.C.; Filatova, T. A conceptual design for a bilateral agent-based land market with heterogeneous economic agents. *Comput. Environ. Urban Syst.* **2008**, *32*, 454–463. [CrossRef]
26. Gilbert, N.; Hawksworth, J.C.; Swinney, P.A. *An Agent-Based Model of the English Housing Market*; Centre for Research in Social Simulation, University of Surrey: Guildford, UK, 2009.
27. Ligmann-Zielinska, A.; Jankowski, P. Exploring normative scenarios of land use development decisions with an agent-based simulation laboratory. *Comput. Environ. Urban Syst.* **2010**, *34*, 409–423. [CrossRef]
28. Farinós Dasi, J.; Olcina Cantos, J.; Rico Amorós, A.; Rodríguez Navarro, C.; Romero Renau, L.D.; Espejo Marín, C.; Vera Rebollo, J.F. Planes estratégicos territoriales de carácter supramunicipal. *Boletín de la Asociación de Geógrafos Españoles* **2005**, *39*, 117–149.
29. Garcia, M. The breakdown of the spanish urban growth model: Social and territorial effects of the global crisis. *Int. J. Urban Reg. Res.* **2010**, *34*, 967–980. [CrossRef]
30. Burriel, E.L. Subversion of land-use plans and the housing bubble in spain. *Urban Res. Pract.* **2011**, *4*, 232–249. [CrossRef]
31. Burriel, E.L. Empty urbanism: The bursting of the spanish housing bubble. *Urban Res. Pract.* **2016**, *9*, 158–180. [CrossRef]
32. Filatova, T.; Parker, D.C.; Van Der Veen, A. Agent-based urban land markets: Agent's princing behavior, land prices and urban land use change. *J. Artif. Soc. Soc. Simul.* **2009**, *12*, 1–3.
33. Ligmann-Zielinska, A.; Church, R.L.; Jankowski, P. Spatial optimization as a generative technique for sustainable multiobjective land-use allocation. *Int. J. Geogr. Inf. Sci.* **2008**, *22*, 601–622. [CrossRef]
34. Haque, A.; Asami, Y. Optimizing urban land use allocation for planners and real estate developers. *Comput. Environ. Urban Syst.* **2014**, *46*, 57–69. [CrossRef]
35. Moglen, G.E.; Gabriel, S.A.; Faria, J.A. A framework for quantitative smart growth in land development. *J. Am. Water Resour. Assoc.* **2003**, *39*, 947–959. [CrossRef]
36. Magliocca, N.R.; Brown, D.G.; McConnell, V.D.; Nassauer, J.I.; Westbrook, S.E. Effects of alternative developer decision-making models on the production of ecological subdivision designs: Experimental results from an agent-based model. *Environ. Plan. B* **2014**, *41*, 907–927. [CrossRef]
37. An, L.; Linderman, M.; Qi, J.; Shortridge, A.; Liu, J. Exploring complexity in a human-environment system: An agent-based spatial model for multidisciplinary and multiscale integration. *Ann. Assoc. Am. Geogr.* **2005**, *95*, 54–79. [CrossRef]
38. Gaube, V.; Remesch, A. Impact of urban planning on household's residential decisions: An agent-based simulation model for vienna. *Environ. Model. Softw.* **2013**, *45*, 92–103. [CrossRef] [PubMed]
39. Ligmann-Zielinska, A.; Jankowski, P. Agent-based models as laboratories for spatially explicit planning policies. *Environ. Plan. B Plan. Des.* **2007**, *34*, 316–335. [CrossRef]
40. Plata Rocha, W.; Gómez Delgado, M.; Bosque Sendra, J. Desarrollo de modelos de crecimiento urbano óptimo para la cm aplicando métodos de evaluación multicriterio y sistemas de información geográfica. *GeoFocus* **2010**, *10*, 103–134.
41. INE. Spanish Statistical National Institute. 2016. Available online: www.ine.es (accessed on 15 November 2017).
42. OSE (Observatory of Sustainability in Spain). *Changes in Land-Use in Spain. Implications for Sustainable Development*; Mundiprensa: Madrid, Spain, 2006.
43. Miralles i Garcia, J.L.; Díaz Aguirre, S.; Altur Grau, V.J. Environmental impact on the mediterranean spanish coast produced by the latest process of urban development. In *The Sustainable City 2012*; Pacetti, M., Passerini, G., Brebbia, C.A., Latini, G., Eds.; WIT Press: Southampton, UK, 2012; Volume 155, pp. 379–390.
44. Gallardo, M.; Martínez-Vega, J. Three decades of land-use changes in the region of madrid and how they relate to territorial planning. *Eur. Plan. Stud.* **2016**, *24*, 1016–1033. [CrossRef]
45. ESPON (European Observation Network for Territorial Development and Cohesion). *European Land Use Patterns (EU-LUPA)-Applied Research 2013/1/8*; European Comission: Luxembourg, 2012.
46. Dijkstra, L.; Garcilazo, E.; McCann, P. The effects of the global financial crisis on european regions and cities. *J. Econ. Geogr.* **2015**, *15*, 935–949. [CrossRef]
47. Cantergiani, C.C.; Gomez Delgado, M. Diseño de un modelo basado en agentes para simular el crecimiento urbano en el corredor del henares (comunidad de madrid). *Boletín de la Asociación de Geógrafos Españoles* **2016**, *70*, 259–283. [CrossRef]

48. Cantergiani, C.C.; Barros, J.; Gómez Delgado, M. How real estate agents behavior affects urban growth: An agent-based model approach. In Proceedings of the Advances in Computational Social Science and Social Simulation Congeference, Barcelona, Spain, 1–5 September 2014; Miguel, F., Amblard, F., Madella, X., Eds.; Universitat Autónoma de Barcelona: Barcelona, Spain, 2014; pp. 491–493.
49. Wilensky, U. *Netlogo*; Center for Connected Learning and Computer-Based Modeling, Northwestern University: Evanston, IL, USA, 1999.
50. Plata Rocha, W.; Gómez Delgado, M.; Bosque Sendra, J. Simulation urban growth scenarios using gis and multicriteria evaluation techniques. Case study: Madrid region, Spain. *Environ. Plan. B* **2011**, *38*, 1012–1031. [CrossRef]
51. Barreira-González, P.; Aguilera-Benavente, F.; Gómez-Delgado, M. Implementation and Calibration of a New Irregular Cellular Automata-Based Model for Local Urban Growth Simulation: The Mugica Model. Available online: http://journals.sagepub.com/doi/abs/10.1177/2399808317709280?journalCode=epbb (accessed on 16 May 2017).
52. Corine Land Cover 2000. Nomenclature at Leval 3 (Spain). Available online: https://www.europeandataportal.eu/data/en/dataset/spaignclc2000_nivel3201307180000 (accessed on 26 April 2011).
53. Qureshi, M.E.; Harrison, S.R.; Wegener, M.K. Validation of multicriteria analysis models. *Agric. Syst.* **1999**, *62*, 105–116. [CrossRef]
54. Truong, M.T.; Amblard, F.; Gaudou, B.; Sibertin-Blanc, C. To calibrate & validate an agent-based simulation model—An application of the combination framework of bi solution & multi-agent platform. In Proceedings of the 6th International Conference on Agents and Artificial Intelligence (ICAART 2014), Angers, France, 6–8 March 2014; pp. 172–183.
55. Acosta-Michlik, L.; Rounsevell, M.D.A.; Bakker, M.; Van Doorn, A.; Gómez Delgado, M.; Delgado, M. An agent-based assessment of land use and ecosystem changes in traditional agricultural landscape of portugal. *Intell. Inf. Manag.* **2014**, *6*, 55–80.
56. Ligtenberg, A.; van Lammeren, R.J.A.; Bregt, A.K.; Beulens, A.J.M. Validation of an agent-based model for spatial planning: A role-playing approach. *Comput. Environ. Urban Syst.* **2010**, *34*, 424–434. [CrossRef]
57. Brown, D.G.; Page, S.; Riolo, R.; Zellner, M.; Rand, W. Path dependence and the validation of agent-based spatial models of land use. *Int. J. Geogr. Inf. Sci.* **2005**, *19*, 153–174. [CrossRef]
58. Li, Y.; Brimicombe, A.J.; Li, C. Agent-based services for the validation and calibration of multi-agent models. *Comput. Environ. Urban Syst.* **2008**, *32*, 464–473. [CrossRef]
59. SIOSE. (Sistema de Información Sobre Ocupación del Suelo de España). Available online: http://www.siose.es/ (accessed on 15 November 2017).
60. Bajracharya, K.; Duboz, R. Comparison of three agent-based platforms on the basis of a simple epidemiological model (WIP). In Proceedings of the Symposium on Theory of Modeling & Simulation—DEVS Integrative M&S Symposium (DEVS 13), San Diego, CA, USA, 7–10 April 2013; Society for Computer Simulation International, Ed.; Association for Computing Machinery: San Diego, CA, USA, 2013.
61. Winfrey, C.M.; Baldwin, B.A.; Cummings, M.A.; Ghosh, P. Osm: An evolutionary system of systems framework for modeling and simulation. In Proceedings of the 2014 Annual Simulation Symposium, Tampa, FL, USA, 13–16 April 2014; Society for Computer Simulation International: San Diego, CA, USA, 2014; pp. 1–8.
62. Crooks, A.; Castle, C.; Batty, M. Key challenges in agent-based modelling for geo-spatial simulation. *Comput. Environ. Urban Syst.* **2008**, *32*, 417–430. [CrossRef]
63. Heppenstall, A.; Malleson, N.; Crooks, A. "Space, the final frontier": How good are agent-based models at simulating individuals and space in cities? *Systems* **2016**, *4*, 9. [CrossRef]
64. Filatova, T.; Verburg, P.H.; Parker, D.C.; Stannard, C.A. Spatial agent-based models for socio-ecological systems: Challenges and prospects. *Environ. Model. Softw.* **2013**, *45*, 1–7. [CrossRef]

environments

MDPI

Article

Assessing Land Use-Cover Changes and Modelling Change Scenarios in Two Mountain Spanish National Parks

Javier Martínez-Vega [1,*], Andrés Díaz [2], José Miguel Nava [2], Marta Gallardo [3] and Pilar Echavarría [1]

[1]   Institute of Economy, Geography and Demography, Spanish National Research Council (IEGD-CSIC), Associated Unit GEOLAB, Albasanz, 26–28, 28037 Madrid, Spain; pilar.echavarria@cchs.csic.es
[2]   Department of Geology, Geography and Environment, University of Alcalá, Colegios 2, 28801 Alcalá de Henares, Spain; andresdiazmartin.87@gmail.com (A.D.); joseminava91@gmail.com (J.M.N.)
[3]   Department of Geography, University of Murcia, Santo Cristo 1, 30001 Murcia, Spain; marta.gallardo@um.es
*   Correspondence: javier.martinez@cchs.csic.es; Tel.: +34-91-602-2395

Received: 29 September 2017; Accepted: 4 November 2017; Published: 7 November 2017

Abstract: Land Use-Cover Changes (LUCCs) are one of the main problems for the preservation of biodiversity. Protected Areas (PAs) do not escape this threat. Some processes, such as intensive recreational use, forest fires or the expansion of artificial areas taking place inside and around them in response to their appeal, question their environmental sustainability and their efficiency. In this paper, we analyze the LUCCs that took place between 1990 and 2006 in two National Parks (NPs) belonging to the Spanish network and in their surroundings: Ordesa and Monte Perdido (Ordesa NP) and Sierra de Guadarrama (Guadarrama NP). We also simulate land use changes between 2006 and 2030 by means of Artificial Neural Networks (ANNs), taking into account two scenarios: trend and green. Finally, we perform a multi-temporal analysis of natural habitat fragmentation in each NP. The results show that the NPs analyzed are well-preserved and have seen hardly any significant LUCCs inside them. However, Socioeconomic Influence Zones (SIZs) and buffers are subject to different dynamics. In the SIZ and buffer of the Ordesa NP, there has been an expansion of built-up areas (annual rate of change = +1.19) around small urban hubs and ski resorts. There has also been a gradual recovery of natural areas, which had been interrupted by forest fires. The invasion of sub-alpine grasslands by shrubs is clear (+2735 ha). The SIZ and buffer of the Guadarrama NP are subject to urban sprawl in forest areas and to the construction of road infrastructures (+5549 ha and an annual rate of change = +1.20). Industrial area has multiplied by 3.3 in 20 years. The consequences are an increase in the Wildland-Urban Interface (WUI), greater risk of forest fires and greater fragmentation of natural habitats (+0.04 in SIZ). In the change scenarios, if conditions change as expected, the specific threats facing each NP can be expected to increase. There are substantial differences between the scenarios depending on whether or not incentives are accepted and legal restrictions are respected.

Keywords: land use-cover changes; change scenarios; artificial neural networks; habitat fragmentation; protected areas; Spain

---

1. Introduction

Protected Areas (PAs) are a key for mitigating climate change, preserving biodiversity, providing ecosystem services and fostering human wellbeing. The declaration of PAs has increased globally because of increased environmental sensitivity [1]; to counter the threats of climate change [2]; land use changes [3]; deforestation [4]; the risk of flooding [5]; the risk of forest fires [6]; habitat fragmentation [7]; the propagation of invasive species [8]; urban pressure [9]; and recreational use [10].

In 1990, PAs covered 8.6% of the Earth's surface. According to the World Data Base of PAs [11], they have expanded from 84,577 in 2003 to 217,155 in 2016. 93% of them occupy 19.8 million km^2, equivalent to 14.7% of the worldwide surface area of terrestrial ecosystems and continental and inland waters, excluding the Antarctic. The remainder are Protected Marine Areas, which cover 14.9 million km^2, 4.12% of the world's oceans and 10.2% of marine and coastal waters under national jurisdiction [11–13]. To reach Aichi Goal 11, the Convention on Biological Diversity recommends that by 2020, at least 17% of terrestrial areas and continental waters be recommended, as well as 10% of coastal and marine areas [14].

In Europe, PAs occupy 13.6% of the land surface and of continental waters [15]. In Spain, from 1990–2013, the number of PAs multiplied by seven, and their surface area tripled [16]. In the worldwide and continental context, Spain plays a relevant role in the preservation of biological diversity. Today, more than 27% of the surface area occupied by terrestrial ecosystems and their inland waters is protected under national, European or worldwide networks. It is the EU country that contributes most territory to the Natura 2000 network.

Amongst the categories defined by the International Union for Conservation of Nature (IUCN), National Parks (NPs) are the figure that is most widely used for PAs of high natural value [17]. In Spain, NPs are covered by a law enacted 100 years ago [18] and recast recently [19]. Since 1918, when the first two national parks were declared, the network has grown at an average annual rate of 23%. Such parks are a key in various Spanish strategies for nature protection (for example, the Spanish Strategic Plan for the Natural Heritage and Biodiversity 2011–2017).

For decades, the scientific community has been showing interest in the spatial and temporal analysis of Land Use-Cover Changes (LUCCs) [20–22]. Cartography of LUCCs is crucial for monitoring ecosystems at different scales [23–25] and for assessing the impact of changes on PAs and biodiversity. Their study allows one to know the size, extent, type and trends of LUCCs and to identify the main factors of change. This is a fundamental previous step for the design of conservation policies in PAs [3], especially in NPs [26]. Such an analysis is also required for assessing the efficiency of PAs [27].

Moreover, increasing importance is being placed by scientists and managers on the LUCCs taking place inside PAs and in their surroundings [28]. External pressure from such transformations has different impacts on the biodiversity of PAs and their Socioeconomic Influence Zones (SIZs), reduces their efficiency and amounts to a threat for environmental sustainability [29,30].

Some aggressive LUCCs—expansion of intensive farming lands and of built areas—lead, amongst others, to the fragmentation of natural habitats [31], the loss of connectivity for habitats of ecological interest [32], a reduction in the floodable area of protected wetlands and poorer water quality in such wetlands [33].

At a local level, in the Pyrenees of the province of Huesca, rising temperatures are leading to invasion by woody vegetation and displacement of forests to higher altitudes, with a decline in grasslands [34]. The reduction in traditional farming activities has been an explanatory factor for recolonization by woody species [35]. The invasion of abandoned farmland and of grassland by more xerophilous woody species [36] increases the risk of forest fires.

In the Sierra de Guadarrama, close to the metropolitan area of Madrid, there is an increased recreational pressure on the PAs [37], which are attractive for their landscape and their environments. There is also great pressure from urban sprawl on ecosystems of high ecological value [38–40].

In order to help in the decision-making of the stakeholders, the scientific community is working on the simulation of various change scenarios using various methods and tools: logistic regression [41], neural networks [42,43], cellular automata [44–46] and, sometimes, combined methods [47,48] or compared [49,50].

The main objective of this research is to analyze the changes in land use that occurred, between 1990 and 2006, in two mountain Spanish National Parks and in their surroundings. In addition, considering these trends of change in recent decades, we intend to simulate different scenarios of change—trend and green—that are expected between 2006 and 2030. For this, we use a simulator based on Artificial Neural Networks (ANNs). In addition, we have analyzed the changes in the fragmentation of natural and semi-natural habitats recorded in the last few decades, as well as those that are expected if the simulated changes are met. We compare the processes that have occurred inside the two NPs with those that have occurred outside of them, in their surroundings, in order to find substantial differences relating to environmental sustainability. The ultimate aim of the research is to provide information that will be of use to the managers of affected NPs and to local administrators in their preventive environmental and territorial decisions. We also aim to increase awareness among those responsible for the Spanish network so that similar methods can be applied in other NPs.

2. Methods

2.1. Study Areas

Mountain ecosystems are the focus of interest in this work. For this reason, we have chosen two mountain NPs that, besides belonging to two different biogeographic regions, had contrasting geographical and socioeconomic characteristics. For this study, we selected a sample of two mountainous NPs belonging to the Spanish network: one of the oldest (Ordesa NP) and the newest (Guadarrama NP); see Figure 1.

Ordesa NP is located in a rural area with poor accessibility, along the frontier with France and far from the main urban centers in Spain. It was declared in 1918 to be representative of high mountain ecosystems and of systems linked to erosion formations and sedimentary rocks in the Pyrenee Province of the EuroSiberian Region [51]. Guadarrama NP is the latest to be declared, in 2013, although several groups have shown interest in its declaration as a national park since the beginning of the 20th century. It is representative of mountain ecosystems of the Mediterranean biogeographical region. It is in a peri-urban zone, less than 50 km from Madrid and its metropolitan area, which has a population of about 6.5 million inhabitants [52]. It is very close to motorways and high-speed railway lines. It is very accessible and is visited by 3,000,000 visitors a year [53]. We intend to know if the processes of change of land uses are significantly different between these two NPs.

In addition, we also selected various areas surrounding them: their Peripheral Protection Zones (PPZs), Socioeconomic Influence Zones (SIZs) and external 5 km-wide buffers plus the extensions of the buffers up to the administrative limits of the municipalities that are totally or partially affected by them. In Table 1, we summarize the main characteristics of the study areas.

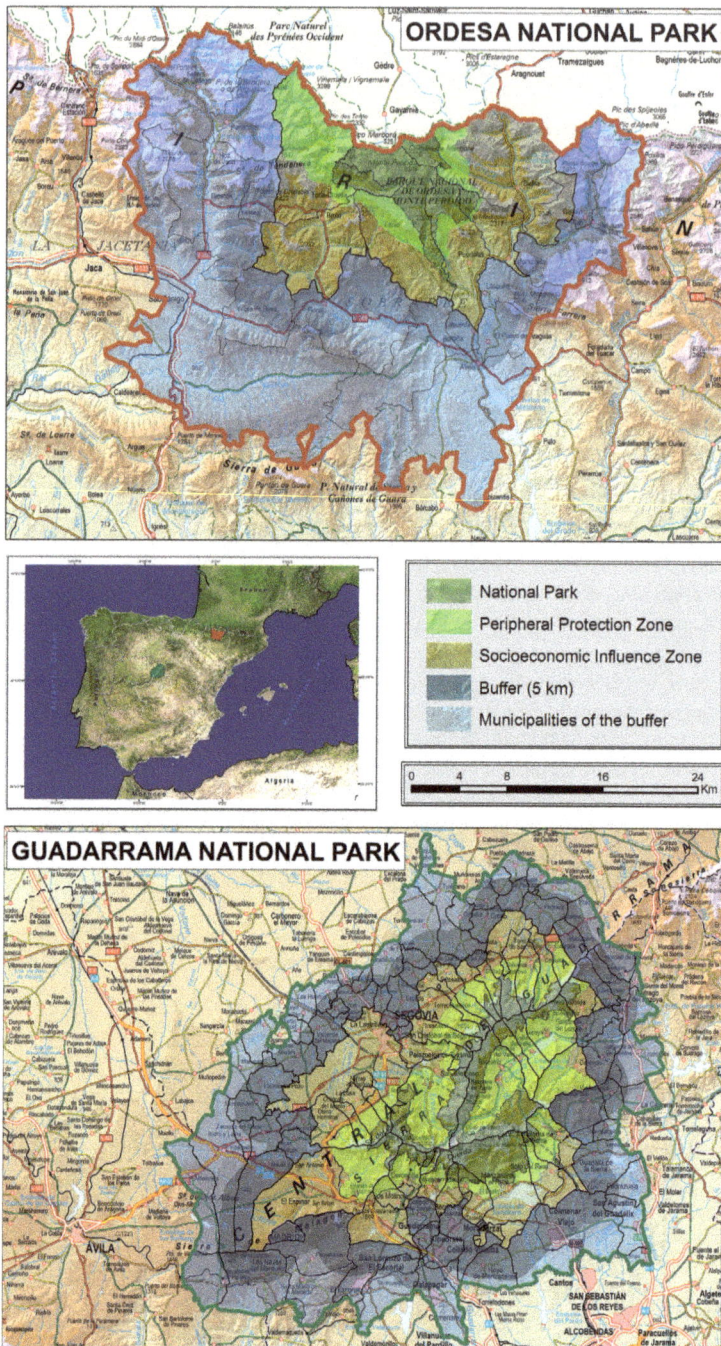

Figure 1. Study areas. Source: Map of Spain 1:500,000. Geodetic Reference System: European Terrestrial Reference System (ETRS89) Peninsula and Balearic Islands, Regente Canarias (REGCAN95) for Canaries. Cartographic projection system: UTM 30 extended.

Table 1. Summary of the main characteristics of the national parks analyzed. NP, National Park.

Characteristics	Ordesa NP	Guadarrama NP
Study Area	300,624 ha	410,000 ha
National Park	15,692 ha	33,960 ha [54]
Peripheral Protection Zone	6164 ha	49,062 ha
Socioeconomic Influence Zone	67,435 ha	77,064 ha
Buffer	62,696 ha	132,620 ha
Natural Systems of Interest	15 [55]	10
Species of Plants	1404	>1000
Endemic Species	50	83
Visitors per Year	600,000	3,000,000
Other Protected Areas within The Study Area	11 Sites of Community Importance 7 Special Protection Areas 2 Regional Parks (RPs) Natural Monuments of Glaciers and Massifs	9 Special Areas of Conservation 8 Special Protection Areas 2 Regional Parks

2.2. Materials

Our research follows the workflow shown in Figure 2. We used ArcGIS v10.3 (ESRI Inc., Redlands, CA, USA) for vector processing of the downloaded data and to draw up the buffer. For LUCC analyses, we used IDRISI-TerrSet (v.18.31, Clark Labs, Clark University, Worcester, MA, USA) [56] and Land Change Modeller (LCM) (Clark Labs, Clark University, Worcester, MA, USA) for designing the scenarios. For the calibration and validation of land use change scenarios, we used the Map Comparison Kit v3.2.3 software (RIKS BV, Maastricht, The Netherlands) [57]. Finally, we used GUIDOS-MSPA (v. 2.4, rev. 2, Joint Research Centre, European Commission, Ispra, Italy) [58] to analyze the spatial landscape pattern. We used three datasets developed by the CORINE Land Cover (CLC) Project [59]: years 1990, 2000 and 2006.

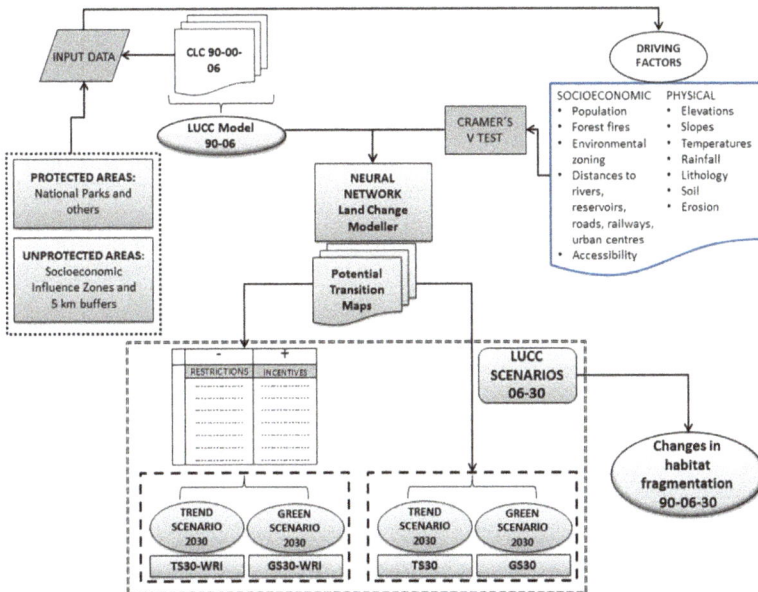

Figure 2. Framework. CLC, CORINE Land Cover; LUCC, Land Use-Cover Change.

2.3. Methods

2.3.1. Land Use Changes between 1990 and 2006

Firstly, although the size of buffers around PAs has been established in various ways in the literature [3,26,60], in this study, we chose to design a 5-km buffer, which suits the characteristics of the two study areas and is also in line with other previous studies [28,31]. A buffer with this size increases the probability of the territories included in these "control areas" having similar geographical characteristics to the "cases" (NPs and their SIZs). On the contrary, if the buffer is too small, the probability of finding differences in land uses-land cover, inside and outside NP, would be small.

Second, we drew up an initial cross-tabulation between the CLC vector maps to identify unusual and unexpected land use changes (Figure A1). Some authors find interpretation and location errors [61,62]. After checking any doubts with the aerial orthophotos provided under the National Plan for Aerial Orthophotography [63], any errors found were corrected. Subsequently, we transformed the CLC vector maps to a raster format with a 50 × 50 m pixel size. From the CLC legend, we made two different groupings. The first is a simplification of Level 3 into six categories: Urban areas (URB), Industrial areas (IND), arable land and permanent crops (Agricultural (AGR)), Heterogeneous agricultural areas (HET), Forests (FOR) and Shrubs and Grasslands (SHR-GRAS). We considered the OTH category (Other: open spaces with little vegetation, wetlands and water bodies) to be stable. Because of the singularity of the Ordesa NP, the CLC Level 3 was simplified by dividing it into six categories: Urban areas (URB), Agricultural areas (AGR), Grasslands (GRAS), Shrubs (SHR), Forests (FOR) and Others (OTH: open spaces with little vegetation, wetlands and water bodies) (Table A1). We used this grouping to simulate future scenarios.

Third, we drew up cross-tabulation matrices [42] to obtain values and maps of changes between 1990 and 2006. We calculated the annual rate of change of each use [43]. We then compared the results with the PAs and with the 5-km buffer. The aim was to find some of the main processes of land use change that had already taken place: built-up land, naturalization and disturbances and exchanges in natural areas [3,64].

2.3.2. Simulated Scenarios of Change in 2030

Firstly, using LCM, we simulated land use in 2030 under different scenarios. In the Ordesa NP, we considered two different scenarios: (a) trend scenario and (b) green scenario. The trend or "business as usual" scenario shows what would happen if the past trend in 1990–2000–2006 were to continue until 2030. The green scenario shows what would happen if there were more active reforestation policies and if greater importance were placed on the natural environment. In both scenarios, we considered two alternatives. Under the first, we considered that land planners will respect the restrictions imposed by law in certain zones (NPs and other PAs, public utility forests or the public domain zones of rivers, roads and railway lines) (Table A2). We also considered the incentives that planners are promoting in certain zones to encourage the naturalization of land for various purposes (for example, to expand habitats of community importance). Under the second alternative, we did not take these restrictions or incentives into account to avoid distorting the simulation, for example excessive preparation for use changes in the location or the elimination of other conditioning factors [65]. We also considered that the stakeholders acting in the territory might not respect legal restrictions on certain land uses or might not take incentives into account. Although such scenarios would be illegal and would not be a sustainable alternative, examples are often found. In the Guadarrama NP, we only took into account a trend scenario, both with and without restrictions and incentives.

In the model simulated with LCM, we related land use and driving factors (Table 2) by means of a Multi-Layer Perceptron neural network (MLP). The MLP classifier uses a Backward Propagation algorithm (BP), one of the most used neural networks. We have taken into account a variable number of input and output nodes depending on the designed scenarios (between 7 (in Ordesa NP) and 28 (in Guadarrama NP)). For each category, the number of pixels per class is randomly divided between training and testing: half of the sample size for training and half for testing. The first are used in the analysis and the second to validate the results. Samples used for the training process are taken from pixels that have and have not undergone the transition being modeled. We selected an automatic and dynamic training to get the proper weights both for the connection between the input and hidden layer and between the hidden and the output layer for the classification of the unknown pixels. We ran the model 13 times, changing the learning rate parameters after obtaining an accuracy rate above 50% for each transition.

Table 2. Summary of the characteristics and sources of the data used for LUCC analysis and simulation.

Name	Description	Source
Numerical Cartographic Base (BCN)	Digital cartographic base in vector format, on scales of 1:100,000 and 1:500,000	Spanish National Centre for Geographic Information (CNIG) http://centrodedescargas.cnig.es/CentroDescargas/index.jsp
Administrative limits	Administrative limits of provinces and municipalities	Nomenclature of Territorial Units for Statistics (NUTS) database, Eurostat
Historical aerial orthophotos	Historical aerial orthophotos with a 0.5 × 0.5 m pixel size for different years (2004–2008)	Spanish National Centre for Geographic Information (CNIG) http://centrodedescargas.cnig.es/CentroDescargas/index.jsp
LANDSAT images	LANDSAT-TM and -ETM images for different years	U.S. Geological Survey (USGS) https://landsat.usgs.gov
Land use-Land cover maps	CORINE Land Cover maps for 1990, 2000 and 2006	Spanish National Centre for Geographic Information (CNIG) http://centrodedescargas.cnig.es/CentroDescargas/index.jsp
Elevations	Elevations resulting from the digital elevation model with a 25 × 25 m pixel size	Spanish National Centre for Geographic Information (CNIG) http://centrodedescargas.cnig.es/CentroDescargas/index.jsp
Slopes	Slopes resulting from the digital elevation model with a 25 × 25 m pixel size	Spanish National Centre for Geographic Information (CNIG) http://centrodedescargas.cnig.es/CentroDescargas/index.jsp
Temperatures	Average annual temperature with a spatial resolution of 200 × 200 m	Digital Climate Atlas for the Iberian Peninsula http://opengis.uab.es/wms/iberia
Precipitation	Average annual precipitation with a spatial resolution of 200 × 200 m	Digital Climate Atlas for the Iberian Peninsula http://opengis.uab.es/wms/iberia
Lithology	Lithological maps for Madrid and Castilla y León on scales of 1:200,000 and 1:100,000	Cartographic Service of the Community of Madrid; Spatial Data Infrastructure of Castilla y León www.idecyl.jcyl.es
Soils	Soil map on scales of 1:1,000,000	Soil Geographical Database of Eurasia, European Soil Datacentre
Erosion states	Map of erosion states on a scale of 1:1,000,000	Nature database http://sig.mapama.es/bdn/visor.html

Table 2. *Cont.*

Name	Description	Source
Public utility forests	Public utility forests	Nature database http://sig.mapama.es/bdn/visor.html
Forest fires	Ignition points, hotspots and/or burned areas	Spanish Ministry of the Environment data base, Moderate Resolution Imaging Spectroradiometer (MODIS) https://reverb.echo.nasa.gov/ Government of Aragon (Research Project GA-LC-042/2011)
Environmental zoning	Ordesa NP and Guadarrama NP management zones	Spanish Autonomous Body for National Parks
Population density	Variation in population density from 1991–2011	Spanish National Statistics Institute www.ine.es

We selected the driving factors by consulting experts and prior studies. We used Cramer's V statistic [66] to test the explanatory power of each variable and selected the most relevant.

We did carry out calibration in order to improve the results in the Ordesa NP. Taking the sequence of maps for 1990–2000 as a base, we simulated a land use model in 2006 and compared it with the real map for 2006. The calibration compared the number of pixels and the spatial location of each land use-land cover with that simulated. Models were tuned by minimizing the disagreement between actual and simulated maps. This was done changing the matrix of transition probabilities, selecting the driving factors and adding or changing the size and/or weight of the neighborhoods.

We used the Kappa simulation statistic (*Kappa Sim*), which tests only the changed areas of the map [67–69]. In addition, we have used *TransLoc* and *Transition*, which evaluate the accuracy in the location and quantity, respectively, of the pixels that experience a land use change.

Second, we drew up cross-tabulation matrices [42] to obtain values and maps of changes between 2006 and 2030. We calculated the annual rate of change of each use [43]. We then compared the results with the PAs and with the 5-km buffer. The aim was to find some of the main processes of land use change that had already taken place and that could be expected in different scenarios: built-up land, naturalization and disturbances and exchanges in natural areas.

2.3.3. Fragmentation of Habitats of Interest

In order to find the dynamics of landscape structure, we took into account Level 1 of the CLC legend. We reclassified the maps in binary format. In the case of Guadarrama, we considered Class 1 (artificial areas) as the background and combined Classes 2, 3, 4 and 5 in a single target category (agricultural and natural areas) linked to the habitats represented in the National Park and the surroundings. In the case of Ordesa, we considered Class 1 (shrubs) as the background and only Class 2 (herbaceous vegetation) as the target category. The managers of this National Park and of other PAs around it are worried about the invasion of alpine and sub-alpine grasslands (focal habitat) by shrubs resulting from the reduction in extensive cattle-breeding and from climate change.

We calculated, at Guadarrama NP, an index for the fragmentation of agricultural and natural habitats and for temporal variations (1990–2006) in terms of their size and spatial pattern. The Morphological Spatial Pattern Analysis (MSPA) algorithm in the GUIDOS software [58] classified each pixel by its geometric position on the matrix being analyzed, distinguishing between seven entities: (1) cores, (2) islets, (3) perforations, (4) edges, (5) loops, (6) bridges and (7) branches. We took into account the following parameters: analysis of pixel connectivity in 8 directions (cardinal and diagonal) = 1; transition pixels = 1; distinction between external and internal edges (perforations) in the core class.

We calculated a Habitat Fragmentation Index (HFI) [6], in our case the sum of agricultural and natural habitats in NP and in PPZ, SIZ and the corresponding buffer. This goes, with continuous values, from 1 (greatest fragmentation) to 2 (least fragmentation). It assigns a different weight to each of the entities mapped in terms of relations between resilience and spatial coherence [70,71]. There is a constant gradation from the core (greatest weight) to the islets (least weight). The index relates the number of pixels in each category or fragmentation entity to their weights. In the Ordesa NP, we used the same method to calculate a fragmentation rate for the alpine and sub-alpine grasslands, which, in this case, are habitats of interest. We studied changes expected between 2006 and 2030, considering the green scenario (GS), without incentives or restrictions (GS30), in order to show changes in the landscape if the PA regulatory instruments are not respected.

3. Results

3.1. Land Use Changes between 1990 and 2006

In general and as expected, there were few changes in the Ordesa NP. This is a rural district in which land uses are strictly regulated because they belong to an NP, two RPs and other sites within the Natura 2000 Network (1.1937). Global persistence is over 98% in the study area (Table 3). The urban areas underwent the greatest annual rate of change (1.1937), and the classes with the greatest net changes were grasslands (about −2800 ha) and shrubs (+2463 ha).

Table 3. Matrix and statistics for land use changes, in % of the total study area, between 1990 (rows) and 2006 (columns) in Ordesa NP and its surroundings.

Land Use-Land Cover	URB	AGR	GRAS	SHR	FOR	OTH	Total	Losses	Persistence	Total Change	Swap	Net Change	Annual Rate of Change
URB	0.28	0.00	0.00	0.00	0.00	0.00	0.28	0.00	0.28	0.18	0.00	0.18	1.193
AGR	0.09	8.13	0.00	0.01	0.00	0.00	8.23	0.10	8.13	0.10	0.20	−0.10	−0.029
GRAS	0.05	0.00	12.11	0.91	0.00	0.00	13.08	0.96	12.11	1.01	1.92	−0.92	−0.175
SHR	0.03	0.00	0.02	20.72	0.25	0.00	21.02	0.31	20.72	1.43	0.61	0.82	0.092
FOR	0.01	0.00	0.02	0.20	40.32	0.02	40.57	0.25	40.32	0.50	0.50	0.00	0.000
OTH	0.00	0.00	0.00	0.00	0.00	16.82	16.83	0.00	16.82	0.02	0.00	0.02	0.003
Total	0.45	8.13	12.16	21.84	40.57	16.85	100		98.38				
Gains	0.18	0.00	0.04	1.13	0.25	0.02							

URB = Urban areas; AGR = Agricultural areas; GRAS = Grasslands; SHR = Shrubs; FOR = Forests; OTH = Others.

Figure 3 shows the spatial distribution of LUCCs taking place between 1990 and 2006. In the NP and in its PPZ, no changes were recorded. Most of them were located inside the buffer and some in its SIZ. The use changes correspond to three different processes: natural regeneration, degradation of the vegetation and increase in artificial surfaces. The most extensive changes correspond to a progression of the vegetation to the climax, from grasslands to shrubs to forest (in green colors). They can be seen in detail in Window C of Figure 4. There are two possible causes for such progressions. The first is the abandonment of agricultural land and, above all, of grasslands because of the abandonment of traditional cattle-farming. Population density is low (<5 inhabitants/km²) and the rate of ageing high (>26%). From an ecological point of view, this could be considered positive. However, it is worrying for land managers because sub-alpine grasslands are focal habitats in the NP. In addition, this transition involves an increase in available potential fuel and, consequently, an increase in the risk of forest fires in an area of high ecological value.

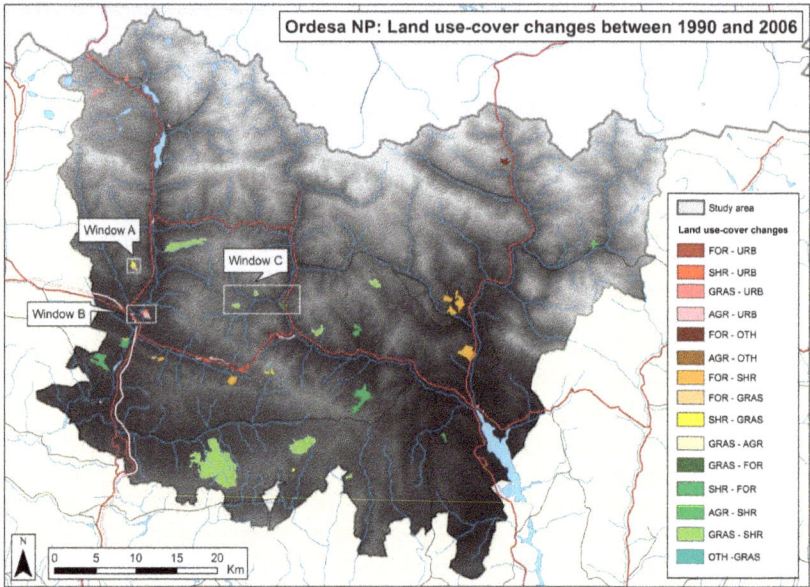

Figure 3. Land use-cover changes that took place between 1990 and 2006 in the Ordesa NP and its surroundings. URB = Urban areas; AGR = Agricultural areas; GRAS = Grasslands; SHR = Shrubs; FOR = Forests; OTH = Others.

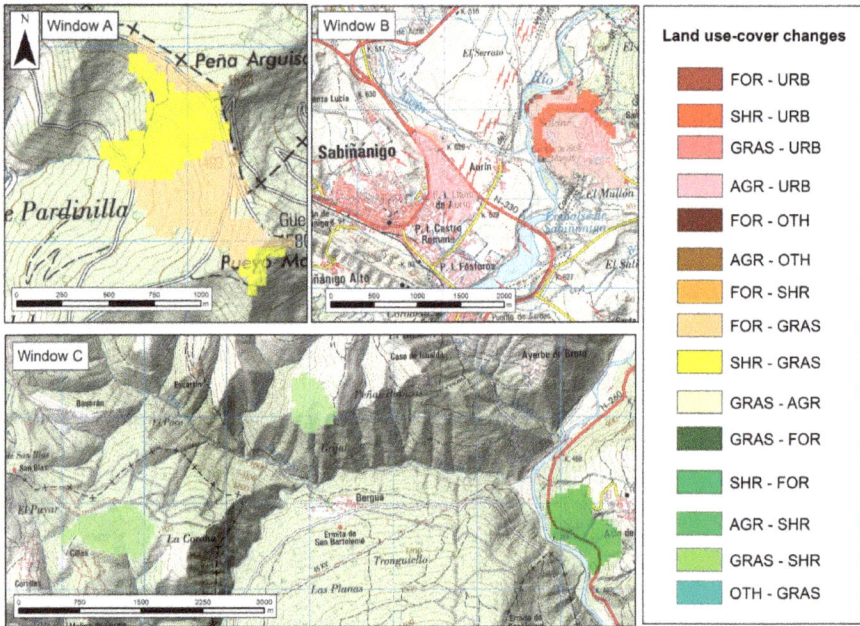

Figure 4. Details of land use changes that took place between 1990 and 2006 in the area of socioeconomic influence and the buffer of the Ordesa NP. URB = Urban areas; AGR = Agricultural areas; GRAS = Grasslands; SHR = Shrubs; FOR = Forests; OTH = Others.

According to the CLC90-06 series and at the landscape scale, there was no invasion of grasslands by shrubs within the NP. Secondly, the progression of vegetation is the consequence of recovery from previous forest fires.

Changes associated with degradation of vegetation (yellow colors) stem from more recent forest fires and can be seen in greater detail in Window A of Figure 4 (transitions from forest or shrubs to grassland). Finally, the expansion of urban zones and infrastructure (red colors) mostly took place in the surroundings of Sabiñánigo, the center of the district with the largest population and industries. Window B of Figure 4 shows a golf course under construction. In the center of Figure 3, certain long patches can be seen, which have changed to artificial use as a result of the widening of the N-260 motorway. In the NW quadrant of the same figure, there are also small grasslands that have been converted into new ski slopes and other infrastructure associated with skiing in Formigal.

In the Guadarrama NP and its surroundings, most of the changes are concentrated in the south, in the region of Madrid and, especially, in the buffer and its SIZ (Figure 5). This is the most dynamic region from the demographic and socio-economic points of view, so it is the region that exerts the greatest pressure on the environment.

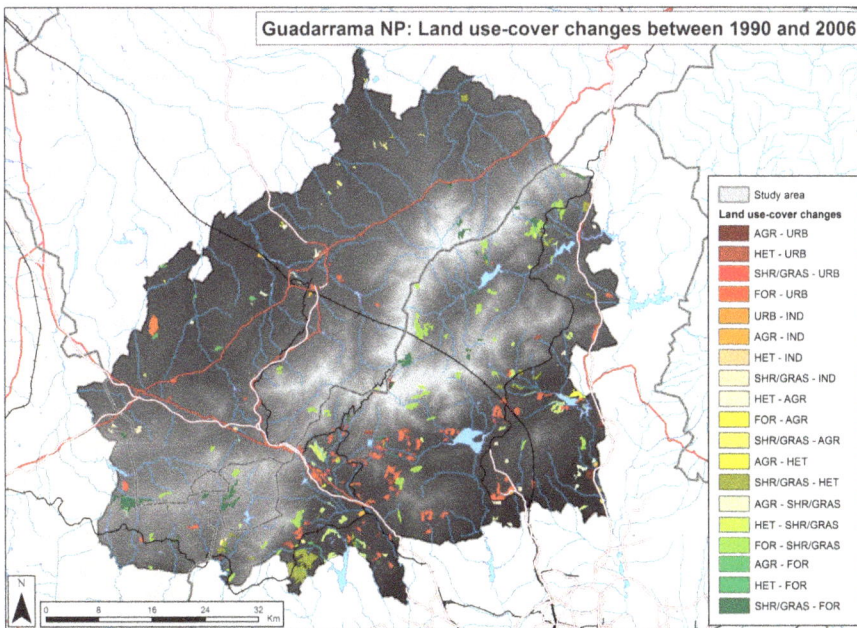

Figure 5. Land use-cover changes that took place between 1990 and 2006 in the Guadarrama NP and its surroundings. URB = Urban areas; IND = Industrial areas; AGR = Agricultural areas; HET = Heterogeneous agricultural areas; FOR = Forests; SHR-GRAS = Shrubs and Grasslands.

In this study area, there were also similar processes, although of different intensity: increase in artificial areas, natural regeneration and degradation of vegetation (Table 4). Improved accessibility to Madrid has facilitated urban development of the towns to the south of the mountains (region of Madrid) and, also, although to a lesser extent, to the north (region of Castilla y León). Artificial areas increased by more than 5500 ha, for both residential (+1.19%) and industrial and commercial uses (+0.18%). Low-density, single-family housing predominates, mostly for use as holiday homes. These changes occurred in former forest areas (shrubs, grassland and forests) and on land occupied by heterogeneous agriculture. This phenomenon did not affect either the NP or the PPZ, but did affect the SIZ, and this amounts to one of the most worrying threats because of its proximity to the NP.

Table 4. Matrix and statistics on land use change, in % of the total study area, between 1990 (rows) and 2006 (columns), in the Guadarrama NP and its surroundings.

Land Use-Land Cover	URB	IND	AGR	HET	FOR	SHR-GRAS	Total	Losses	Persistence	Total Change	Swap	Net Change	Annual Rate of Change
URB	2.41	0.03	0.00	0.00	0.00	0.00	2.43	0.03	2.41	1.22	0.05	1.17	0.945
IND	0.00	0.08	0.00	0.00	0.00	0.00	0.08	0.00	0.08	0.18	0.00	0.18	2.888
AGR	0.06	0.04	14.31	0.06	0.04	0.10	14.61	0.30	14.31	0.41	0.60	−0.19	−0.031
HET	0.11	0.01	0.02	6.55	0.02	0.42	7.13	0.58	6.55	1.13	1.15	−0.02	−0.007
FOR	0.10	0.00	0.01	0.00	17.67	0.77	18.55	0.87	17.67	1.55	1.75	−0.20	−0.026
SHR-GRAS	0.93	0.11	0.09	0.50	0.62	54.97	57.20	2.23	54.97	3.52	4.47	−0.95	−0.040
Total	3.60	0.26	14.42	7.11	18.35	56.25	100.00	4.01	95.99				
Gains	1.19	0.18	0.11	0.56	0.68	1.29	4.01						

URB = Urban areas; IND = Industrial areas; AGR = Agricultural areas; HET = Heterogeneous agricultural areas; FOR = Forests; SHR-GRAS = Shrubs and Grasslands.

There were many interchanges between the other land uses, but they all resulted in negative net change. On the one hand, there was an invasion of former agricultural zones by shrubs (0.52% of the study area). In most cases, this was marginal agricultural land with limited capacity for agricultural use because of important biophysical limitations (slopes, shallow useful soil or stony soil, risk of erosion, bioclimatic limitations, etc.). From the environmental point of view, there are two sides to this phenomenon. On the one hand, it has led to an increase in available fuel, and its extension has led to increased connectivity between forest masses, thus increasing the risk and propagation of forest fires. Conversely, it allows for greater carbon capture. In addition, the regeneration of vegetation is the natural evolution of ecosystems towards the climax. Indeed, it stimulates the structural connectivity between PAs, for example.

Forests have gained by more than 0.68% of the study area, as a result of incentives created by management plans for the PAs located within the study area, including the actual NP. However, the opposite phenomenon was stronger. During the same period, more than 0.87% of the study area occupied by forests was lost. Most of them are now occupied by shrubs and grassland. This transition mainly occurred in the NP, its PPZ and its SIZ. Many of these patches were deforested by fires during this same period [72]. Although there are ample provisions of firefighters and fire-fighting equipment, fires continue to take place and to be propagated throughout this study area. An example was the fire in Abantos which affected 500 ha of the south slope of this mountain in 1999. The dense pine forests have been replaced by grassland and, more recently, by shrubs.

3.2. Simulated Scenarios of Change in 2030

Cramer's V test (Table A3) shows that, in the Ordesa NP, the variable that is most closely associated with all classes of land use is altitude (0.4247). Next come other variables that are useful for the simulation such as those relating to relief (slopes), climate (temperatures and precipitation), accessibility and, finally, changes in population density according to the censuses that are closest to the study period. However, we eliminated the slope variable because it is highly correlated with altitude. In the Guadarrama NP, access to and from Madrid is the factor that is most closely associated with all uses, especially the most intensive: urban, industrial and agricultural. On the other hand, slopes explain more of the changes in marginal agricultural and forestry uses. The distance to reservoirs and accessibility to roads are moderately associated with some land uses. The remaining variables are practically unrelated to land uses, so, since they are not likely to explain any significant change, they were not selected.

In the Ordesa NP, we made various adjustments and calibrations in order to generate the simulated map for 2006 and to compare it with the real map for 2006. Considering CLC90 as the base map of the series, the comparison results in the following statistics: *Kappa Sim* = 0.865, *TransLoc* = 0.957 and

Transition = 0.904. The URB classes (*Kappa Sim* = 0.428) and AGR (*Kappa Sim* = 0.539) score the worst values, while FOR reaches the best values (*Kappa Sim* = 0.912).

Finally, we used two matrices for change probabilities in line with the changes occurring over recent decades and with the two simulated scenarios (Table A4). If the expected conditions prevail, the trend scenario will see few land use changes. However, the green scenario predicts that more agricultural land will be abandoned, with the development of natural ecosystems. In Figure 6, we show land use changes expected between 2006 and 2030, in a Green Scenario with Restrictions and Incentives (GS30-WRI). Again, we can see two opposing trends. On the one hand, there is an increase in the artificial area (reds), with an annual rate of change of +1.2870 (Table 5). If the expected conditions are met, urban expansion will take place in the surroundings of Sabañánigo, mainly on former agricultural land and also on scrubland. Other transitions to artificial areas can be expected in the areas that are closest to the main roads. On the other hand, natural regeneration of vegetation (green colors) is the natural evolution of ecosystems towards the climax. We can expect transitions of agricultural land (about 1000 ha) and of grassland (also about 1000 ha) towards shrubs. In ecological terms, the latter are those that most worry PA managers because focal habitats might be lost. Finally, patches of grassland and shrubs can be expected to evolve towards forest (about 2000 ha).

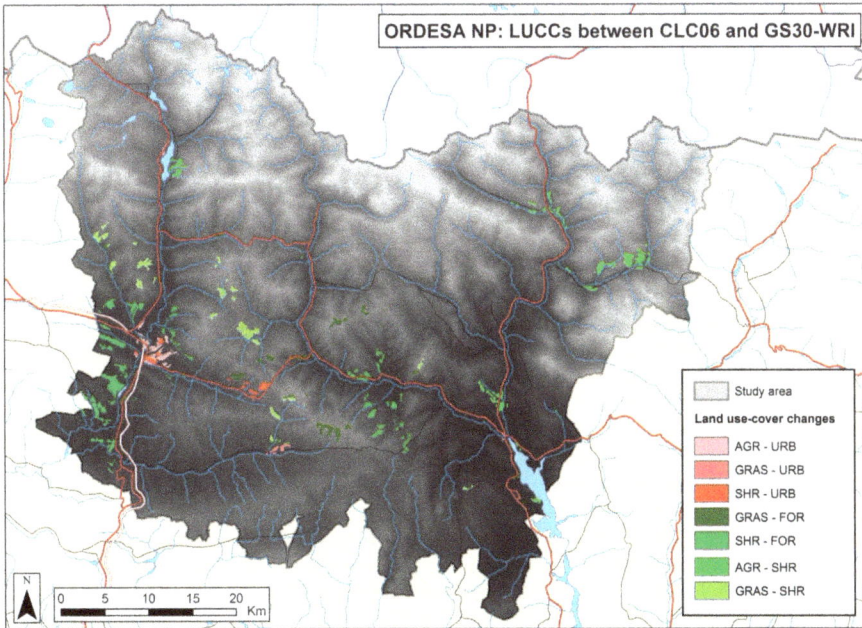

Figure 6. Simulated model of LUCCs between 2006 and 2030 in a Green Scenario with Restrictions and Incentives (GS30-WRI) in the Ordesa NP and its surroundings. URB = Urban areas; AGR = Agricultural areas; GRAS = Grasslands; SHR = Shrubs; FOR = Forests.

Table 5. Matrix and statistics on expected land use changes, in % of the total study area, between 2006 (rows) and 2030 (columns) in the Ordesa NP and its surroundings, considering the Green Scenario with Restrictions and Incentives (GS30-WRI).

Land Use-Land Cover	URB	AGR	GRAS	SHR	FOR	OTH	Total	Losses	Persistence	Total Change	Swap	Net Change	Annual Rate of Change
URB	0.45	0.00	0.00	0.00	0.00	0.00	0.45	0.00	0.45	0.32	0.00	0.32	1.2870
AGR	0.17	7.64	0.00	0.33	0.00	0.00	8.13	0.50	7.64	0.50	1.00	−0.50	−0.1524
GRAS	0.04	0.00	11.57	0.30	0.24	0.00	12.16	0.58	11.57	0.58	1.17	−0.58	−0.1187
SHR	0.11	0.00	0.00	21.08	0.66	0.00	21.84	0.76	21.08	1.39	1.53	−0.14	−0.0150
FOR	0.00	0.00	0.00	0.00	40.57	0.00	40.57	0.00	40.57	0.90	0.00	0.90	0.0529
OTH	0.00	0.00	0.00	0.00	0.00	16.85	16.85	0.00	16.85	0.00	0.00	0.00	0.0000
Total	0.77	7.64	11.57	21.71	41.47	16.85	100.00		98.15	3.69	3.69	0.00	
Gains	0.32	0.00	0.00	0.63	0.90	0.00							

URB = Urban areas; AGR = Agricultural areas; GRAS = Grasslands; SHR = Shrubs; FOR = Forests; OTH = Others.

In Figures 7 and 8, we show the differences between the two simulated change scenarios: GS30-WRI, in which restrictions and incentives are taken into account, and GS30, in which neither restrictions, nor incentives are considered. An example is shown in Windows A. If conditions are as expected, the green patches will be grassland in the GS30-WRI scenario, in line with the policy to promote the maintenance of pastures in PAs, while in GS30, they will probably be shrubs. The transition to shrubs goes against the management plan for the Sierra y Cañones de Guara Regional Park that appears in this Window. In Windows B, G30-WRI predicts a slight increase in the urban area within the town of Aínsa, because of incentives from the Government of Aragon [73], while GS30 simulates these patches as agricultural areas. Windows C represents the difference between the two models on the banks of the river Cinca, the Pineta reservoir and the A13B road. Any use change in these areas of public domain is prohibited. For this reason, GS30-WRI respects its current agricultural use, while GS30 predicts a transition to shrubs.

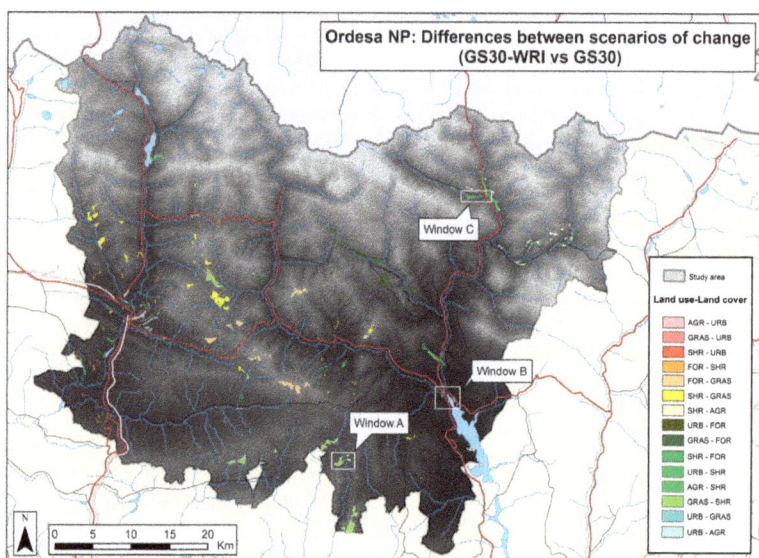

Figure 7. Expected differences in land use in the Ordesa NP and its surroundings according to the simulated scenarios for 2030: Green Scenario with Restrictions and Incentives (GS30-WRI), on the left of the legend, and Green Scenario without restrictions or incentives (GS30), on the right of the legend. URB = Urban areas; AGR = Agricultural areas; GRAS = Grasslands; SHR = Shrubs; FOR = Forests.

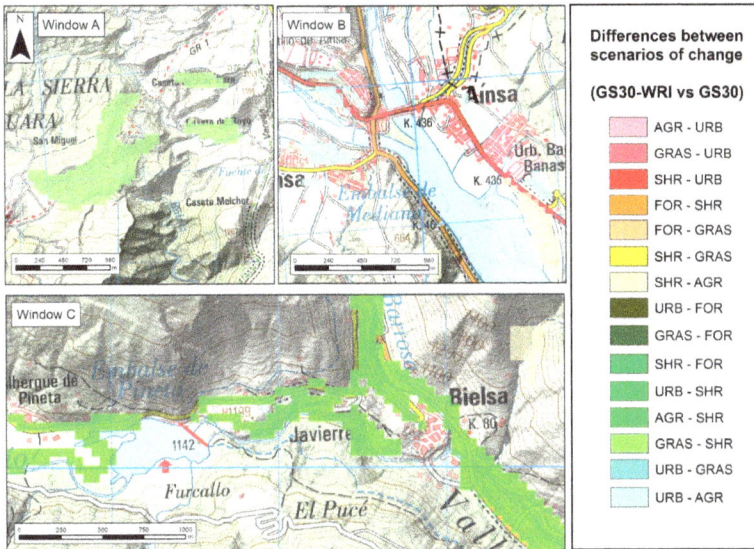

Figure 8. Large-scale Windows from Figure 7 (GS30-WRI, left of the legend, versus GS30, right of the legend): (Window A) incentive for maintaining grassland in the Regional Park of Sierra y Cañones de Guara; (Window B) incentive for urbanization in the town of Aínsa; (Window C) restriction on use change in the area of public domain beside water and along roads. URB = Urban areas; AGR = Agricultural areas; GRAS = Grasslands; SHR = Shrubs; FOR = Forests.

Logically, in the Guadarrama NP, the changes expected between 2006 and the Trend Scenario (TS30) are similar to those that took place between 1990 and 2006, although the change patches are larger. In the Trend Scenario with Restrictions and Incentives (TS30-WRI), artificial areas will probably continue to increase, and new urban zones will mainly be located on the Madrid side of the buffer (Figure 9). The process of the regeneration of vegetation will affect all the zones. In the PPZ and on the northern side of the SIZ, transitions are predicted from agricultural areas to grassland and shrubs because of the abandonment of farming. In the SIZ, the herbaceous and woody vegetation is expected to progress towards forest. In the western buffer, forest patches are expected to become consolidated in response to the incentives offered by several PAs (Campo Azálvaro y Pinares de Peguerinos Special Protection Area (SPA) and the Voltoya River Basin).

In Figure 10, we show the differences between simulated land use maps for 2030 (TS30 vs. TS30-WRI). In the NP and PPZ, small patches of heterogeneous agriculture are expected to be transformed into shrubs and grassland as a consequence of the incentives offered under the NP management plan. In the buffer and in the SIZ, there are differences between the two simulations: in the first model, there are areas that could be urban, while in the second, they will be shrubs and grassland (around the A6 motorway and the Santillana reservoir and in other areas noted in Table A2b, where urbanization is restricted, so transitions from any use to urban are not possible). In the western buffer, forest patches are expected to become consolidated as a result of incentives offered by several PAs (Campo Azálvaro y Pinares de Peguerinos SPA and the Voltoya River Basin).

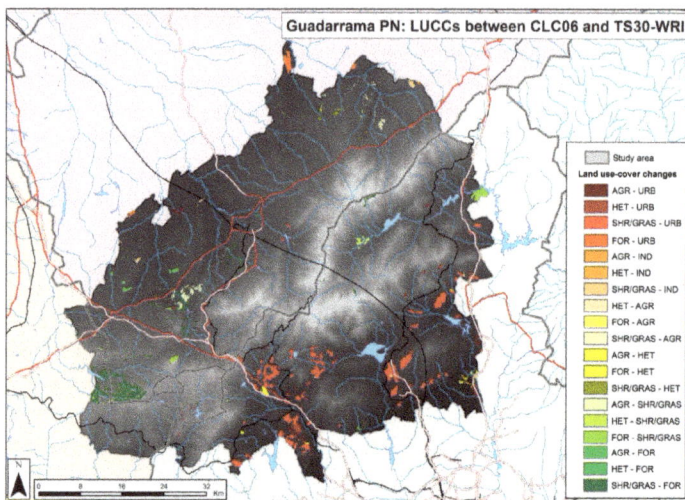

Figure 9. Simulated model of LUCCs between 2006 and 2030 in a Trend Scenario with Restrictions and Incentives (TS30-WRI) in the Guadarrama NP and its surroundings. URB = Urban areas; IND = Industrial areas; AGR = Agricultural areas; HET = Heterogeneous agricultural areas; FOR = Forests; SHR-GRAS = Shrubs and Grasslands.

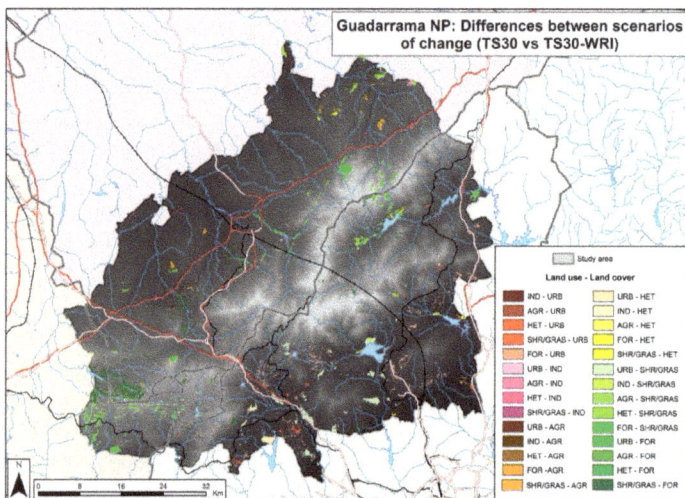

Figure 10. Expected differences in land use in the Guadarrama NP and its surroundings according to simulated scenarios for 2030: Trend Scenario without restrictions or incentives (TS30), on the left of the legend, and Trend Scenario with Restrictions and Incentives (TS30-WRI), on the right of the legend. URB = Urban areas; IND = Industrial areas; AGR = Agricultural areas; HET = Heterogeneous agricultural areas; FOR = Forests; SHR-GRAS = Shrubs and Grasslands.

3.3. Fragmentation of Habitats of Interest

At the landscape scale, alpine and sub-alpine grasslands are hardly fragmented by shrubs (HFI = 1.84, Table 6) in the Ordesa NP. From the NP towards its periphery, there is a gradient of increasing fragmentation (1.21 in the buffer). If the conditions predicted in the GS30 scenario prevail, the fragmentation can be expected to increase slightly in all areas, except in the NP.

Table 6. Expected evolution of the fragmentation index for habitats of interest (alpine and sub-alpine grasslands) in the Ordesa NP and its surroundings between 2006 and 2030, taking into account the Green Scenario (GS30).

Year	NP	PPZ	SIZ	RP	SPA-SCI	Buffer
2006	1.84	1.62	1.39	1.37	1.32	1.21
2030	1.84	1.61	1.38	1.33	1.30	1.20

NP = National Park, PPZ = Peripheral Protection Zone, SIZ = Socioeconomic Influence Zone, RP = Regional Park; SPA = Special Protection Area, SCI = Sites of Community Importance.

There is also a clear difference between PA networks. NPs are more effective for maintaining grassland ecosystems than Regional Parks. Of special interest are Sites of Community Importance (SCIs) and SPAs, which see values very close to those of unprotected areas. This may be the result of a lack of planning and of efficient management. In Figure 11, we show an example of the retreat of grasslands seen between 1990 and 2006 and predicted between 2006 and 2030. It is clear that the core, edges, branches and bridges of grasslands are disappearing and being invaded by shrubs (in grey colors). This trend is creating increasingly serious ecological and environmental problems, against the wishes of naturalists and PA managers.

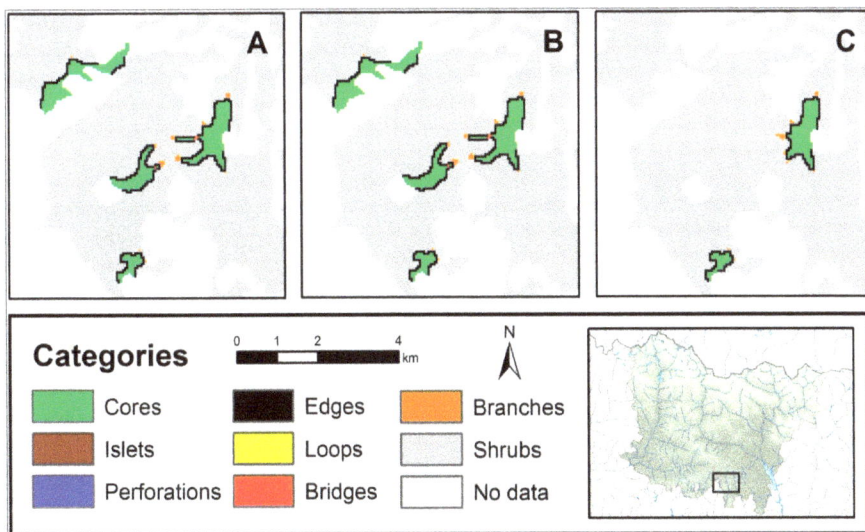

Figure 11. Trend of the fragmentation of grasslands in a window of the Regional Park of Cañones y Sierra de Guara (SE of the Ordesa NP study area) in three years: 1990 (**A**), 2006 (**B**) and 2030 (**C**).

In the case of the Guadarrama NP, urban zones and artificial infrastructure have not fragmented the agricultural and natural ecosystems of the NP and of its PPZ (Table 7). They have done so slightly in the surroundings, but this process has had a greater impact on the SIZ, especially its southern slopes, than on the buffer. Urban sprawl and newly-built infrastructure (high-speed train and AP61 motorway) have fragmented ecosystems of interest in the SIZ (Figure 12), causing negative visual impacts and obstructing movement for terrestrial mammals. Although this land does not fall under their authority, the managers of the NPs should therefore carry out special monitoring of these threats and draw up corrective and protective measures together with the regional and local land planning authorities.

Table 7. Trend in the fragmentation index of natural and semi-natural habitats in the Guadarrama NP and its surroundings between 1990 and 2006.

Year	NP	PPZ	SIZ	Buffer
1990	2.00	2.00	1.94	1.96
2006	2.00	2.00	1.90	1.95

NP = National Park, PPZ = Peripheral Protection Zone, SIZ = Socioeconomic Influence Zone.

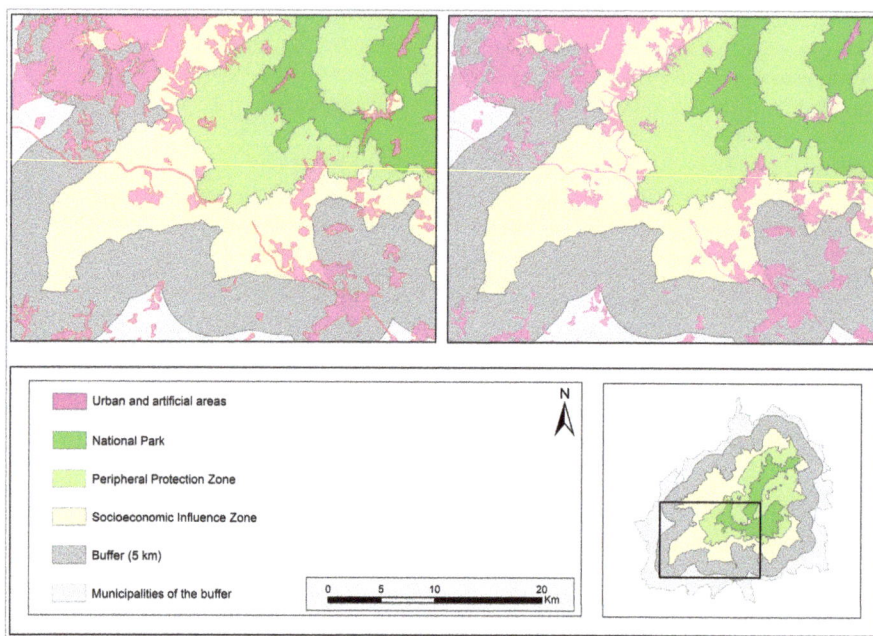

Figure 12. Trend in fragmentation of semi-natural and natural ecosystems in a window for the SW of the Guadarrama NP and its surroundings between 1990 (**left**) and 2006 (**right**). The pink patches are urban and artificial (infrastructure) areas.

4. Discussion

There have been few attempts at guiding future planning of PAs [2,60,73,74]. PANORAMA [75] is an online platform that shows examples, in PAs of the world, of planning based on participation between local communities and managers. Our research is in line with this objective. The simulated scenarios and other variants may be a good starting-point for discussion and agreements between local communities and managers.

From a methodological point of view, the main discussion revolves around the spatial and temporal dimensions of the data used as inputs in the simulations. In spite of errors [61,62], we consider that CLC is a standard source of information that is available at the European level, allowing studies to be replicated in other locations. In addition, it has a historical series from 1990. The main drawbacks are the scale, which offers limited detail for local studies and, recently, the change in method in the 2012 series. We explored the Spanish Land Occupation Information System (in Spanish SIOSE) [76], which has a larger scale (1:25,000), but its historical sequence is short (2005–2012), its legend is complex, and each polygon is defined by more than a single class [77]. We also considered the possibility of using the new Spanish Forest Map on a scale of 1:50,000 [78], but its time sequence is also short (2009–2012). It might be assumed that in order to calibrate the model better, and therefore achieve a better simulation, it would be necessary to use the longest historical sequence of data available. In this case, we could use the historical aerial photographs taken in 1945 to build a land use map for that year. Instead of using just two dates (for the start and the finish), Paegelow proposes that all available dates be used, in irregular periods to cover non-linear trends during calibration [79]. However, Candau [80] and Clarke [81] show that the use of an excessive number of data amounts to an unjustified effort in processing, introduces various change trends that often counter each other, makes the simulation more difficult and also introduces sources of error in the model resulting from the different methods and sources used to draw up the maps in the series.

From the point of view of simulating change scenarios, other authors have compared different analytical techniques [49]. We chose the ANN technique over others for several reasons. Operationally, it is easy to use, and the results are similar to those obtained using logistic regression techniques [31]. In order to avoid conditioning the predictive model, some Cellular Automata (CA) models do not pay much attention to zoning [65], which contains legal restrictions and incentives, among other aspects. However, other CA models take into account the zoning criteria [82].

From the point of view of territorial management, this is a key aspect that we think should not be ignored. Our simulation models simultaneously consider the geographic [83], multivariable and multitemporal models. Our simulations do not take into account only variables related to the neighborhood factor, but also other biophysical and socioeconomic variables. We worked with static variables. We were only able to take into account current legislation, without considering possible changes in it. Additionally, we were unable to consider the construction of future infrastructure that might change accessibility. Our models do not predict changes in the metropolis or in its behavior. This would have required knowing infrastructure and urban plans and including them in the simulated models as dynamic variables. Such information, which may be unknown or politically sensitive, is not always available. Finally, in spite of the importance of forest fires for modelling the landscape and for many land use changes, they should not be included in simulations because we cannot know where fires are going to take place. It is only possible to know where there is a greater probability or risk of forest fires [84].

From the point of view of calibrating the models and validating the resulting maps, some authors show that the probability of being correct is fairly limited [85]. In calibration, LCM does not contain as many changeable stochastic parameters as other techniques such as CA [68] or Conversion of Land Use and its Effects (CLUE) [41], which allow for a better fit. Another problem is to simulate a scenario with restrictions considering past trends since 2006, prior to the declaration of the Guadarrama NP and approval of its management plan. Finally, it is not possible to validate the results. Obviously, it is not yet possible to compare any of the simulated models with the real land use map for 2030. Some authors [86] do not validate their results for similar reasons, although they defend their usefulness as a starting point for discussion among stakeholders. Simulations do not provide a single solution for end users, but they do facilitate communication and debate among different stakeholder groups [81]. Moreover, for simulations to be useful to end users, Barredo argues that models have to differentiate between a larger number of land uses, even distinguishing between land use and land cover [87]. It also seems essential to include sectoral policies and plans in models, however complex they may be. Finally, Gómez-Delgado points to the difficulty of simulating the distribution of land uses of different types, based on different driving factors that compete simultaneously in a single territory [88].

The results obtained in the Guadarrama NP are in line with previous studies in nearby or similar study areas [62,65,89]. The trend scenarios built are in line with the results obtained by other authors [60]. These researchers modelled trend scenarios in 1260 PAs in the USA and concluded that the greatest threat for them and for their buffers is urban expansion. The main land use changes taking place in Ordesa NP are similar to the findings of other studies in the same area [34,36,90] and in other mountain systems on the planet [91,92]. From the point of view of landscape structure, the results are similar to those obtained by other authors [93] from two different approaches: fragmentation versus connectivity. We confirmed differences in the habitat fragmentation rates among, and within, PA networks [31].

We also aim to increase awareness among those responsible for the Spanish network so that similar methods can be applied in other NPs. To achieve this objective, we have invited their managers to a Workshop on Monitoring and Evaluation of Sustainability in National Parks. The expert panel included managers of national parks and of other Pas, as well as regional managers of territorial planning, mayors and representatives of local action groups. We consulted their opinions to find out their assessment of the method used in this work and the feasibility of its implementation as a tool to help make spatial decisions. Globally, they appreciated the models as a starting point for the debate.

5. Conclusions

We can confirm that the cores of both NPs are not subject to significant LUCCs on an intermediate scale of analysis. This fact is a partial indicator of efficiency. However, their areas of influence and buffers are subject to pressure and threats that might affect the sustainability of the NPs in the future, especially those that are located close to large cities. Managers should carry out constant monitoring in order to minimize the impact of LUCCs and of visitors, in the form of soil sealing, fragmentation of natural habitats, lower water quality, increased risk of forest fires or the concentration of visitors in a few hotspots, such as visitor centers and the most frequently-used paths. Some of these are irreversible processes.

Simulations of change scenarios provide knowledge of potential transitions in land use. They indicate which zones are most likely to change, and such information is useful for land planning. Simulations may be a starting-point enabling stakeholders to share opinions and reach agreements on the future management of their resources. On a continental scale, habitat fragmentation models, inside and outside NPs, may be of great interest for the managers of the Natura 2000 Network because fragmentation may affect connectivity in the PAs that belong to it.

Acknowledgments: This study was funded by the Spanish Ministry of Economy, Industry and Competitiveness in the framework of the DISESGLOB project (CSO2013-42421-P). We thank the anonymous reviewers for their helpful comments.

Author Contributions: Javier Martínez-Vega and Marta Gallardo conceived of and designed the experiments. Andrés Díaz and José M. Nava performed the experiments, analyzed the data and wrote a provisional draft. Pilar Echavarría was in charge of the final elaboration of the maps. Javier Martínez-Vega wrote the paper.

Conflicts of Interest: The authors declare no conflict of interest.

Appendix A

Code	111	112	121	122	133	142	211	212	222	231	241	242	243	311	312	313	321	322	323	324	331	332	333	334	335	412	511	512	Code
111	0	2	2	1	2	2	3	3	3	3	3	3	3	3	3	3	3	3	3	3	3	3	3	3	3	3	3	3	111
112	1	0	1	1	1	1	3	3	3	3	3	3	3	3	3	3	3	3	3	3	3	3	3	3	3	3	3	3	112
121	1	1	0	1	1	1	3	3	3	3	3	3	3	3	3	3	3	3	3	3	3	2	3	3	3	3	3	3	121
122	1	1	1	0	1	1	3	3	3	3	3	3	3	3	3	3	3	3	3	3	3	2	3	3	3	3	3	3	122
133	1	1	1	1	0	1	2	2	2	2	2	2	2	2	2	2	2	2	2	2	2	2	2	3	3	3	2	1	133
142	1	1	1	1	1	0	3	3	3	3	3	3	3	3	3	3	3	3	3	2	3	3	3	3	3	3	2	2	142
211	1	1	1	1	1	1	0	1	1	1	1	1	1	1	1	1	1	1	1	1	3	3	3	3	3	3	2	1	211
212	1	1	1	1	1	1	1	0	1	1	1	1	1	1	1	1	1	1	1	1	3	3	3	3	3	3	2	1	212
222	1	1	1	1	1	1	1	1	0	1	1	1	1	1	1	1	1	3	1	1	3	3	2	3	3	3	2	1	222
231	1	1	1	1	1	1	1	1	1	0	1	1	1	1	1	1	2	3	1	1	2	3	2	3	3	2	2	1	231
241	1	1	1	1	1	1	1	1	1	1	0	2	2	1	1	1	1	3	1	1	2	3	2	3	3	3	2	1	241
242	1	1	1	1	1	1	1	1	1	1	2	0	1	1	1	1	1	3	1	1	2	3	2	3	3	3	2	1	242
243	1	1	1	1	1	1	1	1	1	1	2	1	0	1	1	1	1	3	1	1	2	3	2	3	3	3	2	1	243
311	1	1	1	1	1	1	1	2	1	1	1	1	1	0	2	1	1	3	1	1	2	3	2	1	3	2	2	1	311
312	1	1	1	1	1	1	1	2	1	1	1	1	1	2	0	1	1	3	1	1	2	3	2	1	3	2	2	1	312
313	1	1	1	1	1	1	1	2	1	1	1	1	1	1	1	0	1	3	1	1	2	3	2	1	3	2	2	1	313
321	1	1	1	1	1	1	1	1	1	2	1	1	1	1	1	1	0	3	1	1	1	2	1	2	2	2	2	1	321
322	1	1	1	1	1	1	1	3	1	1	1	1	1	1	1	1	1	0	3	2	1	3	2	1	3	2	2	1	322
323	1	1	1	1	1	1	1	1	1	1	1	1	1	1	1	1	1	3	0	1	1	1	1	3	3	2	2	1	323
324	1	1	1	1	1	1	1	1	1	1	1	1	1	1	1	1	1	3	1	0	2	2	1	3	3	2	2	1	324
331	1	1	1	1	1	1	2	2	2	2	2	2	2	1	1	1	1	1	2	1	0	2	1	3	3	2	2	1	331
332	1	1	1	1	1	2	3	3	2	3	2	2	2	3	3	3	2	3	2	2	2	0	1	3	1	3	3	3	332
333	1	1	1	1	1	1	2	2	2	2	2	2	2	2	2	2	2	2	2	2	1	1	0	2	1	3	2	1	333
334	1	3	3	3	3	3	1	1	2	3	3	3	3	1	1	1	1	1	1	1	3	3	1	0	3	1	2	1	334
335	3	3	3	3	3	3	3	3	3	3	3	3	3	3	3	3	2	3	3	3	3	1	1	3	0	3	2	2	335
412	3	2	1	1	1	2	1	3	3	1	2	2	2	1	1	1	1	2	3	1	3	3	2	1	3	0	2	1	412
511	3	2	2	2	1	2	2	2	2	2	2	2	2	2	2	2	2	2	2	2	3	1	3	3	3	2	0	1	511
512	3	2	2	2	1	2	2	2	2	2	2	2	2	2	2	2	2	2	2	2	2	2	1	3	3	1	2	0	512
Code	111	112	121	122	131	142	211	212	222	231	241	242	243	311	312	313	321	322	323	324	331	332	333	334	335	412	511	512	Code

Figure A1. Matrix for probability of change between two dates, in line with Level 3 of CLC (CORINE Land Cover): Persistence (0, grey), Possible change (1, green), Strange change (2, orange), Unexpected change (3, red). In the rows, the CLC classes for the starting year (t1, for example, 1990). In columns, the CLC classes for the final year (t2, for example, 2006). The meaning of the codes for the CLC classes can be consulted in Bossard et al., 2000 [59]. Changes in types 2 and 3 suggest revision by field work or consultation of aerial orthophotographs with a greater spatial resolution.

Table A1. Reclassification of CLC Level 3 classes in the Ordesa and Guadarrama NPs for analysis of Land use-cover change and simulation of scenarios.

Ordesa NP	CLC Classes, Level 3	Guadarrama NP
	111 Continuous urban fabric	Urban areas (URB)
	112 Discontinuous urban fabric	
	121 Industrial or commercial units	Industrial areas (IND)
Urban areas (URB)	122 Road and rail networks and associated land	Urban areas (URB)
	123 Port areas	
	124 Airports	
	131 Mineral extraction sites	
	132 Dump sites	
	133 Construction sites	
	141 Green urban areas	
	142 Sport and leisure facilities	
	211 Non-irrigated arable land	Arable land and permanent crops (AGR)
	212 Permanently irrigated land	
Agricultural areas (AGR)	213 Rice fields	
	221 Vineyards	
	222 Fruit trees and berry plantations	
	223 Olive groves	

Table A1. *Cont.*

Ordesa NP	CLC Classes, Level 3	Guadarrama NP
	231 Pastures 241 Annual crops associated with permanent crops 242 Complex cultivation patterns 243 Land principally occupied by agriculture, with significant areas of natural vegetation 244 Agro-forestry areas	Heterogeneous agricultural areas (HET)
Forests (FOR)	311 Broadleaved forests 312 Coniferous forests 313 Mixed forests	Forests (FOR)
Grasslands (GRAS)	321 Natural grasslands	Shrubs and grasslands (SHR-GRAS)
Shrubs (SHR)	322 Moors and heathland 323 Sclerophyllous vegetation 324 Transitional woodland-scrub	
Others (OTH)	331 Beaches, dunes, sands 332 Bare rocks 333 Sparsely vegetated areas 334 Burnt areas 335 Glaciers and perpetual snow 411 Inland marshes 412 Peat bogs 421 Salt marshes 422 Salines 423 Intertidal flats 511 Water courses 512 Water bodies 521 Coastal lagoons 522 Estuaries 523 Sea and ocean	Others (OTH)

Notes: CLC = CORINE Land Cover; NP = National Park.

Table A2. Land uses-land cover and transitions with restrictions or incentives in the Ordesa NP (**a**) and Guadarrama NP (**b**) and in their surroundings by management areas.

(a)			
Regulations	**Zone**	**Restrictions**	**Incentives**
Management plan for the Ordesa NP	NP and PPZ	Any transition from FOR or GRAS. Any transition to URB or AGR.	
	SIZ	Any transition from GRAS	From SHR to GRAS
Management plan for the Sierra y Cañones de Guara Regional Park	Between 1000 and 1600 masl.	Any transition from GRAS	From SHR to GRAS
Management plan for Posets Maladeta Regional Park	Between 1000 and 1600 masl.	Any transition from GRAS	From SHR to GRAS
Protection plan for the Natural Monument of the Pyrenean Glaciers		Any LUCC	
Act 1/2001 on water	100 m. restricted zone	Any LUCC	
Act 39/2003 on the railway sector	70 m. from the outside edges of levelled ground	Any LUCC	
Act 8/1998 on roads in Aragon	50 m. on both sides of motorways and highways; 18 m. on conventional roads	Any LUCC	
Act 21/2015 on forests	Burnt areas	From OTH to URB or AGR	
Landscape map of the district of Sobrabe [73]	Town of Aínsa		From AGR to URB
(b)			
Regulation	**Zone**	**Restrictions**	**Incentives**
Natural resource plan for the Guadarrama NP (region of Madrid)	Reserve	URB, IND, AGR, HET, SHR-GRAS	FOR
	Maximum protection	URB, IND, AGR, HET, SHR-GRAS	
	Organised use of natural resources	URB, IND, AGR, HET, SHR-GRAS	
	Conservation and maintenance of traditional uses	URB, IND, AGR, HET, SHR-GRAS	HET
	Traditional settlements and special areas	URB, IND, AGR, HET, SHR-GRAS	
	Transition zones	SHR-GRAS	URB-IND
	Abantos protected landscape	URB-IND	

Table A2. *Cont.*

Regulation	Zone	Restrictions	Incentives
Natural resource plan for the Guadarrama NP (region of Castilla y León)	Special plan	URB, IND, AGR, HET, FOR, SHR-GRAS	
	Compatible use A		AGR, HET, FOR
	General use		URB, IND
	Limited use (common, peaks and special interest)	URB, IND, AGR, HET, SHR-GRAS	FOR
Management plan for the summit, cirque and lagoons of Peñalara Regional Park	Maximum reserve	URB, IND	FOR
	Special protection	URB, IND	FOR
	Educational interest	URB, IND	FOR
	Buffer and preservation	URB, IND	FOR
	Forest and cattle	URB, IND, AGR, SHR-GRAS	FOR
	Recreational use	URB, IND, HET	FOR
	Special use	URB, IND, AGR, HET, FOR, SHR-GRAS	FOR
Management plan for the Cuenca Alta del Manzanares Regional Park	Natural reserve	URB, IND	HET, SHR-GRAS
	Agricultural District park (protection, production and regeneration)	URB, IND	HET, SHR-GRAS
	Urban planning areas and transition		HET, SHR-GRAS
Management plan for the Cuenca del Río Manzanares SAC			FOR
Management plan for the River Guadarrama Regional Park	Maximum protection	URB, IND and from FOR to HET	FOR
	Protection and improvement	URB, IND and from FOR to HET	
	Countryside	URB, IND and from FOR to HET	
Management plan for the Cuenca del Río Guadalix SAC			FOR
Management plan for the Cuenca del Río Lozoya y Sierra Norte SAC			SHR-GRAS
Management plan for the Cuenca de los Ríos Alberche y Cofio SAC-SPA	Priority conservation A and B	URB, IND, AGR and from FOR to HET	FOR
	General use		FOR
Management plan for Campo Azálvaro y Pinares de Peguerinos SAC-SPA			HET, FOR, SHR-GRAS
Management plan for Encinares de los Ríos Adaja y Voltoya SAC-SPA			HET, SHR-GRAS
Management plan for Sabinares de Somosierra SAC			HET, SHR-GRAS
Management plan for Lagunas de Cantalejo SAC-SPA			FOR
Management plan for Valles del Voltoya y Zorita SAC-SPA			FOR
Management plan for Pinares del bajo Alberche SAC-SPA			FOR
Act 21/2015 on forests	Burned areas	From OTH to URB or AGR	
Act 16/1995 on forests and nature protection in Madrid	Public utility forests	From FOR to HET	
Act 1/2001 on water	100 m. restricted zone	Any LUCC	
Act 7/1990 on protection of reservoirs and wetlands in Madrid	50 m. peripheral zone	URB, IND, AGR, HET, SHR-GRAS	
Act 39/2003 on the railway sector	70 m. from the outside edges of levelled land	Any LUCC	
Act 3/1991 on roads in Madrid	50 m. on both sides of motorways and highways; 25 m. on conventional roads	Any LUCC	
Act 2/1990 on roads in Castilla y León	25 m. on both sides of motorways and highways; 8 m. on conventional roads	Any LUCC	

Notes: NP = National Park; PPZ = Peripheral Protection Zone; SIZ = Socioeconomic Influence Zone; LUCC = Land Use-Cover Change; URB = Urban areas; IND = Industrial areas; AGR = Agricultural areas; HET = Heterogeneous agricultural areas; FOR = Forests; GRAS = Grasslands; SHR = Shrubs; OTH = Others.

Table A3. Cramer's V test in the Ordesa NP (**a**) and Guadarrama NP (**b**) and their surroundings.

(a)

Driving Factors	URB	AGR	GRAS	SHR	FOR	OTH	Overall
Altitudes	0.0621	**0.3648**	**0.4757**	**0.2263**	**0.4515**	**0.7075**	**0.4247**
Slopes	0.0000	0.0811	**0.2072**	0.0850	**0.4150**	0.1436	**0.2081**
Temperatures	0.0000	0.0960	**0.4259**	**0.4738**	**0.3285**	**0.2444**	**0.3795**
Precipitation	0.0000	0.0446	**0.1684**	**0.2485**	**0.2006**	**0.2404**	**0.2090**
Ranker	0.0000	0.0081	0.1070	0.0003	0.0427	0.0568	0.0975
Rendzina	0.0000	0.0229	0.0232	0.0845	0.0862	0.0437	0.0508
Calcaric Fluvisol	0.0000	0.0550	0.0206	0.0673	**0.1909**	0.0107	0.0845
Dystrict Lithosol	0.0000	0.0108	0.0595	0.0471	0.0403	0.0510	0.0584
Calcaric Lithosol	0.0000	0.0101	0.0543	0.0641	0.0224	0.0469	0.0401
Chromic Vertisol	0.0000	0.0107	0.0219	0.0193	0.0091	0.0230	0.0149
Calcaric Cambisol	0.0000	0.0117	0.0993	**0.1640**	0.0906	0.0136	**0.1201**
Humic Cambisol	0.0000	0.0024	0.1096	**0.1612**	0.0449	0.0888	0.0937
Forest fires 86-05	0.0000	0.0060	0.0568	0.0288	0.0276	**0.1515**	0.0620
Population density 91-11	0.0000	0.0394	**0.2271**	**0.2674**	**0.1739**	**0.2236**	**0.1904**
Distance to roads	0.0000	0.1104	**0.2472**	**0.2620**	**0.3484**	0.1066	**0.2310**
Distance to reservoirs	0.0000	0.0649	**0.1879**	0.0942	**0.1679**	0.1258	**0.1367**
Distance to rivers	0.0000	0.0327	0.1014	0.0914	**0.1630**	0.0395	**0.1001**
Accesibility to Huesca	0.1527	**0.2891**	**0.2415**	0.1429	**0.2693**	**0.4076**	**0.2639**
Accesibility to Sabiñánigo	0.0000	0.1348	**0.2671**	**0.2089**	**0.3148**	0.1099	**0.2256**
Accesibility to railway stations	0.0000	0.1287	**0.2713**	**0.2117**	**0.3089**	0.1029	**0.2247**

(b)

Driving Factors	URB	IND	AGR	HET	FOR	SHR-GRAS	Overall
Slopes	0.0000	0.1015	0.0309	**0.3619**	**0.1693**	**0.3642**	**0.1713**
Sedimentary rocks	0.0000	0.0534	0.0081	**0.5988**	0.0174	0.0919	**0.1283**
Granite rocks	0.0000	**0.1642**	0.0224	**0.2045**	0.0222	0.0782	0.0819
Igneous rocks	0.0000	0.0156	0.0123	0.0398	0.0105	0.0266	0.0175
Metamorphic rocks	0.0000	0.0808	0.0281	**0.3430**	0.0292	**0.1559**	**0.1062**
Distance to reservoirs	0.0000	0.1005	0.0465	**0.5632**	0.1178	0.1243	**0.1587**
Distance to rivers	0.0000	0.0315	0.0678	**0.1614**	0.0686	0.0529	0.0637
Accesibility to roads	0.0000	**0.1600**	0.0504	**0.1749**	**0.1588**	**0.1981**	**0.1237**
Accesibility to Madrid	0.2519	**0.1797**	**0.2925**	0.1346	**0.1679**	**0.1684**	**0.1992**

Note: Factors with greater explanatory power regarding land uses are in bold print (values > 0.15).

Table A4. Markov probability matrices, calculated in the Ordesa and Monte Perdido National Park and its surroundings, in a trend scenario (**a**) and a green scenario (**b**).

(a)

T1/T2	URB	AGR	GRAS	SHR	FOR	OTH
URB	1.0000	0.0000	0.0000	0.0000	0.0000	0.0000
AGR	0.0001	0.9996	0.0000	0.0003	0.0000	0.0000
GRAS	0.0017	0.0000	0.9979	0.0003	0.0001	0.0000
SHR	0.0000	0.0000	0.0000	0.9998	0.0002	0.0000
FOR	0.0000	0.0000	0.0000	0.0002	0.9998	0.0000
OTH	0.0000	0.0000	0.0000	0.0000	0.0000	1.0000

(b)

T1/T2	URB	AGR	GRAS	SHR	FOR	OTH
URB	1.0000	0.0000	0.0000	0.0000	0.0000	0.0000
AGR	0.0210	0.8590	0.0500	0.0400	0.0300	0.0000
GRAS	0.0030	0.0050	0.9470	0.0250	0.0200	0.0000
SHR	0.0050	0.0050	0.0000	0.9600	0.0300	0.0000
FOR	0.0000	0.0000	0.0000	0.0000	1.0000	0.0000
OTH	0.0000	0.0000	0.0000	0.0000	0.0000	1.0000

Notes: URB = Urban areas; AGR = Agricultural areas; GRAS = Grasslands; SHR = Shrubs; FOR = Forests; OTH = Others; T1 = 2006 (in rows); T2 = 2030 (in columns).

References

1. Wandersee, S.M.; An, L.; López-Carr, D.; Yang, Y. Perception and decisions in modeling coupled human and natural systems: A case study from Fanjingshan National Nature Reserve, China. *Ecol. Model.* **2012**, *229*, 37–49. [CrossRef]
2. Ruiz-Mallén, I.; Corbera, E.; Calvo-Boyero, D.; Reyes-García, D. Participatory scenarios to explore local adaptation to global change in biosphere reserves: Experiences from Bolivia and Mexico. *Environ. Sci. Policy* **2015**, *54*, 398–408. [CrossRef]
3. Martínez-Fernández, J.; Ruiz-Benito, P.; Zavala, M.A. Recent land cover changes in Spain across biogeographical regions and protection levels: Implications for conservation policies. *Land Use Policy* **2015**, *44*, 62–75. [CrossRef]
4. Food and Agriculture Organization of the United Nations. *Global Forest Resources Assessment 2010*; FAO: Rome, Italy, 2010.
5. Saraswati, G. Development Directives In Disaster-Prone Areas Based on Identification Level Vulnerability Using Geographical Information System Applications in Bogor Regency. *Procedia Soc. Behav. Sci.* **2014**, *135*, 112–117. [CrossRef]
6. Chuvieco, E.; Martínez, S.; Román, M.V.; Hantson, S.; Pettinari, M.L. Integration of ecological and socio-economic factors to assess global vulnerability to wildfire. *Glob. Ecol. Biogeogr.* **2014**, *23*, 245–258. [CrossRef]
7. Dantas de Paula, M.; Groeneveld, J.; Huth, A. Tropical forest degradation and recovery in fragmented landscapes. Simulating changes in tree community, forest hidrology and carbon balance. *Glob. Ecol. Conserv.* **2015**, *3*, 664–677. [CrossRef]
8. Lei, C.; Lin, Z.; Zhang, Q. The spreading front of invasive species in favorable habitat or unfavorable habitat. *J. Differ. Equ.* **2014**, *257*, 145–166. [CrossRef]
9. McDonald, R.I. Implications of Urbanization for Conservation and Biodiversity Protection. In *Encyclopedia of Biodiversity*, 2nd ed.; Levin, S.A., Ed.; Academic Press: Amsterdam, The Netherlands, 2013; pp. 231–244.
10. López Lambas, M.E.; Ricci, S. Planning and management of mobility in natural protected areas. *Procedia Soc. Behav. Sci.* **2014**, *162*, 320–329. [CrossRef]
11. UNEP-World Conservation Monitoring Centre. World Database on Protected Areas. 2012. Available online: https://www.protectedplanet.net/c/world-database-on-protected-areas (accessed on 29 September 2017).
12. Venter, O.; Fuller, R.A.; Segan, D.B.; Carwardine, J.; Brooks, T.; Butchart, S.H.M.; Di Marco, M.; Iwamura, T.; Joseph, L.; O'Grady, D.; et al. Targeting Global Protected Area Expansion for Imperiled Biodiversity. *PLoS Biol.* **2014**, *12*, e1001891. [CrossRef] [PubMed]
13. Juffe-Bignoli, D.; Burgess, N.D.; Bingham, H.; Belle, E.M.S.; de Lima, M.G.; Deguignet, M.; Bertzky, B.; Milam, A.N.; Martinez-Lopez, J.; Lewis, E.; et al. *Protected Planet Report 2014*; UNEP-WCMC: Cambridge, UK, 2014.
14. Convention on Biological Diversity. COP 10 Decision X/2: X/2. Strategic Plan for Biodiversity 2011–2020. Available online: https://www.cbd.int/decision/cop/?id=12268 (accessed on 12 September 2017).
15. Deguignet, M.; Juffe-Bignoli, D.; Harrison, J.; MacSharry, B.; Burgess, N.; Kingston, N. *United Nations List of Protected Areas*; UNEP-WCMC: Cambridge, UK, 2014.
16. EUROPARC-España. *Anuario 2013 del Estado de las Áreas Protegidas en España*; Fundación Fernando González Bernáldez: Madrid, Spain, 2014.
17. Dudley, N. *Guidelines for Applying Protected Area Management Categories*; IUCN: Gland, Switzerland, 2008.
18. Spanish Government. Ley de 7 de Diciembre de 1916, de Parques Nacionales de España. Gaceta de Madrid. 1916. Available online: https://www.boe.es/datos/pdfs/BOE/1916/343/A00575-00575.pdf (accessed on 19 September 2017).
19. Spanish Government. Ley 30/2014, de 3 de Diciembre, de Parques Nacionales. Boletín Oficial del Estado. 2014. Available online: https://www.boe.es/boe/dias/2014/12/04/pdfs/BOE-A-2014-12588.pdf (accessed on 19 September 2017).
20. Lambin, E.F.; Turner, B.L.; Geist, H.J.; Agbola, S.B.; Angelsen, A.; Folke, C.; Bruce, J.W.; Coomes, O.T.; Dirzo, R.; George, P.S.; et al. The causes of land-use and land-cover change: Moving beyond the myths. *Glob. Environ. Chang.* **2001**, *11*, 261–269. [CrossRef]

21. Romero-Calcerrada, R.; Perry, G.L.W. The role of land abandonment in landscape dynamics in the SPA Encinares del río Alberche y Cofio, Central Spain, 1984–1999. *Landsc. Urban Plan.* **2004**, *66*, 217–232. [CrossRef]
22. Viedma, O.; Moity, N.; Moreno, J.M. Changes in landscape fire-hazard during the second half of the 20th century: Agriculture abandonment and the changing role of driving factors. *Agric. Ecosyst. Environ.* **2015**, *207*, 126–140. [CrossRef]
23. Lambin, E.F.; Geist, H.J. *Land-Use and Land-Cover Change. Local Processes and Global Impacts*; Springer-Verlag: Berlin, Germany, 2006.
24. Dong, J.; Xiao, X.; Sheldon, S.; Biradar, C.; Duong, N.; Hazarika, M. A comparison of forest cover maps in mainland Southeast Asia from multiple sources: PALSAR, MERIS, MODIS and FRA. *Remote Sens. Environ.* **2012**, *127*, 60–73. [CrossRef]
25. Yu, W.; Zhou, W.; Qiana, Y.; Yan, J. A new approach for land cover classification and change analysis: Integrating backdating and an object-based method. *Remote Sens. Environ.* **2016**, *177*, 37–47. [CrossRef]
26. Hewitt, R.; Pera, F.; Escobar, F. Cambios recientes en la ocupación del suelo de los parques nacionales. *Cuadernos Geográficos* **2016**, *55*, 46–84.
27. Rodríguez-Rodríguez, D.; Martínez-Vega, J. What should be evaluated from a manager's perspective? Developing a salient protected area effectiveness evaluation system for managers and scientists in Spain. *Ecol. Indic.* **2016**, *64*, 289–296. [CrossRef]
28. Spracklen, B.D.; Kalamandeen, M.; Galbraith, D.; Gloor, E.; Spracklen, D.V. A global analysis of deforestation in moist tropical forest protected areas. *PLoS ONE* **2015**, *10*, e0143886. [CrossRef] [PubMed]
29. Mcdonald, R.I.; Forman, R.T.T.; Kareiva, P.; Neugarten, R.; Salzer, D.; Fisher, J. Urban effects, distance, and protected areas in an urbanizing world. *Landsc. Urban Plan.* **2009**, *93*, 63–75. [CrossRef]
30. Radeloff, V.C.; Stewart, S.I.; Hawbaker, T.J.; Gimmi, U.; Pidgeon, A.M.; Flather, C.H.; Hammer, R.B.; Helmers, D.P. Housing growth in and near United States protected areas limits their conservation value. *PNAS* **2010**, *107*, 940–945. [CrossRef] [PubMed]
31. Gallardo, M.; Martinez-Vega, J. Future Land Use Change Dynamics in Natural Protected Areas—Madrid Region Case Study. In Proceedings of the 3rd International Conference on Geographical Information Systems Theory, Applications and Management (GISTAM 2017), Porto, Portugal, 27–28 April 2017; Ragia, L., Rocha, J.G., Laurini, R., Eds.; SCITEPRESS–Science and Technology Publications: Setúbal, Portugal, 2017; pp. 370–377.
32. Gurrutxaga, M.; Rubio, L.; Saura, S. Key connectors in protected forest area networks and the impact of highways: A transnational case study from the Cantabrian Range to the Western Alps (SW Europe). *Landsc. Urban Plan.* **2011**, *101*, 310–320. [CrossRef]
33. Pascual-Aguilar, J.; Andreu, V.; Gimeno-García, E.; Picó, Y. Current anthropogenic pressures on agro-ecological protected coastal wetlands. *Sci. Total Environ.* **2015**, *503–504*, 190–199. [CrossRef] [PubMed]
34. Gartzia, M.; Pérez-Cabello, F.; Bueno, C.B.; Alados, C.L. Physiognomic and physiologic changes in mountain grasslands in response to environmental and anthropogenic factors. *Appl. Geogr.* **2016**, *66*, 1–11. [CrossRef]
35. Alados, C.L.; Errea, P.; Gartzia, M.; Saiz, H.; Escós, J. Positive and Negative Feedbacks and Free-Scale Pattern Distribution in Rural-Population Dynamics. *PLoS ONE* **2014**, *9*, e114561. [CrossRef] [PubMed]
36. Gartzia, M.; Alados, C.L.; Pérez-Cabello, F. Assessment of the effects of biophysical and anthropogenic factors on woody plant encroachment in dense and sparse mountain grasslands based on remote sensing data. *Prog. Phys. Geogr.* **2014**, *38*, 201–217. [CrossRef]
37. Gómez-Limón, F.J.; de Lucio, J.V. Recreational activities and loss of diversity in grasslands in Alta Manzanares Natural Park, Spain. *Biol. Conserv.* **1995**, *74*, 99–105. [CrossRef]
38. Palomino, D.; Carrascal, L.M. Urban influence on birds at regional scale: A case study with the avifauna of northern Madrid province. *Landsc. Urban Plan.* **2006**, *77*, 276–290. [CrossRef]
39. Palomino, D.; Carrascal, L.M. Habitat associations of a raptor community in a mosaic landscape of Central Spain under urban development. *Landsc. Urban Plan.* **2007**, *83*, 268–274. [CrossRef]
40. García-Ureta, A. Wilderness protection in Spain. In *Wilderness Protection in Europe: The Role of International European and National Law*; Bastmeijer, K., Ed.; Cambridge University Press: Cambridge, UK, 2016; pp. 455–481.

41. Verburg, P.H.; Overmars, K.P. Combining top-down and bottom-up dynamics in land use modeling: Exploring the future of abandoned farmlands in Europe with the Dyna-CLUE model. *Landsc. Ecol.* **2009**, *24*, 1167–1181. [CrossRef]
42. Pontius, R.G., Jr.; Huffaker, D.; Denman, K. Useful techniques of validation for spatially explicit land-change models. *Ecol. Model.* **2004**, *179*, 445–461. [CrossRef]
43. Rodríguez Eraso, N.; Armenteras-Pascual, D.; Retana Alumbreros, J. Land use and land cover change in the Colombian Andes: Dynamics and future scenarios. *J. Land Use Sci.* **2013**, *8*, 154–174. [CrossRef]
44. Soares-Filho, B.S.; Pennachin, C.L.; Cerqueria, G. DINAMICA—A stochastic cellular automata model designed to simulate the landscape dynamics in an Amazonian colonization frontier. *Ecol. Model.* **2002**, *154*, 217–235. [CrossRef]
45. Walsh, S.J.; Entwisle, B.; Rindfuss, R.R.; Page, P.H. Spatial simulation modelling of land use/land cover change scenarios in northeastern Thailand: A cellular automata approach. *J. Land Use Sci.* **2006**, *1*, 5–28. [CrossRef]
46. Mancosu, E.; Gago-Silva, A.; Barbosa, A.; de Bono, A.; Ivanov, E.; Lehmann, A.; Fons, J. Future land-use change scenarios for the Black Sea catchment. *Environ. Sci. Policy* **2015**, *46*, 26–36. [CrossRef]
47. Pontius, R.G., Jr.; Boersma, W.; Castella, J.-C.; Clarke, K.; De Nijs, T.C.M.; Dietzel, C.; Duan, Z.; Fotsing, E.; Goldstein, N.; Kok, K.; et al. Comparing the input, output and validation maps for several models of land change. *Ann. Reg. Sci.* **2008**, *42*, 11–37. [CrossRef]
48. Nogueira-Terra, T.; Ferreira dos Santos, R.; Cortijo, D. Land use changes in protected areas and their future: The legal effectiveness of landscape protection. *Land Use Policy* **2014**, *38*, 378–387. [CrossRef]
49. Díaz-Pacheco, J.; Hewitt, R. Modelado del cambio de usos del suelo urbano a través de Redes Neuronales Artificiales. Comparación con dos aplicaciones de software. *Geofocus* **2013**, *14*, 1–22.
50. Gallardo, M.; Gómez, I.; Vilar, L.; Martínez-Vega, J.; Martín, M.P. Impacts of future land use/land cover on wildfire occurrence in the Madrid region. *Reg. Environ. Chang.* **2016**, *16*, 1047–1061. [CrossRef]
51. Spanish Government. Declaración del Parque Nacional del Valle de Ordesa. Gaceta de Madrid. 1918. Available online: http://www.mapama.gob.es/es/red-parques-nacionales/nuestros-parques/ordesa/_creacion_tcm7-257426.pdf (accessed on 19 September 2017).
52. Instituto Nacional de Estadística (INE). INEbase. Demografía y Población. Padrón. Población por Municipios. Cifras Oficiales de Población de los Municipios Españoles: Revisión del Padrón Municipal. Cifras Oficiales de Población Resultantes de la Revisión del Padrón Municipal a 1 de enero. 2017. Available online: http://www.ine.es/dynt3/inebase/index.htm?padre=525 (accessed on 3 September 2017).
53. Ministerio de Agricultura, Pesca, Alimentación y Medio Ambiente (MAPAMA). Memoria de la Red de Parques Nacionales. 2015. Available online: http://www.mapama.gob.es/es/red-parques-nacionales/divulgacion/memoria-2015_tcm7-454259.pdf (accessed on 3 September 2017).
54. Spanish Government. Ley 7/2013, de 25 de junio, de declaración del Parque Nacional de la Sierra de Guadarrama. Boletín Oficial del Estado. 2013. Available online: https://www.boe.es/boe/dias/2013/06/26/pdfs/BOE-A-2013-6900.pdf (accessed on 19 September 2017).
55. Ministerio de Agricultura, Pesca, Alimentación y Medio Ambiente (MAPAMA). Memoria de la Red de Parques Nacionales. 2014. Available online: http://www.mapama.gob.es/es/red-parques-nacionales/divulgacion/memoria-2014_tcm7-454256.pdf (accessed on 29 September 2017).
56. Eastman, J.R. *IDRISI Terrset Manual*; Clark Labs-Clark University: Worcester, MA, USA, 2016.
57. Map Comparison Kit 3, User Manual. Research Institute for Knowledge Systems. Available online: https://www.dropbox.com/s/aybrp1f7fwgv2nr/MCK_Manual_3_2_1.pdf (accessed on 30 October 2017).
58. Soille, P.; Vogt, P. Morphological segmentation of binary patterns. *Pattern Recognit. Lett.* **2009**, *30*, 456–459. [CrossRef]
59. Bossard, M.; Feranec, J.; Otahel, J. *CORINE Land Cover Technical Guide—Addendum 2000*; European Environment Agency: Copenhagen, Danmark, 2000.
60. Martinuzzi, S.; Radeloff, V.C.; Joppa, L.N.; Hamilton, C.M.; Helmers, D.P.; Plantinga, D.P.; Lewis, D.J. Scenarios of future land use change around United States' protected areas. *Biol. Conserv.* **2015**, *184*, 446–455. [CrossRef]
61. Catalá Mateo, R.; Bosque Sendra, J.; Plata Rocha, W. Análisis de posibles errores en la base de datos CORINE Land Cover (1990–2000) en la Comunidad de Madrid. *Estud. Geogr.* **2008**, *49*, 81–104.

62. Díaz-Pacheco, J.; Gutiérrez, J. Exploring the limitations of CORINE Land Cover for monitoring urban land-use dynamics in metropolitan areas. *J. Land Use Sci.* **2013**, *9*, 243–259. [CrossRef]

63. Centro Nacional de Información Geográfica (CNIG). Centro de Descargas. Available online: http://centrodedescargas.cnig.es/CentroDescargas/index.jsp (accessed on 29 September 2017).

64. Stellmes, M.; Röder, A.; Udelhoven, T.; Hill, J. Mapping syndromes of land change in Spain with remote sensing time series, demographic and climatic data. *Land Use Policy* **2013**, *30*, 685–702. [CrossRef]

65. Díaz-Pacheco, J. Ciudades, Autómatas Celulares y Sistemas Complejos: Evaluación de un Modelo Dinámico de Cambio de Usos de Suelo urbano de Madrid. Ph.D. Thesis, Universidad Complutense de Madrid, Madrid, Spain, 10 September 2015.

66. Cramér, H. *Mathematical Methods of Statistics*; Princeton University Press: Princeton, NJ, USA, 1946.

67. Pontius, R.G., Jr.; Millones, M. Death to Kappa: Birth of quantity disagreement and allocation disagreement for accuracy assessment. *Int. J. Remote Sens.* **2011**, *32*, 4407–4429. [CrossRef]

68. Van Vliet, J.; Bregt, A.K.; Hagen-Zanker, A. Revisiting Kappa to account for change in the accuracy assessment of land-use change models. *Ecol. Model.* **2011**, *222*, 1367–1375. [CrossRef]

69. Van Vliet, J.; Naus, N.; van Lammeren, R.J.; Bregt, A.K.; Hurkens, J.; van Delden, H. Measuring the neighbourhood effect to calibrate land use models. *Comput. Environ. Urban* **2013**, *41*, 55–64. [CrossRef]

70. Opdam, P.; Verboom, J.; Pouwels, R. Landscape cohesion: An index for the conservation potential of landscapes for biodiversity. *Landsc. Ecol.* **2003**, *18*, 113–126. [CrossRef]

71. Opdam, P.; Steingrover, E.; Rooij, S.V. Ecological networks: A spatial concept for multi-actor planning of sustainable landscapes. *Landsc. Urban Plan.* **2006**, *75*, 322–332. [CrossRef]

72. Gómez, I.; Martín, M.P.; Salas, F.J.; Gallardo, M. Análisis del régimen histórico de incendios forestales en la Comunidad de Madrid (1985–2010) y su relación con los cambios de usos del suelo. In Proceedings of the 15th Congreso Nacional de Tecnologías de la Información Geográfica, Madrid, Spain, 19–21 September 2012; Martínez-Vega, J., Martín, M.P., Eds.; AGE: Madrid, Spain, 2012; pp. 71–82.

73. Dolz, P.A.; Cáncer, L. *Mapa de Paisaje de la Comarca de Sobrarbe. Mapa de Prospectiva del Paisaje*; Gobierno de Aragón: Zaragoza, Spain, 2009.

74. Hewitt, R.; van Delden, H.; Escobar, F. Participatory land use modelling, pathways to an integrated approach. *Environ. Model. Softw.* **2014**, *52*, 149–164. [CrossRef]

75. IUCN. PANORAMA: Solutions for a Healthy Planet. 2017. Available online: https://www.iucn.org/theme/protected-areas/our-work/protected-area-solutions/panorama-solutions-healthy-planet (accessed on 12 September 2017).

76. Valcarcel, N.; Villa, G.; Arozarena, A.; Garcia-Asensio, L.; Caballero, M.E.; Porcuna, A.; Domenech, E.; Peces, J.J. SIOSE, a Successful Test Bench Towards Harmonization and Integration of Land Cover/Use Information As Environmental Reference Data. *Remote Sens. Spat. Inf. Sci.* **2009**, *37*, 1159–1164.

77. García-Álvarez, D.; Camacho Olmedo, M.T. Changes in the methodology used in the production of the Spanish CORINE: Uncertainty analysis of the new maps. *Int. J. Appl. Earth Obs.* **2017**, *63*, 55–67. [CrossRef]

78. Ministerio de Agricultura, Pesca, Alimentación y Medio Ambiente (MAPAMA). Banco de Datos de la Naturaleza. Available online: http://sig.mapama.es/bdn/visor.html (accessed on 29 September 2017).

79. Paegelow, M. Land Change Modeling Handling with Various Training Dates. In Proceedings of the 3rd International Conference on Geographical Information Systems Theory, Applications and Management (GISTAM 2017), Porto, Portugal, 27–28 April 2017; Ragia, L., Rocha, J.G., Laurini, R., Eds.; SCITEPRESS–Science and Technology Publications: Setúbal, Portugal, 2017; pp. 350–356.

80. Candau, J. Calibrating a cellular automaton model of urban growth in a timely manner. In Proceedings of the 4th International Conference on Integrating GIS and Environmental Modeling (GIS/EM4), Banff, AB, Canada, 2–8 September 2000.

81. Clarke, K. Improving SLEUTH Calibration with a Genetic Algorithm. In Proceedings of the 3rd International Conference on Geographical Information Systems Theory, Applications and Management (GISTAM 2017), Porto, Portugal, 27–28 April 2017; Ragia, L., Rocha, J.G., Laurini, R., Eds.; SCITEPRESS–Science and Technology Publications: Setúbal, Portugal, 2017; pp. 319–326.

82. Metronamica. Version 4.2. Available online: http://www.metronamica.nl/models_zoning.php (accessed on 30 October 2017).

83. Tobler, W.R. A computer movie simulating urban growth in the Detroit region. *Econ. Geogr.* **1970**, *46*, 234–240. [CrossRef]

84. Chuvieco, E.; Aguado, I.; Jurdao, S.; Pettinari, M.L.; Yebra, M.; Salas, J.; Hantson, S.; de la Riva, J.; Ibarra, P.; Rodrigues, M.; et al. Integrating geospatial information into fire risk assessment. *Int. J. Wildl. Fire* **2012**, *23*, 606–619. [CrossRef]

85. Pascual, V.; Aguilera, F.; Plata, W.; Gómez-Delgado, M.; Bosque, J. Simulación de modelos de crecimiento urbano: Métodos de comparación con los mapas reales. In *Tecnologías de la Información Geográfica: La Información Geográfica al Servicio de los Ciudadanos*; Ojeda, J., Pita, M.F., Vallejo, I., Eds.; Universidad de Sevilla: Sevilla, Spain, 2010; pp. 1000–1013.

86. Antoni, J.P.; Judge, V.; Vuidel, G.; Klein, O. Using Constraint Cellular Automata to Simulate Urban Development in a Cross-border Area. In Proceedings of the 3rd International Conference on Geographical Information Systems Theory, Applications and Management (GISTAM 2017), Porto, Portugal, 27–28 April 2017; Ragia, L., Rocha, J.G., Laurini, R., Eds.; SCITEPRESS–Science and Technology Publications: Setúbal, Portugal, 2017; pp. 366–369.

87. Barredo, J.I.; (Institute for Environment and Sustainability of the EC Joint Research Centre, Ispra, Italy). Round Table, Geomatic approaches for modelling land change scenarios. Personal communication, 2017.

88. Gómez-Delgado, M.; (University of Alcalá, Alcalá de Henares, Spain). Round Table, Geomatic approaches for modelling land change scenarios. Personal communication, 2017.

89. Díaz-Pacheco, J.; García-Palomares, J.C. A highly detailed land-use vector map for Madrid region based on photo-interpretation. *J. Maps* **2014**, *10*, 424–433. [CrossRef]

90. Lasanta-Martínez, T.; Vicente-Serrano, S.M.; Cuadrats, J.M. Mountain Mediterranean landscape evolution caused by the abandonment of traditional primary activities: A study of the Spanish Central Pyrenees. *Appl. Geogr.* **2005**, *25*, 47–65. [CrossRef]

91. Sitko, I.; Troll, M. Timberline changes in relation to summer farming in the western Chornohora (Ukrainian Carpathians). *Mt. Res. Dev.* **2008**, *28*, 263–271. [CrossRef]

92. Stueve, K.M.; Isaacs, R.E.; Tyrrell, L.E.; Densmore, R.V. Spatial variability of biotic and abiotic tree establishment constraints across a treeline ecotone in the Alaska Range. *Ecology* **2011**, *92*, 496–506. [CrossRef] [PubMed]

93. Gartzia, M.; Fillat, F.; Pérez-Cabello, F.; Alados, C.L. Influence of Agropastoral System Components on Mountain Grassland Vulnerability Estimated by Connectivity Loss. *PLoS ONE* **2016**, *11*, e0155193. [CrossRef] [PubMed]

environments

MDPI

Article

Socioeconomic Indicators for the Evaluation and Monitoring of Climate Change in National Parks: An Analysis of the Sierra de Guadarrama National Park (Spain)

Iván López [1],* and Mercedes Pardo [2]

[1] Department of Psychology and Sociology, University of Zaragoza, c/Pedro Cerbuna, 12, 50009 Zaragoza, Spain
[2] Department of Social Analysis, University Carlos III of Madrid, c/Madrid, 126, 28903 Getafe, Spain; mercedes.pardo@uc3m.es
* Correspondence: ivalopez@unizar.es; Tel.: +34-976-8765-54578

Received: 27 November 2017; Accepted: 8 February 2018; Published: 12 February 2018

Abstract: This paper analyzes the importance of assessing and controlling the social and economic impact of climate change in national parks. To this end, a system of indicators for evaluation and monitoring is proposed for the Sierra de Guadarrama National Park, one of the most important in Spain. Based on the Driving forces-Pressure-State-Impact-Response (DPSIR) framework, the designed system uses official statistical data in combination with data to be collected through ad hoc qualitative research. The result is a system of indicators that monitors the use of natural resources, the demographic evolution, economic activities, social interactions, and policies. Adapted to different contexts, these indicators could also be used in other national parks and similar natural protected areas throughout the world. This type of indicator system is one of the first to be carried out in Spain's national parks. The result is a system that can be useful not only in itself, but also one that can catalyze climate change planning and management of national parks.

Keywords: socioeconomic indicators; national parks; climate change; sustainable development; Sierra de Guadarrama; Spain

---

1. Introduction

Anthropogenic climate change, which is produced by greenhouse gas emissions from human activities added to natural climate variability [1], is one of the most serious problems of global environmental change faced by contemporary societies [2].

The need to identify the current and foreseeable impacts of climate change as well as its mitigation and adaptation presents challenges in scientific, political, economic and social spheres [2]. Among these challenges is addressing the potential impacts on national parks [3].

National parks are privileged spaces for monitoring climate change impacts [4–7]. As they are protected spaces in their biophysical characteristics and limited in their socioeconomic activities, they are easier to control than other spaces that are subjected to social and economic dynamics. In addition, high mountain areas—as is the case of the Sierra de Guadarrama National Park—are a good indicator of the possible effects of climate change on other parts of the planet, as they are particularly sensitive to global environmental changes [8,9].

Consequently, the identification, evaluation and monitoring of the impact of climate change on the park values (biological, cultural, etc.) is an important task for science and for identifying appropriate management actions [3,6,10,11].

There is already experience in monitoring systems with indicators related to biophysical conservation and evaluation of conservation management [12–16] as well as the impact of global environmental change on national parks [4,6,7,17]. However, monitoring the social systems that are both producing climate change and being impacted by climate change in national parks is much scarcer [3,18–21].

There are fifteen national parks in Spain, and for only two of these—Picos de Europa and Sierra de Guadarrama—has a system of indicators for the assessment and monitoring of the socioeconomic impact of climate change been developed. Given the recent creation of these monitoring systems, they have not yet collected enough time-series of data to detect trends in any socioeconomic indicators.

In this paper, we present the system of indicators developed for the Sierra de Guadarrama National Park. We first describe the special biophysical and cultural characteristics of the Sierra de Guadarrama. Secondly, we highlight the relevance of a system of socioeconomic indicators to evaluate and monitor climate change in the Sierra de Guadarrama National Park. Then, we explain the methodology used to develop the indicator system. Finally, we present the selected indicators, the conclusions, and some lines of discussion.

1.1. Sierra de Guadarrama: Object of Desire for Kings, Nobles, Clergymen and Novelists, Since the Middle Ages

The Sierra de Guadarrama National Park occupies 33,960 hectares, and is located in the mountain range of the Central System (Figure 1), forming part of the natural division between the northern and southern plateaus that make up the center of the Iberian Peninsula (Spain and Portugal). In addition, its peripheral protection zone is 62,687.26 hectares (this has its own legal regime, designed to promote the values of the park in its surroundings and to minimize the ecological or visual impact of the exterior over the interior of the park), and its legal area of socioeconomic influence is 175,593.40 hectares (Figure 2)—the total area of the municipalities where the National Park and its Peripheral Protection Zone are located [22].

Figure 1. Sierra de Guadarrama National Park location.

The Sierra de Guadarrama has been present in Spanish literature [23] since the Middle Ages: the Archpriest of Hita (1283–1350), Cervantes (Don Quixote) (1547–1616), Lope de Vega (1562–1635), Tirso de Molina (1579–1648), Zorrilla (1607–1648), Pío Baroja (1872–1956), Cela (a Nobel Prize winner) (1916–2002), Sanchez Ferlosio (1927–2004), and Vicente Aleixandre (1898–1984), are some of the authors who have referred to it. This is not surprising, as the Sierra de Guadarrama offers grandiose and majestic scenery, and thoroughly enigmatic settings [23] (p. 24).

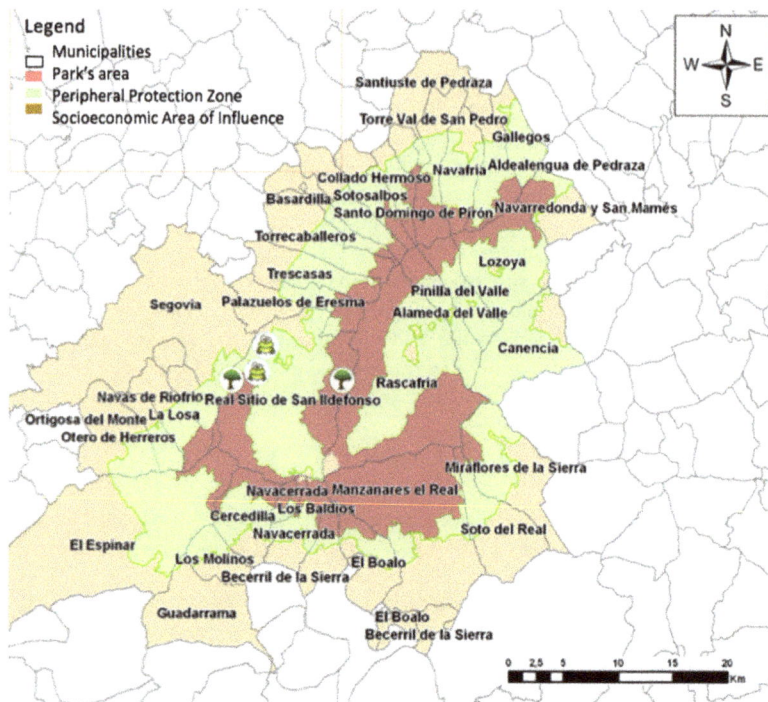

Figure 2. The park territorial limits, its zone of protection, and its area of socioeconomic influence.

The natural riches of the environs of Sierra de Guadarrama attracted the interest of kings, nobles and clergymen, who chose this area to build their palaces, fortresses, monasteries and churches, resulting in a wealth of heritage. Many of these attractions are inside or around the park, and are an addition to the park's appeal. Highlights include the Monastery of El Paular, the Castle of Manzanares, and the Royal Site of San Ildefonso [23] (p. 25).

This park is a representative sample of the natural systems of high Mediterranean mountains (Peñalara is the highest at 2428 m), as are its alpine grasslands and pastures, pine and Pyrenean oak forests, peatlands, with glacier and periglacial modeling, and the presence of unique reliefs and geological elements. The main ecosystems of the park are *Pinus sylvestris* pine trees on siliceous soils; high mountain lakes and wetlands; formations and reliefs of mountains and high mountains; the geomorphology of granite rock that distinguishes the shape of the unique relief and landscape; gall-oak and Pyrenean oak groves; supraforestal thickets, high mountain pastures, high, woody, gravelly steppes; and forests of pine, savin juniper and juniper [24].

Its biophysical values have been internationally recognized. The park, besides being a national park, has, totally or partially, other forms of international protection. It is a Special Protection Area for Birds (SPA), parts of the park are included in two Biosphere Reserves (BR) (Cuenca Alta del Manzanares BR; Real Sitio de San Ildefonso-El Espinar BR), it is included in the International Ramsar List, and is designated a Site of Community Importance (SCI) with 25 habitats of interest, four of which are of priority. Spain occupies second place in the European Union's habitats of interest ranking and third in that of priority habitats. Sierra de Guadarrama is also characterized [25] by its floristic richness and contains a large number of threatened and/or endemic species. Its special climatic conditions and its location in the transition zone between the Eurosiberian and Mediterranean regions have favored the processes of endemism. For example, in relation to flora, 40 species of interest have been cataloged; 4 are on the Red List of Spanish vascular flora, 35 are in the catalog of protected flora of the region of Madrid, and 10 are in the catalog of the region of Castile and Leon. Also found here are 83 endemic plants of the Iberian Peninsula, some of them exclusive to the Central System and others to Sierra de Guadarrama.

In addition, the park has cultural values, such as the remains of traditional socioeconomic activities and trades (transhumant pastoralists, cowherds, stonecutters, oxen, charcoal workers, carters, neighbors, etc.), remnants of pastoral pastures on the top of the sierra, the ruins of shearing ranches or the brick chimneys of old sawmills, among others. These remains bring us closer to a world of traditions that influenced the local culture for centuries and shaped the territory. It is also worth mentioning the Roman road that crosses the Park, and several drovers' roads and cattle routes dating from the Middle Ages to displace the transhumant herds—millions of Merino sheep of good wool to market to other parts of the world.

Today, most of these activities have been lost, with cattle still kept for meat production. Tourism, based on the landscape, values of nature and cultural heritage, has become one of the main economic sectors in the area.

Despite its natural values, the area was not declared a national park until 2013 [24]. In order to meet the criteria to reach category II of the IUCN, this law was modified in 2014 [24]. The process to acquire this category is still ongoing. The first National Park in Spain dates back to 1916 (Covadonga National Park).

The Park belongs to two autonomous communities (regional governments). Sixty-four percent of its area corresponds to the Autonomous Community of Madrid and a little over 36 per cent belongs to the province of Segovia, in the Autonomous Community of Castile and Leon. There are 28 municipalities included in the geographical limits of the Park.

The aforementioned natural and cultural values, as well as the park's proximity—35 km—to the Madrid metropolitan area, tend to attract large numbers of people (3.8 million visits in 2014 [24]). This mass tourism produces one of the main challenges faced by the Sierra de Guadarrama National Park: the tension between the conservation of the park and the economic interests of the municipalities within the park or in the protection area surrounding it—28 included plus 34 in its area of socioeconomic influence.

Such a conflict became more visible in 2013 when the Sierra de Guadarrama was declared a National Park. Some argue that an excessive touristic focus was given to the detriment of the conservation objective [26], and that there is a lack of coordination between protection efforts and the pursuit of traditional activities [27].

The valuable ecosystem of the Sierra de Guadarrama National Park is under threat. On the one hand by global warming, to which the park is particularly vulnerable. On the other, by an existing tendency to prioritize the economic interests of local communities instead of the conservation of the park. Both pressures have the potential to interact: for example, changed land use by humans could exacerbate the effects of climate change on the natural and cultural resources of the park. Despite these, there is very little evaluation and monitoring, mitigation and adaptation to climate change [2] of the Sierra de Guadarrama National Park [28]. The current process of drafting the obligatory Master Plan for the Use and Management of the Park could be an opportunity to address climate change, particularly its socioeconomic dimensions, more directly.

1.2. The System of Socioeconomic Indicators for the Evaluation and Monitoring of Climate Change in the Sierra de Guadarrama National Park

The aim of designing and operating a system of socioeconomic indicators for the evaluation and monitoring of climate change in the Sierra de Guadarrama National Park responds to the need to have a sufficient set of data to monitor the short, medium and long-term effects of climate change in the social and economic sphere of this protected natural space.

In the face of climate change, such monitoring is crucial for the development and implementation of plans [29–31], to conserve natural resources and to the living conditions of the communities dependent on these resources.

Those plans need to be based on an approach of seeking to increase the resilience of natural and social systems [32–38] in the face of considerable uncertainty about the specific changes that might occur and their timing and magnitude [29,39].

To do this successfully requires consideration of a potentially wide range of valued assets, whose vulnerability to different aspects of climate change will vary, and a range of interacting biophysicial and social processes at a range of spatial scales [40].

Hence, an appropriate set of indicators needs to be carefully chosen to be able to track changes in the most important elements of these complex systems over relevant timescales and spatial scales. The collected data will enable park managers to efficiently respond to a complex and changing natural and social environment [39,41].

However, managers and planners still have little guidance or training on how to address the social aspects of vulnerability to climate change in their management and planning [42,43]. This deficit can jeopardize the management strategies of these areas as well as the public support for them.

Thus, the objectives of the research presented here have been (1) the definition of a system of indicators for the evaluation and monitoring of the impact of climate change in the social and economic environment of the Sierra de Guadarrama National Park, specifying those that may be generalizable to other national parks of similar characteristics; (2) the design and development of an updated database.

2. Methodology

Following Land and Spilerman [44], the indicators refer to those parameters (statistics, data and all forms of evidence) that allow us to evaluate where we are and where we are going, in relation to the objectives set. The variables and indices that have the characteristic of indicators are those sensitive to changes, whether they are of social or physical nature, and trends of natural or social origin. As a whole, the system of indicators should show the relationships between the elements of the system studied and the underlying interactions [45].

To address the research objectives, we drew on a range of relevant existing sources of information relating to conservation and management of national parks and natural resources in Spain. First of all, the legal framework on which the general objectives for national parks in Spain are based; these focus primarily on the protection of their biogeophysical values [46]. However, the sustainable development of the municipalities situated within the park's area of socioeconomic influence, is also considered by law [46]. With regard to monitoring, the National Parks Network of Spain [47] proposes to develop and maintain a monitoring and evaluative system for the ecological, socioeconomic and functional aspects of each park and the Network as a whole. In addition, we have considered both the criteria and indicators for sustainable forest management in Spanish forests [48] and the evaluation of public use of national parks in Spain by the Autonomous National Parks Organization [49], the System of Indicators for the Evaluation and Monitoring of the Socio-economic Impact of the Impact of Global Change in the Picos de Europa National Park, as well as the system proposed for the Integrated Assessment of Protected Areas of the region of Madrid [16], among other sources.

Then, taking all of these aspects and arguments into account, the indicators selected were based on the following criteria. Firstly, they were selected according to the socioeconomic characteristics of the park's municipalities and the area of influence and the availability of the data [50].

Secondly, we considered the functions that the indicators should fulfill [51]: the continuous recording of the dynamics of the socioeconomic system and the analysis of the trends of change, either by natural or social causes; the improvement of the knowledge of the system, through the compilation or generation of new information regarding the social and the economic impact of climate change on the national park; the forecast for specific and/or global changes in the system, especially alterations or damage due to unexpected events; the identification, where appropriate, of the effects of management practices on the dynamics of social systems, and detection of undesirable effects. To do this, the research team took into consideration literature analysis, existing accessible statistical information, the park management office's annual reports, and those indicators that may be more sensitive to change.

Finally, the focus was on the concordance of the preceding two criteria with the overall goal of progressing towards the sustainable development of the communities that influence or are influenced by the National Park [52–54], according to the United Nations sustainable development goals.

The indicators developed here are the result of a selection from the many possibilities resulting from the great complexity of the natural and social systems that intertwine in protected natural spaces. This selection, made using rigorous and explicit criteria, has been necessary in order to obtain a number of indicators not too large in order to maximize the information and minimize the cost. To this end, we considered the extent to which the indicators are specific and unequivocal, easy to interpret, accessible, significant and relevant, sensitive to change, valid, verifiable and reproducible, and, above all, useful tools for action.

A balance has also been sought between the indicators of general use relating to protected natural areas and those developed for the particular case of the Sierra de Guadarrama National Park. The use of general-purpose indicators allows comparison between different protected areas and their integration into larger monitoring projects, and therefore the achievement of relevant time series.

Different indicator systems use alternative frameworks for impact analysis and sustainability [55]. In this case, we have used the Driving forces-Pressure-State-Impact-Response (DPSIR) framework, developed by the European Environment Agency (EEA) [56], which is the one used by the Spanish Ministry of Agriculture, Fishing, Food and Environment to elaborate the Water Indicators System [57].

The EEA defines "Driving forces" as "the social, demographic and economic developments in societies and the corresponding changes in lifestyles, overall levels of consumption and production patterns" [56] (p. 8). "Pressure" indicators describe the "developments in release of substances (emissions), physical and biological agents, the use of resources and the use of land" [56] (p. 9).

Pressure indicators are outside the scope of this study. Climate change is a global process that is barely affected by the activities taking place in the park, and the focus of our indicator framework is on impacts and adaptation. Therefore we do not consider it necessary to develop indicators to monitor factors (emissions of CO_2 and other greenhouse gases) that cause climate change. Nor do we consider it necessary to create additional indicators to monitor climate change itself (for example changes in temperature and rainfall), as the park has weather stations with continuous meteorological meters installed and annual reports are kept.

We have focused instead on the identification of indicators for (1) the "State" category, a description of the quantity and quality of socioeconomic phenomena in the studied area; (2) the "Impacts", the changes in the social, economic and environmental dimensions, which are caused by changes in the "State" of the system; and (3) the society's "Response" to change the pressures and the state of the environment for the solution of the problem in question, as illustrated in Figure 3.

"Impact" indicators will provide data about change in the "State", but it will not be possible to establish, a priori, a causal relation, since the park's socio-ecologic system is affected by other factors as well. As the Intergovernmental Panel on Climate Change IPCC (2014) concludes, "many processes and mechanisms are well understood, but others are not. Complex interactions among multiple climatic and non-climatic influences changing over time lead to persistent uncertainties, which in turn lead to the possibility of surprises" [2] (p. 151). Further, the impact of important socioeconomic factors could emerge in the medium or long term [58], depending as well on the adopted mitigation and adaptation measures.

Even so, "for most economic sectors, the impacts of drivers such as changes in population, age structure, income, technology, relative prices, lifestyle, regulation, and governance are projected to be large relative to the impacts of climate change" [58] (p. 19). This emphasizes the importance of the evaluative and monitoring systems of climate change, in this case the socio-economic impact regarding the national park.

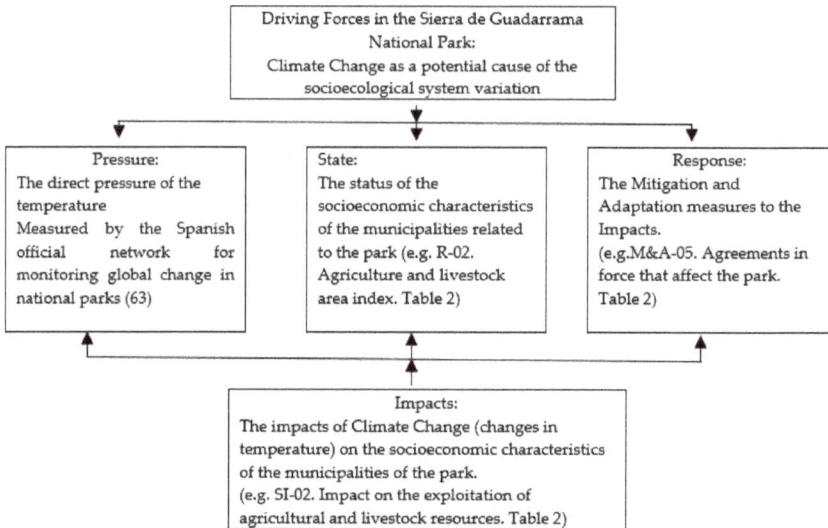

Figure 3. Example of Driving forces-Pressure-State-Impact-Response (DPSIR) scheme.

We propose a system based on a basic chain-of-causality among the indicators and their mutual dependence. This is achieved by indicating on each indicator what other indicators we consider has a relation with. The starting assumption is that the object of evaluation and monitoring is a system, formed by a series of elements interrelated to each other by different processes [59,60].

Niemeijer and Groot [61] consider it important to advance the development of indicators from causal chains to causal networks, that is to say, including all systemic interrelationships between indicators. This approach enriches but also complicates the issue. In any case, it is a question of finding the appropriate balance of indicators to identify relevant trends of change for policy-making and explain the overall functioning of the system and its remoteness or approximation to sustainability [46]. This approach also makes it feasible to inform civil society and support communication with societies [62,63]. Finally, it must be taken into account that the use of the selected indicators requires continuous revision. The indicator here proposed are just the beginning of a monitoring system that will enable adjusting the model to better address its multi-causal dimension.

A final methodological issue regards the information used to elaborate the Sierra de Guadarrama National Park indicators. To a large extent, data has been gathered from official statistical sources. This is a limitation, as the collected data lacked in some cases the level of disaggregation necessary for some of the indicators. Even so, they have been maintained for their role in the whole system. It is expected that the information will be provided in the future.

3. Results

The indicators here presented have been elaborated to fit the socio-economic conditions of the Sierra de Guadarrama National Park. However, they could be adapted and used in other protected natural areas.

The category "State" has been labelled as "Receptor Environment" (RE) in this indicator system. Taking into account the socio-economic characteristics of the Sierra de Guadarrama National Park, the following categories have been proposed, differentiating "group" and "subgroup":

1. Use of natural resources:
    a. Land use
    b. Agrarian resources use
    c. Water use
    d. Energy use
    e. Waste treatment

2. Demography:
    a. Population and its characteristics
    b. Activity, occupation and unemployment

3. Economy:
    a. Employment in productive activities
    b. Tourist activity
    c. Public investments
    d. Income and transfers

4. Society:
    a. Education
    b. Health
    c. Quality and living conditions

The indicators of the "Impact" (SI) are those of the future "State", which is to say, considering the changes within the time taken into consideration.

Finally, the indicators of the "Response" are those including mitigation and adaptation (M&A) to climate change. Two levels have been differentiated: "group" and "subgroup".

1. Governance
    a. Regulated
    b. Not regulated; informal
    c. National park management

2. Social and research instruments
    a. Information and communication
    b. Social perception
    c. Training, qualification and participation
    d. Social research

Table 1 has been designed for each of the indicators and includes: the name of the indicator; the frame of reference; the "group" and "subgroup"; the objectives it pursues; its justification; the measurement parameters or variables that define it; the data source; the scope and period to which they refer; and the relation with other indicators. All this is part of the necessary monitoring protocol to ensure its quality.

As a result, we have developed seventy-nine indicators altogether, which are listed on the table below. It contains thirty indicators regarding both the biophysical and socio-economic means (State) that could be affected by the impact of global and climate change (RE); twenty indicators regarding the future socioeconomic "Impact" of climate change (SI); twenty-nine indicators regarding the measures (Response) to mitigate and to adapt to climate change (M&A). All these indicators are available on the internet and can be accessed following the links provided in the Supplementary Materials at the end of this article.

Table 2 shows a list of all the indicators developed and the indicators with which they are related.

Table 1. Example of an indicator.

Indicator Name	Agricultural and Livestock Area Index	
Frame of reference	Receptor Environment	
Group of indicators	Natural resources	Reference number RE-02
Subgroup	Uses of the territory	
	Characteristics of the Indicator	
Objective, definition and justification of the indicator	It comprises the agricultural and livestock exploitations within the territory, which include the strata of agricultural crops, scrub, pasture, and grassland of the National Forest Inventory. It seeks to reflect uses of the territory by uses that do not entail an irreversible transformation of the national park.	
Measurement parameters	Percentage of the agricultural and livestock area respect to the total area of the park.	
Calculation formula	Agrarian area multiplied by 100 divided by the total area.	
Unit of measurement	Percentage rate, result of dividing hectares by hectares.	
Possible disaggregations	By municipalities of the park.	
Source of information	III National Forest Inventory. Data for the Sierra de Guadarrama National Park.	
Referred area	Territory included within the delimitation of the national park.	
Data availability	Upon request on the Management Office of the national park.	
Measurement periodicity	The corresponding to the update of the National Forest Inventory.	
Responsibility for the veracity of the data	Ministry of Agriculture, Food and Environment.	
Indicators to which is related	RE-01, RE-03, RE-04. (Table 2)	
Reference values	Other national parks.	
	Values of the Indicator for the Different Areas and Periods	
Year	Municipalities of the national park in Segovia	Municipalities of the national park in Madrid	Total municipalities of the Sierra de Guadarrama National Park
2015			
2014			
2013			

Comments: Model of the Indicator Data-Sheet. Its quantification would require an ad-hoc investigation, not included in the scope of this work.

Table 2. Indicators for the Socio-Economic Monitoring and Evaluation System of Climate Change of the Sierra de Guadarrama National Park.

Receptor Environment Indicators (RE)			Socioeconomic Impact Indicators (SI)			Mitigation and Adaptation Indicators (M&A)		
N°	Indicator Name	Related indicators	N°	Indicator Name	Related indicators	N°	Indicator Name	Related M&A Indicators
RE-01	Wooded forest index	RE-02, RE-03, RE-04	SI-01	Impact on wooded forest	RE-01, RE-02, RE-03, RE-04, IS 02, SI-03	M&A-01	Meetings held by the governing and social participatory bodies of the park	M&A-02
RE-02	Agricultural and livestock area index	RE-01, RE-03, RE-04	SI-02	Impact on the exploitation of agricultural and livestock resources	RE-02, RE-01, RE-03, RE-04, SI-01, SI-3, SI-04	M&A-02	Agreements reached by the governing and social participatory bodies of the national park	M&A-01
RE-03	Agricultural and livestock forest index	RE-01, RE-02, RE-04	SI-03	Impact on the agricultural and livestock forest index	RE-03, RE-01, RE-02, RE-04, SI-02, SI-04	M&A-03	General legislation development	M&A-04, M&A-05
RE-04	Scrubland index	RE-01, RE-02, RE-03	SI-04	Impact on the development of scrubland	RE-04, RE-01, RE-02, RE-03, SI-02, SI-03	M&A-04	Level of development of park planning	M&A-03, M&A-05
RE-05	Water consumption of supply network	RE-06, R-21, RE-23	SI-05	Impact on water consumption supplied	RE-05	M&A-05	Agreements in force that affect the park	M&A-03, M&A-04
RE-06	Water treated by purification systems	RE-05	SI-06	Impact on energy consumption	RE-07	M&A-06	Records of sanction of activities processed	M&A-07
RE-07	Energy consumption	RE-08, R-21, RE-23	SI-07	Impact on generated waste	RE-09	M&A-07	Records of authorization of activities processed	M&A-06
RE-08	Energy production	RE-07	SI-08	Impact on demographic dependency ratio	RE-12, RE-11, RE-13, SI-09	M&A-08	Unregulated governance activities	With all M&A
RE-09	Generated waste	RE-10, R-21, RE-23	SI-09	Impact on the aging index	RE-13, RE-11, RE-12, SI-08	M&A-09	Cleared areas for fire protection and for improvements of uses of pasture	M&A-10
RE-10	Separate waste collection	RE-09	SI-10	Impact on the immigration rate	RE-14, RE-11, RE-13, RE-15, RE-16, RE-17, SI-11, SI-12	M&A-10	Treated areas for fire prevention and for improvement of uses of pasture	M&A-09
RE-11	Demographic pyramid of population	RE-12, R-13, RE-14, RE-15	SI-11	Impact on the active population rate	RE-15, RE-16, RE-17, SI-12	M&A-11	Damage to wildlife	With all M&A
RE-12	Demographic dependency rate	RE-11, RE-13	SI-12	Impact on the unemployment rate	RE-17, RE-15, RE-11, SI-11	M&A-12	Participants in the park's environmental education program	M&A-11, M&A-27, M&A-28, M&A-29
RE-13	Aging index	RE-11, RE-12	SI-13	Impact on the agrarian productivity base rate	RE-18, RE-19, SI-14	M&A-13	Participants in the park's volunteer program	M&A-11, M&A-12, M&A-27, M&A-28, M&A-29
RE-14	Immigration rate	RE-11, RE-13, RE-15, RE-16, RE-17	SI-14	Impact on the service economy rate	RE-19, RE-18, SI-13	M&A-14	Waste removed from the park	M&A-05
RE-15	Active population rate	RE-16, RE-17	SI-15	Impact on tourist accommodation capacity	RE-20, RE-21, RE-22, RE-23, SI-14	M&A-15	Areas affected by forest fires	M&A-16
RE-16	Occupied population rate	RE-15, RE-17	SI-16	Impact on annual Park visits	RE-21, RE-20, RE-22, RE-23, SI-15	M&A-16	Investment in prevention and extinction of forest fires in the park	M&A-15, M&A-17

Table 2. *Cont.*

Receptor Environment Indicators (RE)			Socioeconomic Impact Indicators (SI)			Mitigation and Adaptation Indicators (M&A)		
RE-17	Registered unemployment	RE-15, RE-11	SI-17	Impact on the seasonality of visits	RE-22, RE-20, RE-21, RE-23, SI-16	M&A-17	Public investments in the park	M&A-15, M&A-16
RE-18	Agrarian productivity base rate	RE-19	SI-18	Impact on the touristic uses of housing	RE-23, RE-20, RE-21, RE-22, SI-16	M&A-18	Grants given in the municipalities of the park	M&A-16
RE-19	Service economy rate	RE-18	SI-19	Impact on public investment per capita	RE-24, RE-25, RE-26, SI-20	M&A-19	Compensation for the cease of activities incompatible with the regime of the park	M&A-17
RE-20	Tourist accommodation capacity	RE-21, RE-22, RE-23	SI-20	Impact on health infrastructure	RE-24, RE-25, RE-28, RE-11, RE-13, SI-19	M&A-20	People attended at the visitor centers	M&A-20, M&A-21
RE-21	Park visits	RE-20, RE-22, RE-23				M&A-21	School group visits to the park	M&A-22, M&A-23
RE-22	Seasonality of Park visits	RE-20, RE-21, RE-23				M&A-22	Non-school group visits to the park	M&A-20
RE-23	Secondary uses of housing	RE-20, RE-21, RE-22				M&A-23	Brochures and other information formats edited by the park	M&A-24
RE-24	Public investment per capita	RE-25, RE-26				M&A-24	Specific publications related to global and climate change edited by the park	M&A-25
RE-25	Municipal investment per capita	RE-24, RE-26				M&A-25	Social perception of global and climate change; Social participation	M&A-11, M&A-12, M&A-26, M&A-27, M&A-28, M&A-29
RE-26	Municipal indebtedness per capita	RE-24, RE-25				M&A-26	Training and enabling activities about global and climate change	M&A-12, M&A-13, M&A-28, M&A-30
RE-27	Rate of university graduates	RE-11				M&A-27	Research on the impact of global and climate change on the biogeophysical environment of the park	M&A-28
RE-28	Health infrastructure index	RE-11, RE-13, RE-24, RE-25				M&A-28	Research on the impact of global and climate change on the social and economic environment of the park	M&A-29
RE-29	Home equipment	RE-23				M&A-29	Visits to the website of the National Parks Autonomous Body within the section monitoring global change in the network of national parks	M&A-12, M&A-13, M&A-27, M&A-28
RE-30	Elderly population living alone	RE-13						

4. Discussion

The first general conclusion is that the corpus of scientific and empirical knowledge on the social and economic impact of climate change on national parks is scant. However, the study of the socioeconomic impact of climate change in national parks is relevant because climate change is one of the most important challenges faced by today's society. Moreover, climate impacts on people in and around parks, and people's response to such impacts could also affect the natural values of the parks.

Thus, it is considered necessary to evaluate and monitor these impacts, with a systematic scientific approach aimed at understanding parks interconnected biophysical and social systems. This requires the development of more ad hoc theoretical and methodological tools for national parks. This work is oriented in that sense, although limited to indicators and data that nevertheless need to be tested and adjusted in the future.

For diverse reasons, the indicators elaborated for the Sierra de Guadarrama National Park vary in their detail, including the lack of sufficiently disaggregated statistical information and the necessary primary research that qualitative indicators require. This primary research is particularly important in order to extend the system to process indicators, limited in this work as well as in many of these features. Such indicators of processes make it possible to examine social phenomena, like relationships between social groups or social perception of trends on sustainability in the area of study. The system of indicators we propose will have to be adjusted in the future, as the processes of interaction of the biophysical and social systems of the Sierra de Guadarrama National Park are better known. The same will be also necessary in the other national parks with similar characteristics.

A system of indicators for the monitoring and evaluation of the social and economic impacts of climate change has the potential to go beyond simple reporting. It can provide information on whether the situation improves or worsens, recedes or progresses, increases or decreases.

Socio-ecological systems generally are multi-causal and different depending the characteristics of the area. Thus, it cannot be a priori determined whether such changes have been caused by climate change or by any other factor or combination of factors. Nonetheless, continuous evaluation will allow to deepen the understanding of the causal relationships between changes in climatic conditions, changes in the socio-ecological system and changes in natural and cultural values of the park. For example, climate change is a direct cause of drought termed "meteorological drought", in addition to human action or "hydric drought" (water infrastructures, responsible water human use or consumption, etc.), which could also have a relevant impact on the economy (agricultures, industry, tourism, etc.), environment (evolution of fauna and flora, territory, etc.) and population's living conditions (consumption, transport, live styles, etc.) and environmental values and attitudes (response dimension). However, this needs to be tested over time.

Ideally, the information provided by a system of socio-economic indicators relating to climate change can be integrated into management planning of national parks to improve decision-making. Moreover, a study on indicators could be the catalyst for the development of comprehensive climate change adaptation plans for individual national parks and protected area networks, which is still limited or non-existent in most national parks in Spain and many other parts of the world.

5. Conclusions

In conclusion, the interpretation of changes in the monitoring process of complex climate-related changes in protected areas is a major challenge. The evaluation of the interactions between climate change and the socio-ecological changes in the park and its area of influence requires a holistic approach and a sufficient time-series of data. The set of socio-economic indicators we have developed provides a framework for collecting and interpreting such, and so will help to inform adaptation planning for the Sierra de Guadarrama National Park. The approach we have taken could also be applied in other similar national parks.

Supplementary Materials: The seventy-nine indicators are available online at: http://portal.uc3m.es/portal/page/portal/grupos_investigacion/sociologia_cambio_climatico/Sociology_of_Climate_Change_and_Sustainable_Development/Receptor%20Environment%20Indicators.%20Sierra%20de%20Guadarrama%20Na.pdf; http://portal.uc3m.es/portal/page/portal/grupos_investigacion/sociologia_cambio_climatico/Sociology_of_Climate_Change_and_Sustainable_Development/Impact%20Indicators.%20Sierra%20de%20Guadarrama%20National%20Park.pdf; http://portal.uc3m.es/portal/page/portal/grupos_investigacion/sociologia_cambio_climatico/Sociology_of_Climate_Change_and_Sustainable_Development/Mitigation%20and%20Adaptation%20Indicators.%20Sierra%20de%20Guadarra.pdf.

Acknowledgments: This research was funded by Fundación Biodiversidad. We did not received funds for covering the costs to publish in open access.

Author Contributions: Both authors directed the research project. Iván López and Mercedes Pardo have contributed to the draft of the paper.

Conflicts of Interest: The author declares no conflict of interest.

References

1. UN. *United Nations Framework Convention on Climate Change*; United Nations: New York, NY, USA, 1992.
2. IPCC. *Climate Change: Synthesis Report*; WMO: Geneva, Switzerland, 2014; p. 151. ISBN 978-92-9169-143-2.
3. Rannow, S.; Macgregor, N.A.; Albrecht, J.; Crick, H.Q.; Förster, M.; Heiland, S.; Janauer, G.; Morecroft, M.D.; Neubert, M.; Sarbu, A.; et al. Managing protected areas under climate change: Challenges and priorities. *Environ. Manag.* **2014**, *54*, 732–743. [CrossRef] [PubMed]
4. Zamora, R. Las Áreas protegidas como Observatorios del Cambio Global. *Ecosistemas* **2010**, *19*, 1–4.
5. Fancy, S.G.; Bennetts, R.E. Institutionalizing an effective long-term monitoring program in the U.S. National Park Service. In *Design and Analysis of Long-Term Ecological Monitoring Studies*; Gitzen, R.A., Millspaugh, J.J., Cooper, A.B., Licht, D.S., Eds.; Cambridge University Press: Cambridge, UK, 2012; pp. 481–497.
6. Hansen, A.J.; Piekielek, N.; Davis, C.; Haas, J.; Theobald, D.M.; Gross, J.E.; Monahan, W.B.; Olliff, T.; Running, S.W. Exposure of U.S. National Parks to land use and climate change 1900–2100. *Ecol. Appl.* **2014**, *24*, 484–502. [CrossRef] [PubMed]
7. Hansen, A.J.; Monahan, W.; Theobald, D.M.; Olliff, S.T. *Climate Change in Wildlands: Pioneering Approaches to Science and Management*; Island Press: Washington, DC, USA, 2016; ISBN 9781610917124.
8. Löeffler, J.; Anschlag, K.; Baker, B.; Finch, O.D.; Diekkrueger, B.; Wundram, D.; Schröder, B.; Pape, R.; Lundberg, A. Mountain ecosystem response to global change. *Erdkunde* **2011**, *65*, 189–213. [CrossRef]
9. Tiwari, P.C.; Joshi, B. Global Change and Mountains: Consequences, Responses and Opportunities. In *Impact of Global Changes on Mountains: Responses and Adaptation*; Grover, V.I., Borsdorf, A., Breuste, J.H., Tiwari, P.C., Frangetto, F., Eds.; CRC Press: London, UK; New York, NY, USA, 2014; pp. 79–136. ISBN 978-1-4822-0890-0.
10. Lockwood, M.; Worboys, G.; Kothari, A. (Eds.) *Managing Protected Areas: A Global Guide*; Routledge: Milton Park, UK, 2012; ISBN 978-1-84407-303-3.
11. Cook, C.N.; Mascia, M.B.; Schwartz, M.W.; Possingham, H.P.; Fuller, R.A. Achieving conservation science that bridges the knowledge-action boundary. *Conserv. Biol.* **2013**, *27*, 669–678. [CrossRef] [PubMed]
12. Kremen, C. Assessing the Indicator Properties of Species Assemblages for Natural Areas Monitoring. *Ecol. Appl.* **1992**, *2*, 203–217. [CrossRef] [PubMed]
13. Noss, R.F. Assessing and monitoring forest biodiversity: A suggested framework and indicators. *For. Ecol. Manag.* **1999**, *115*, 135–146. [CrossRef]
14. Hockings, M.; Stolton, S.; Leverington, F.; Dudley, N.; Courrau, J. *Evaluating Effectiveness: A Framework for Assessing the Management of Protected Areas*, 2nd ed.; IUCN: Gland, Switzerland; Cambridge, UK, 2006; ISBN 978-2-8317-0939-0.
15. Leverington, F.; Costa, K.L.; Pavese, H.; Lisle, A.; Hockings, M. A global analysis of protected area management effectiveness. *Environ. Manag.* **2010**, *46*, 685–698. [CrossRef] [PubMed]
16. Rodríguez-Rodríguez, D.; Martínez-Vega, J. Results of the implementation of the System for the Integrated Assessment of Protected Areas (SIAPA) to the protected areas of the Autonomous Region of Madrid (Spain). *Ecol. Indic.* **2013**, *34*, 210–220. [CrossRef]

17. Fancy, S.G.; Gross, J.E.; Carter, S.L. Monitoring the condition of natural resources in US national parks. *Environ. Monit. Assess.* **2009**, *151*, 161–174. [CrossRef] [PubMed]
18. Cutter, S.L.; Boruff, B.J.; Shirley, W.L. Social vulnerability to environmental hazards. *Soc. Sci. Q.* **2003**, *84*, 242–261. [CrossRef]
19. Adger, W.N.; Brooks, N.; Bentham, G.; Agnew, M.; Eriksen, S. *New Indicators of Vulnerability and Adaptive Capacity*; Technical Report 7; Tyndall Centre for Climate Change Research: Norwich, UK, 2004.
20. Mitchell, R.; Parkins, J. The challenge of developing social indicators for cumulative effects assessment and land use planning. *Ecol. Soc.* **2011**, *16*, 29:1–29:14. [CrossRef]
21. Petrova, S. *Communities in Transition: Protected Nature and Local People in Eastern and Central Europe*; Routledge: Milton Park, UK, 2014; ISBN 978-1-13-825130-4.
22. Ley 7/2013, de 25 de junio, de Declaración del Parque Nacional de la Sierra de Guadarrama. Available online: https://www.boe.es/diario_boe/txt.php?id=BOE-A-2013-6900 (accessed on 11 January 2018).
23. Parque Nacional Sierra de Guadarrama. *Mundo del Agrónomo* **2015**, *29*, 24–26. Available online: http://www.agronomoscentro.org/images/mda/mda29.pdf (accessed on 24 August 2017).
24. Parque Nacional Sierra de Guadarrama. Available online: http://www.parquenacionalsierraguadarrama.es/en/ (accessed on 25 August 2017).
25. Ramos de Armas, F.J. El Parque Nacional de la Sierra de Guadarrama. *Ambienta: La revista del Ministerio de Medio Ambiente* **2013**, *103*, 4–9.
26. Nieto, N.; Díez, R. Sierra de Guadarrama, un Parque Nacional liberal: Más marca turística que conservación. *El Ecologista* **2014**, *82*, 28–29.
27. Campos Palacín, P.; Carrera Troyano, M. Crítica de la exclusión del aprovechamiento de recursos naturales en los parques nacionales españoles. *Principios de Economía Política* **2007**, *8*, 39–58.
28. Rodríguez-Rodríguez, D.; Martínez-Vega, J. What should be evaluated from a manager's perspective? Developing a salient protected area effectiveness evaluation system for managers and scientists in Spain. *Ecol. Indic.* **2016**, *64*, 289–296.
29. Doak, D.F.; Estes, J.A.; Halpern, B.S.; Jacob, U.; Lindberg, D.R.; Lovvorn, J.; Monson, D.H.; Tinker, M.T.; Williams, T.M.; Wootton, J.T.; et al. Understanding and predicting ecological dynamics: Are major surprises inevitable. *Ecology* **2008**, *89*, 952–961. [CrossRef] [PubMed]
30. Millar, C.I.; Stephenson, N.L.; Stephens, S.L. Climate change and forests of the future: Managing in the face of uncertainty. *Ecol. Appl.* **2007**, *17*, 2145–2151. [CrossRef] [PubMed]
31. Baron, J.S.; Allen, C.D.; Fleishman, E.; Gunderson, L.; Mckenzie, D.; Meyerson, L.; Oropeza, J.; Stephenson, N. National parks. In *Preliminary Review of Adaptation Options for Climate-Sensitive Ecosystems and Resources. A Report by the U.S. Climate Change Science Program and the Subcommittee on Global Change Research*; Julius, S.H., West, J.M., Eds.; Environmental Protection Agency: Washington, DC, USA, 2008; pp. 1–68.
32. Naughton-Treves, L.; Holland, M.B.; Brandon, K. The role of protected areas in conserving biodiversity and sustaining local livelihoods. *Annu. Rev. Environ. Resour.* **2005**, *30*, 219–252. [CrossRef]
33. West, P.; Igoe, J.; Brockington, D. Parks and peoples: The social impact of protected areas. *Annu. Rev. Anthropol.* **2006**, *35*, 251–277. [CrossRef]
34. Berkes, F. Community-based conservation in a globalized world. *Proc. Natl. Acad. Sci. USA* **2007**, *104*, 15188–15193. [CrossRef] [PubMed]
35. Budruk, M.; Phillips, R. (Eds.) *Quality-of-Life Community Indicators for Parks, Recreation and Tourism Management*; Springer: London, UK, 2011; Volume 43, ISBN 978-94-007-3445-6.
36. Holling, C.S. Resilience and stability of ecological systems. *Annu. Rev. Ecol. Syst.* **1973**, *4*, 1–23. [CrossRef]
37. Welsh, M. Resilience and responsibility: Governing uncertainty in a complex world. *Geogr. J.* **2014**, *180*, 15–26. [CrossRef]
38. Tanner, T.; Lewis, D.; Wrathall, D.; Bronen, R.; Cradock-Henry, N.; Huq, S.; Lawless, C.; Nawrotzki, R.; Prasad, V.; Rahman, M.A.; et al. Livelihood resilience in the face of climate change. *Nat. Clim. Chang.* **2015**, *5*, 23–26. [CrossRef]
39. Baron, J.S.; Gunderson, L.; Allen, C.D.; Fleishman, E.; McKenzie, D.; Meyerson, L.A.; Oropeza, J.; Stephenson, N. Options for national parks and reserves for adapting to climate change. *Environ. Manag.* **2009**, *44*, 1033–1042. [CrossRef] [PubMed]
40. Glick, P.; Stein, B.; Edelson, N.A. *Scanning the Conservation Horizon: A Guide to Climate Change Vulnerability Assessment*; National Wildlife Federation: Washington, DC, USA, 2010.

41. Williamson, T.B.; Price, D.T.; Beverly, J.L.; Bothwell, P.M.; Parkins, J.R.; Patriquin, M.N.; Pearce, C.; Stedman, R.C.; Volney, W.J.A. *A Framework for Assessing Vulnerability of Forest-Based Communities to Climate Change*; Natural Resources Canada, Canadian Forest Service, Northern Forestry Centre: Edmonton, AB, Canada, 2007; ISBN 978-0-662-47044-1.
42. Rodríguez-Rodríguez, D.; Martínez-Vega, J. Proposal of a system for the integrated and comparative assessment of protected areas. *Ecol. Indic.* **2012**, *23*, 566–572. [CrossRef]
43. Fischer, A.P.; Paveglio, T.; Carroll, M.; Murphy, D.; Brenkert-Smith, H. Assessing social vulnerability to climate change in human communities near public forests and grasslands: A framework for resource managers and planners. *J. For.* **2013**, *111*, 357–365. [CrossRef]
44. Land, K.C.; Spilerman, S. (Eds.) *Social Indicator Models*; Russel Sage Foundation: New York, NY, USA, 1975; ISBN 978-0-87154-505-3.
45. Force, J.E.; Machlis, G.E. The human ecosystem part II: Social indicators in ecosystem management. *Soc. Nat. Resour.* **1997**, *10*, 369–382. [CrossRef]
46. Ley 30/2014, de 3 de Diciembre, de Parques Nacionales. Available online: https://www.boe.es/boe/dias/2014/12/04/pdfs/BOE-A-2014-12588.pdf (accessed on 25 August 2017).
47. Royal Decree 389/2016, de 22 de octubre, Plan Director de la Red de Parques Nacionales. Available online: https://www.boe.es/boe/dias/2016/10/24/pdfs/BOE-A-2016-9690.pdf (accessed on 29 August 2017).
48. MAGRAMA. Available online: http://www.mapama.gob.es/es/biodiversidad/publicaciones/informe_castellano_criterios_indicarores_gestion_forestal_sostenible_bosques_2012_tcm7-260632.pdf (accessed on 24 August 2017).
49. Muñoz Santos, M.; Benayas Del Álamo, J. *El uso Público en la Red de Parques Nacionales de España: Una Propuesta de Evaluación*; Organismo Autónomo Parques Nacionales: Madrid, Spain, 2012; ISBN 978-84-8014-827-6.
50. Instituto Nacional de Estadística. Lista Completa de Operaciones. 2016. Available online: http://www.ine.es/dyngs/INEbase/listaoperaciones.htm (accessed on 11 January 2018).
51. Jones, D.A.; Hansen, A.J.; Bly, K.; Doherty, K.; Verschuyl, J.P.; Paugh, J.I.; Carle, R.; Story, S.J. Monitoring land use and cover around parks: A conceptual approach. *Remote Sens. Environ.* **2009**, *113*, 1346–1356. [CrossRef]
52. Beckley, T.; Parkins, J.; Stedman, R. Indicators of forest-dependent community sustainability: The evolution of research. *For. Chron.* **2002**, *78*, 626–636. [CrossRef]
53. Gough, A.D.; Innes, J.L.; Allen, S.D. Development of common indicators of sustainable forest management. *Ecol. Indic.* **2008**, *8*, 425–430. [CrossRef]
54. Akamani, K.A. Community resilience model for understanding and assessing the sustainability of forest-dependent communities. *Hum. Ecol. Rev.* **2012**, *19*, 99–109.
55. Singh, R.K.; Murty, H.R.; Gupta, S.K.; Dikshit, A.K. An overview of sustainability assessment methodologies. *Ecol. Indic.* **2012**, *15*, 281–299. [CrossRef]
56. Smeets, E.; Weterings, R. *Environmental Indicator: Typology and Overview*; Technical Report 25; European Environment Agency: Copenhagen, Denmark, 1999.
57. Ministerio de Agricultura, Pesca, Alimentacion y Medio Ambiente. Water Indicators System. Available online: http://www.mapama.gob.es/es/agua/temas/planificacion-hidrologica/sia-/indicadores.aspx (accessed on 29 January 2018).
58. IPCC. Summary for policymakers. In *Climate Change 2014: Impacts, Adaptation, and Vulnerability. Part A: Global and Sectoral Aspects. Contribution of Working Group II to the Fifth Assessment Report of the Intergovernmental Panel on Climate Change*; Field, C.B.V.R., Barros, D.J., Dokken, K.J., Mach, M.D., Mastrandrea, T.E., Bilir, M., Chatterjee, K.L., Ebi, Y.O., Estrada, R.C., Genova, B., et al., Eds.; Cambridge University Press: Cambridge, UK; New York, NY, USA, 2014; pp. 1–32.
59. Machlis, G.E.; Force, J.E.; Burch, W.R., Jr. The human ecosystem part I: The human ecosystem as an organizing concept in ecosystem management. *Soc. Nat. Resour.* **1997**, *10*, 347–367. [CrossRef]
60. Weber, M.; Krogman, N.; Antoniuk, T. Cumulative effects assessment: Linking social, ecological, and governance dimensions. *Ecol. Soc.* **2012**, *17*, 22:1–22:7. [CrossRef]
61. Niemeijer, D.; De Groot, R.S. Framing environmental indicators: Moving from causal chains to causal networks. *Environ. Dev. Sustain.* **2008**, *10*, 89–106. [CrossRef]

62. OECD. *Environmental Indicators: Towards Sustainable Development*; OECD: Paris, France, 2001.
63. Red de Seguimiento del Cambio Global en la Red de Parques Nacionales. Ministerio de Agricultura y Pesca, Alimentación y Medio Ambiente. Available online: http://www.mapama.gob.es/es/red-parques-nacionales/red-seguimiento/resultados-2015-rscg_tcm7-449653.pdf (accessed on 25 November 2017).

environments

MDPI

Article

Fine-Tuning of a Protected Area Effectiveness Evaluation Tool: Implementation on Two Emblematic Spanish National Parks

David Rodríguez-Rodríguez [1,2,*], **Paloma Ibarra** [3], **Javier Martínez-Vega** [1], **Maite Echeverría** [3] and **Pilar Echavarría** [1]

1   Institute of Economy, Geography and Demography, Spanish National Research Council (IEGD-CSIC), Associated Unit GEOLAB. C/Albasanz, 26–28, 28037 Madrid, Spain; javier.martinez@cchs.csic.es (J.M.-V.); pilar.echavarria@cchs.csic.es (P.E.)
2   European Topic Centre-Universidad of Malaga, Andalucía Tech, University of Malaga, 29010 Malaga, Spain
3   Aragonese University Research Institute on Environmental Science, Department of Geography and Territorial Management, University of Zaragoza, 50009 Zaragoza, Spain; pibarra@unizar.es (P.I.); mtecheve@unizar.es (M.E.)
*   Correspondence: david.rodriguez@csic.es or davidrgrg@yahoo.es; Tel.: +34-916-022-322 or +34-951-953-102

Received: 3 August 2017; Accepted: 21 September 2017; Published: 26 September 2017

Abstract: As global biodiversity trends worsen, protected area (PA) environmental effectiveness needs to be assessed to identify strengths and areas to improve. Through a participatory process including PA managers and scientists, we refined the System for the Integrated Assessment of Protected Areas (SIAPA), in order to increase its legitimacy, credibility and salience to end users in Spain. Then, we tested the optimised version of the SIAPA on two emblematic Spanish national parks (NPs): Ordesa y Monte Perdido NP (Ordesa NP) and Sierra de Guadarrama NP (Guadarrama NP). PA managers and scientists largely coincided in the ratings of SIAPA's indicators and indices. Collaboration with Ordesa NP's managers was regular, allowing a nearly complete evaluation of the NP. However, greater collaboration between PA managers and scientists remains a priority in Guadarrama NP. Results show that potential effectiveness is moderate for Ordesa NP and low for Guadarrama NP, according to the indicators that could be evaluated. For Ordesa NP, lack of data on focal habitats and other focal features determined a deficient valuation of its conservation state, although the remaining indicators in that category showed adequate or moderate values. The compilation of those data should be overriding in the NP. In contrast, only climate change posed a serious threat in that NP. The social perception and valuation of both NPs was good, suggesting broad support from local populations and eased management.

Keywords: reserve; environmental sustainability; assessment; indicator; index; SIAPA

---

1. Introduction

Global protected area (PA) coverage is constantly expanding to presently cover 14.7% of terrestrial and freshwater ecosystems and 4.1% of marine ecosystems [1]. However, the status of global biodiversity continues to deteriorate [2]. As a result, increased focus is being put on assessing the effectiveness of PAs as the main global policy to reduce biodiversity loss [1,3–5]. Dozens of PA assessment systems have been developed worldwide [6] and in Europe [7]. RAPPAM (Rapid Assessment and Prioritization of Protected Areas Management) [8] and METT (Management Effectiveness Tracking Tool) [9] are the most broadly and frequently used, especially in contexts of limited availability of data [6]. Nevertheless, issues regarding accurateness and precision of both rapid, opinion-based systems have been raised [10,11]. Some more objective PA assessment and evaluation systems chiefly based on secondary data have since been developed to estimate environmental effectiveness of PAs comprehensively. One of these was the System for the Integrated Assessment of Protected Areas (SIAPA, [12]). The SIAPA evaluates PAs based on 43 indicators that were highly valued by experts and for which common legal, scientific, technical or logical thresholds were established. SIAPA indicators were integrated in six partial effectiveness indices: State of Conservation, Planning, Management, Socioeconomic context, Social perception and valuation, and Threats to conservation. Later, these partial indices were also integrated in an overall PA Effectiveness Index, which allows comparison among PAs that belong, ideally, to the same PA network.

The SIAPA was implemented to evaluate the potential effectiveness of the PAs of the Autonomous Region of Madrid, in Spain [13]. Later on, the methodology underpinning the SIAPA was presented to the main potential users in Spain, including PA network managers, scientists, environmental NGOs (non-governmental organizations) and other stakeholders in a national workshop. There was limited interest in implementing the SIAPA or a similar tool to regularly and consistently assess PA effectiveness among PA managers in Spain [14]. One of the reasons was that the SIAPA had been developed for implementation in the Autonomous Region of Madrid which is little representative of the environmental and socioeconomic characteristics of the whole country. Thus, some indicators and valuation thresholds were not found useful for other Spanish regions. Another reason was exclusion of important indicators in the original SIAPA, such as 'focal ecosystems' extent'. An improved version of SIAPA was produced after valuation of the 43 original SIAPA indicators and 6 indices by workshop participants and consideration of their remarks [15]. However, comments were made on the reduced participation of scientists in this improved SIAPA compared to PA managers, which might limit its legitimacy, credibility and salience, according the Knowledge Systems for Sustainable Development Framework [16].

In this study, we: (1) increased scientists' participation to produce an optimised version of the SIAPA in which the views of two key stakeholder groups, PA managers and scientists, are more widely and equally represented; (2) assessed the relatedness in indicator and index valuations between both stakeholder groups to identify evaluation priorities.; and (3) tested the optimised version of the SIAPA to evaluate the effectiveness of two emblematic national parks (NPs) in Spain: Ordesa NP and Guadarrama NP.

2. Methods

2.1. SIAPA Optimisation

Four researchers from the DISESGLOB project [17] rated each of the 43 original SIAPA indicators according to the Likert scale (1 to 5 points) used in the SIAPA improvement workshop [15], according to their relevance for defining the partial effectiveness indices in which they were included. Those researchers also rated the six partial indices on the same scale according to their importance for the overall environmental effectiveness of a PA. We combined the responses of those scientists ($n = 4$) with those of the PA managers ($n = 12$) and scientists ($n = 3$) from the SIAPA workshop, to have a more balanced representation of both key stakeholder groups in Spain. Responses corresponded to staff from different institutions from the workshop's participants'. The complete responses and institutions from all ratters (workshop participants plus new scientists) are shown in Supplementary 1.

To make the final indicator selection for the optimised version of SIAPA, every indicator was ranked according to their decreasing coefficient of variation (CV; i.e., their standard deviations divided by their means) and the upper tier (percentile 33) was selected. CVs have been used to prioritise indicator selection in previous studies [14,15]. The total indicators' and indices' mean values from the workshop participants and additional scientists ($n = 19$) were used to assign weights for integrating them in their respective partial indices or super-index, respectively (Table 1). The original scales of each indicator and conforming variables were standardised to an ordinal 0, 1, 2 point scale showing 'adequate', 'moderate' or 'deficient' valuation, respectively. The two valuation thresholds dividing the three previous values [18,19] were established for each original scale according to: values established by law, values commonly used by specialized agencies, values found in the literature, or logical, empirical or experience-based values based on the precautionary principle [20].

Detailed profiles for each selected indicator were developed or adapted from [12] (Supplementary 2). On interpretation and comparison grounds, we used the same Effectiveness Index's cut-off values based on the Precautionary Principle as in [12].

Table 1. Calculation, valuation and interpretation of the optimised SIAPA's indices.

Index	Number of Common Indicators	Calculation Formula	Value (Interpretation)
State of Conservation	6		
Planning	3		$_wI \geq 1.5 \rightarrow 2$ points (Adequate)
Management	4	$wI = \sum\limits_{i=1}^{n} x_i \cdot k_i / \sum\limits_{i=1}^{k} k_i$	$1 \leq {}_wI < 1.5 \rightarrow 1$ point (Moderate)
Social and Economic Context	2	where:	$_wI < 1 \rightarrow 0$ points (Deficient)
Social Perception and Valuation	2	wI = partial index	
		x_i = indicator value (0; 1; 2)	$_tI \leq 0.5 \rightarrow 0$ points (Adequate) *
Threats to Conservation ($_tI$)	5	k_i = weighting factor (3.3 to 5.0)	$0.5 < {}_tI < 1 \rightarrow 1$ point (Moderate) *
			$_tI \geq 1 \rightarrow 2$ points (Deficient) *
		$EI = \sum\limits_{i=1}^{n} x_i \cdot k_i / \sum\limits_{i=1}^{k} k_i$	$EI \geq 1.2 \rightarrow 2$ points (Adequate)
Effectiveness (*EI*)	22	where:	$0.8 \leq EI < 1.2 \rightarrow 1$ point (Moderate)
		x_i = index value (0; $\pm$1; $\pm$2)	$EI < 0.8 \rightarrow 0$ points (Deficient)
		k_i = weighting factor (2.8 to 4.7)	

* The values of the Threats to conservation Index and their interpretations are opposite to the other partial indices that positively add to protected area effectiveness.

2.2. Index and Indicator Valuation Comparison by PA Managers and Scientists

The degree of relatedness between each group's index and average indicator valuations was analysed using Spearman rank order correlation test ($\alpha = 0.05$), after checking the non-normality of the original and log-transformed variables. Differences in index and indicator valuations by stakeholder group were then analysed via Kruskall-Wallis test ($\alpha = 0.05$) using SPSS version 23 software (IBM, New York, NY, USA).

2.3. Optimised SIAPA Testing

Two highly symbolic PAs were selected to test the optimised SIAPA: Ordesa NP and Guadarrama NP. Ordesa NP was designated nearly one century ago [21]. It was reclassified and expanded in 1982 to cover its current 15,608 ha that extend over six municipalities [22]. It is a high-mountain NP located in the northern Spanish province of Huesca, in the Pyrenees. Biogeographically, it is located in the Alpine region [23] (Figure 1). It contains 15 of the 27 'natural systems' (natural ecosystems and landscapes defined by their representative vegetation) in the Spanish Law on National Parks [24] and 1404 plant species, including approximately 50 Pyrenean endemic species [25]. Guadarrama NP is the most recent NP in Spain. It was designated in June 2013 over 33,960 ha of 35 municipalities [26]. It is also a medium and high-mountain NP located in the Central Mountain Range, between the provinces of Madrid and Segovia (Figure 1). Biogeographically, it belongs to the Mediterranean Region [23]. It protects 10 natural systems in the Spanish Law on National Parks [24] and more than 1000 flora species, of which 83 are endemic to the Iberian Peninsula [25].

Though from an environmental point of view the two NPs are similar, both representing mountainous Iberian biodiversity, they are socioeconomically very distinct. Ordesa NP is a peripheric, rural NP of difficult accessibility with less than 2000 local residents around it [27] in which tertiary economic activities are predominant but where primary activities are still relevant. It receives approximately 600,000 visitors every year [28]. In contrast, Guadarrama NP is a peri-urban, easily-accessible NP at only 40 min from Madrid by car. Local population around the NP is approximately 150,000, but as many as six million people live within an hour drive from the NP [27]. This results in very high visitation levels, with around 3,000,000 visitors per year [28]. Tertiary economic activities are predominant, whereas primary economic activities are residual.

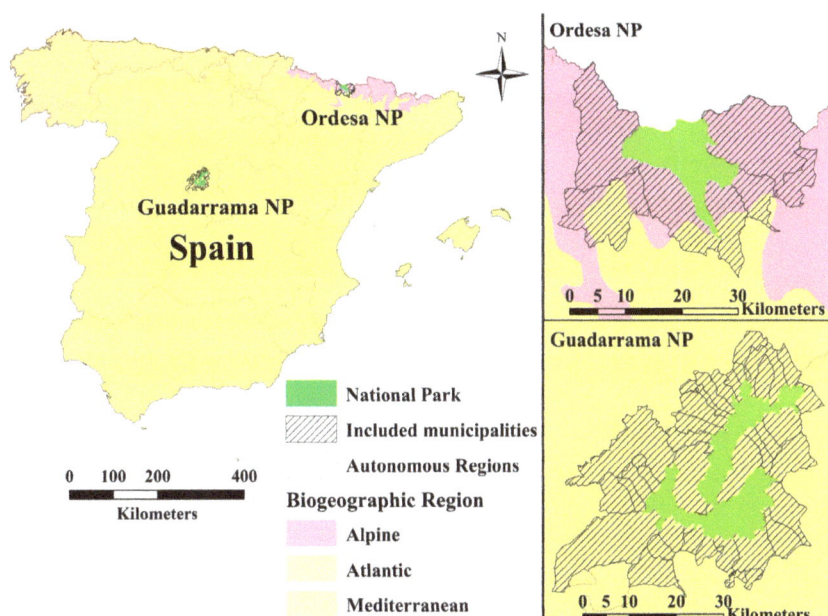

Figure 1. Location of both National Parks in the Spanish regional and biogeographic map (**left**). Perimeters of both National Parks and their conforming municipalities (**right**). Please note that the Canary Islands Region is not showing on the regional map.

Additionally to the resulting common SIAPA indicators, the evaluated NPs' managers were given the chance to identify other case-specific indicators that were most relevant to their respective PAs. Regular phone and/or face-to-face contacts with both NPs' managers were made since 2015 to ensure data provision for the calculation of each indicator. Information exchange with Ordesa NP's managers was frequent, which allowed a nearly complete implementation of the optimised SIAPA in this NP. In contrast, Guadarrama NP's managers stated interest in participating in the project but did not provide us with the required data for evaluation. As a result, only 12 indicators for which other secondary data sources could be retrieved could be evaluated. Thus, the indices of both NPs are mostly calculated from different indicators and compared on the basis of available information for each NP. Evaluations took place between June of 2016 and June of 2017.

3. Results

3.1. SIAPA Optimisation

The templates of the final selection of indicators ($n = 22$) according to the ratings of the complete set of PA managers and scientists (Supplementary 1) is shown in Supplementary 2. Ordesa NP's managers suggested incorporating to the evaluation of the NP the indicator: 'pasture encroachment by woody vegetation' (Supplementary 2).

3.2. Stakeholder Group Valuation of SIAPA's Indicators and Indices

The SIAPA indices' ratings by PA managers were not correlated with the average valuation of the SIAPA indicators within each index. In contrast, there was a weak, significant correlation between both by the group of scientists ($r_{s(37)} = 0.34$; $p = 0.03$). There were no statistically significant differences in index ratings or average indicator ratings between both groups, for the complete set of indices and indicators, and by index. Scientists rated, however, the indicators within State of conservation almost significantly higher than PA managers ($\chi^2_{(1)}$ = 3.73; p = 0.05). The mean valuation of the six partial indices of SIAPA by both stakeholder groups is shown in Table 2.

Table 2. Valuation of the SIAPA indices by protected area managers and scientists (means and standard deviations).

SIAPA Indices	Managers (Mean ± sd)	Scientists (Mean ± sd)
State of conservation	4.91 ± 0.30	4.29 ± 1.25
Planning	4.00 ± 1.00	4.00 ± 1.00
Management	4.27 ± 1.01	4.14 ± 0.90
Socioeconomic context	2.91 ± 1.30	2.71 ± 1.25
Social perception and valuation	3.45 ± 0.93	2.71 ± 1.11
Threats to conservation	4.00 ± 1.10	4.14 ± 1.07

3.3. SIAPA Implementation Results

Summary results on the integrated assessment of Ordesa NP (Table 3) and Guadarrama NP (Table 4) are shown below. Specific results for each indicator and NP can be retrieved from Supplementary 3. 'Pasture encroachment by woody vegetation' was evaluated for Ordesa NP (Supplementary 3) but it was not included in the final effectiveness score of the NP due to its original valuation scale and also on comparison grounds.

Table 3. Summary results from the application of the optimised SIAPA to Ordesa NP.

Ordesa National Park				
NP Area (ha): 15,608 Designation Date: 1918 (1982 Re-Classified)		PPZ [1] Area (ha): 19,679 Evaluation Date: 2016–2017		SIZ [2] Area (ha): 89,341 Evaluation: 1st
Index/Indicator	Value	State	Trend	Evaluation Period
STATE OF CONSERVATION	0	🙁		
Population trends of endangered species or sub-species	2	🙂	NA [3]	2012–2015
Changes in the extent of focal habitats	0	🙁	NA	2013
Changes in the features for which the PA was designated	0	🙁	NA	2012–2015
Visual impact	1	😐	NA	2010
Surface water quality	2	🙂	↔	2014–2015
Health of vegetation [4]	1	😐	↓	2012; 2013; 2015
PLANNING	2	🙂		
Appropriateness of protection regulation	1	😐	NA	2017
Existence of updated management plan	2	🙂	NA	2017
Existence of updated socioeconomic plan	2	🙂	NA	2017
MANAGEMENT	1	😐		
Degree of fulfilment of management objectives Effectiveness of public participation bodies	2	🙂	↔	2012–2015
Existence of sufficient management staff	1	😐	↔	2014–2015
Existence of environmental education and volunteering activities	2	🙂	↔	2014–2015

Table 3. *Cont.*

Ordesa National Park				
NP Area (ha): 15,608 Designation Date: 1918 (1982 Re-Classified)	PPZ [1] Area (ha): 19,679 Evaluation Date: 2016–2017		SIZ [2] Area (ha): 89,341 Evaluation: 1st	
Index/Indicator	Value	State	Trend	Evaluation Period
SOCIOECONOMIC CONTEXT	0	☹		
Local population density	0	☹	↓	2015–2016
Land use changes	1	😐	NA	2006; 2012
SOCIAL PERCEPTION AND VALUATION	2	🙂		
Degree of knowledge of the PA	2	🙂	NA	2016
Personal importance	2	🙂	NA	2016
THREATS TO CONSERVATION	0	🙂		
Fragmentation	0	🙂	↔	2006; 2012
Density of alien invasive species	0	🙂	NA	2016
Density of visitors	1	😐	↓	2014–2015
Activities performed by visitors	0	🙂	NA	2016
Climate change	2	☹	NA	1976–2016
Pasture encroachment by woody vegetation [5]	0	🙂	NA	2006; 2012
EFFECTIVENESS	1	😐		

[1] Peripheral Protection Zone. [2] Socioeconomic Influence Zone. [3] Non-applicable. [4] Trend was calculated comparing the last available data (2015) and the two previous available data (2012–2013). [5] NP's specific indicator.

Table 4. Summary results from the application of the optimised SIAPA to Guadarrama NP [1].

Guadarrama National Park				
NP Area (ha): 33,960 Designation Date: 2013	PPZ Area (ha): 62,687 Evaluation Date: 2016–2017		SIZ Area (ha): 173,632 Evaluation: 1st	
Index/Indicator	Value	State	Trend	Evaluation Period
STATE OF CONSERVATION	1	😐		
Population trends of endangered species or sub-species Changes in the extent of focal habitats Changes in the features for which the PA was designated				
Visual impact	0	☹	NA	2010
Surface water quality	2	🙂	↔	2014–2015
Health of vegetation [2]	1	😐	↑	2014–2015
PLANNING	1	😐		
Appropriateness of protection regulation	2	🙂	NA	2017
Existence of updated management plan	0	☹	NA	2017
Existence of updated socioeconomic plan	2	🙂	NA	2017

Table 4. *Cont.*

Guadarrama National Park				
NP Area (ha): 33,960 Designation Date: 2013	PPZ Area (ha): 62,687 Evaluation Date: 2016–2017			SIZ Area (ha): 173,632 Evaluation: 1st
Index/Indicator	Value	State	Trend	Evaluation Period
MANAGEMENT	1	😐		
Degree of fulfilment of management objectives				
Effectiveness of public participation bodies	1	😐	↔	2015
Existence of sufficient management staff				
Existence of environmental education and volunteering activities	2	🙂	NA	2014
SOCIOECONOMIC CONTEXT	2	🙂		
Local population density	2	🙂	↑	2015–2016
Land use changes				
SOCIAL PERCEPTION AND VALUATION	2	🙂		
Degree of knowledge of the PA	2	🙂	NA	2016
Personal importance	2	🙂	NA	2016
THREATS TO CONSERVATION	2	☹️		
Fragmentation Density of alien invasive species				
Density of visitors	2	☹️	↓	2014–2015
Activities performed by visitors Climate change				
EFFECTIVENESS	0	☹️		

[1] Please, note that effectiveness comparison between both national parks is not straightforward and should be made with care, as different number and type of indicators were often evaluated for each site. Valid comparisons can be made at indicator level and at partial index level (if both indices have the same indicators evaluated). [2] Trend available for plots in the province of Madrid only.

4. Discussion

4.1. SIAPA Optimisation

The 22 indicators in the optimised SIAPA are the most relevant ones to assess PA effectiveness by a relatively highly representative sample of key stakeholders on PAs in Spain. Stakeholder participation in the making of this optimised version of SIAPA is much wider than most similar global initiatives [29]. Eleven of the seventeen regional governments' representatives and the two representatives of the national bodies with some competencies on PAs participated on the managers' side, whereas seven different scientific institutions also provided input. The optimised SIAPA includes all the eight priority indicators from its improved version [15] and 15 of the 28 indicators of the original SIAPA's simplified model, aimed at increasing evaluation efficiency [12]. Fifteen of the 22 indicators of the optimised SIAPA can be currently evaluated using secondary data external to PA administrations in Spain, although some key indicators, such as species', habitats' or other focal features' status cannot. To overcome such evaluation challenge, collaboration between PA managers and scientists is essential in terms of raw data provision but also in terms of processed result return [30,31].

Despite these limitations, our results suggest that a legitimate, credible and salient PA evaluation system can be established in Spain. All the three fundamental criteria to facilitate bridging the science-implementation gap [16] have been substantially improved since the SIAPA's original version [12] by increasing key stakeholder participation. Different PA assessment tools are available in the country [12,15,32,33]. Nevertheless, the fact that PA evaluation is not considered a legal obligation in Spain, with the exception of NPs, and that other more pressuring managerial priorities exist, such as the drafting of management plans for Natura 2000 sites, are likely to still limit the salience of this and any other PA assessment tool [14]. Additionally, insufficient basic data to undertake evaluations, limited institutional interest, reluctance to assessments, and lack of culture on transparency and accountability will also probably hamper the implementation of any sort of external, regular and sound 'environmental audits' in Spanish PAs for some time [13,14].

The different versions of the SIAPA were developed in Spain with participation of Spanish stakeholders. Thus, their implementation will be most salient in Spain and other countries or regions with similar environmental and socioeconomic characteristics (i.e., Euro-Mediterranean countries). In contrast, in countries or regions from different contexts (e.g., tropical or developing countries) some different indicators and/or valuation thresholds may probably be needed according to their own characteristics. Those tailored versions of the SIAPA will increase their legitimacy, credibility and salience by using similar (or, where possible, broader) participatory processes as SIAPAs'.

4.2. Stakeholder Group Valuation Comparison

PA managers and scientists largely agreed on the most and least relevant indices for overall PA effectiveness. For both, State of conservation was the paramount index to assess PA environmental effectiveness, which aligns with mainstream claims [11]. However, they both rated the Socioeconomic context index lowly, in contrast to previous suggestions on its importance for effective conservation [7,13].

Scientists seem to provide more consistent index and indicator ratings or have closer alignment with SIAPA's indicator classification procedures within indices than PA managers. In contrast, the very weak correlation between index and indicator ratings by PA managers suggests either less consistency in their valuations or greater divergence on indicator selection or classification within the SIAPA indices. Lack of agreement on SIAPA indicator classification in indices [15] points to limited consistency in PA managers' ratings of SIAPA's indicators and indices.

4.3. Optimised SIAPA Testing: Effectiveness of Ordesa NP and Guadarrama NP

Ordesa NP scored deficiently in state of conservation and socioeconomic context. However, the poor value of the State of Conservation Index in Ordesa NP is not due to poor indicator values but to absence of data on key conservation features: geomorphological features, air quality and focal habitats. Actually, a crucial state of conservation indicator such as endangered species' population trends shows adequate valuation. Additionally, it is likely that actual values of missing indicators are good, but data are needed to provide evidence of that. Lack of data or existence of inconsistent, poor or outdated data often hamper the fulfilment of PA conservation objectives and the evaluation of PA effectiveness [34,35]. Regarding socioeconomic context, though coarse scale [36] land use-land cover changes remain stable, low and decreasing local population density is changing land management practices and ecosystem composition at finer scales [37,38]. In contrast, Ordesa NP scored positively in threats to conservation, with only two threats showing moderate or high importance. Of those, climate change is likely having the greatest impact on biodiversity in the medium term [39,40] and, opposite to visitor numbers that can be easily regulated, it is a largely unmanageable threat at local or even regional scale [41].

Guadarrama NP scored adequately on socioeconomic context and poorly on threats to conservation, though only one indicator in these indices could be evaluated. Both assessed variables showed opposing trends. Whereas local population density showed low and decreasing figures [27], visitation figures to this peri-urban NP are high and increasing [26,28]. These opposing trends suggest less residential use and rising tourist use in the area, as shown by the large proportion of holiday homes, ranging from 77% to 14% and averaging 51% in the municipalities of Guadarrama NP [42]. Guadarrama mountains have historically been a popular place for recreation [43–45]). Numerous sport and leisure activities were performed in the area before its designation as a NP [43,46,47]. It seems that since its designation as a NP, Guadarrama mountains are attracting more visitors, as it happens elsewhere [48]. Additionally, some massive sport events, such as cross-country marches or bike contests are being organised [49]. Both facts are likely to result in diverse impacts on biodiversity, cultural and geomorphological features [50], and more challenging management, especially in a PA without a management plan to officially regulate such activities yet.

Social perception and valuation was very positive in both NPs. It shows a high identification of residents with each NP which should favour conservation and ease of management [1]. Comparing the effectiveness of both NPs is risky, as different indicators were evaluated for each one, but with the necessary precautions and as guidance, results on effectiveness were better for Ordesa NP than for Guadarrama NP, according to their un-standardised EI's values. This was expected due to the different socioeconomic and managerial characteristics of both NPs. Ordesa NP is a peripheral, geographically isolated NP in which active management has been implemented for a long time [21,22]. It is currently managed by the Environmental Ministry of the Region of Aragon. In contrast, Guadarrama NP is a peri-urban NP with comparatively high residential population density and very high visitation figures [28] in which management of the whole area is very recent and divided between the two regional administrations that share managerial competencies on the NP: Madrid and Castille and Leon. Intra- and inter-administrative inefficiencies are common in Spain and likely affect PA effectiveness [51]. Its peri-urban nature also makes visual impacts in and around this NP more abundant.

Accurate, updated and regularly compiled data on the status of protected biodiversity and other relevant features is a priority task in both NPs, as in any PA [52,53]. In highly pressured Guadarrama NP [47,49], the passing and implementation of a management plan should also be an immediate priority. PA managers and territorial planners could enhance use of existing incentives for municipalities in NPs [24] to help to maintain local population and traditional activities that favour biodiversity [54,55], especially in Ordesa NP. An update of the designation norm of Ordesa NP would be advisable, although the recently passed management plan (2015) facilitates that management is performed according to current information, conservation criteria and practices.

4.4. Study Limitations

Data for evaluating Guadarrama NP were scarce. Collaboration between scientists and PA managers is complex and improvable in most places, making environmental evaluations challenging [16,56]. Spain is not different [13,14]. Thus, the results of applying the SIAPA to this PA were incomplete for four of the six partial indices, and thus its overall effectiveness value can only be regarded as a partial estimation.

Moreover, secondary data used for evaluation are assumed to be sound. Some authors highlight the need to validate raw data to ensure quality for assessments [34]. However appropriate that recommendation is, validating the very high volume and diversity of data used in this or similar evaluations seems beyond the timeframe and cost of usual projects and even beyond the scientific or technical capabilities of the evaluators.

Future studies should explore the use of finer-scale remote sensing data on land uses-land covers, as CORINE Land Cover data [36], despite its pros, depicts too coarse a scale that is insufficient to detect ecologically-relevant, fine scale land use-land cover changes. This limitation likely resulted in unchanging results for some indicators, such as 'land use changes' or 'pasture encroachment by woody vegetation'. Some alternative, finer scale data sources could be SIOSE [57], Spain's Forest Map [58] or SIGPAC [59] for Spain. In Germany, EU's Integrated Administration and Control Systems (IACS) data was used [60]. Bastin et al. [61] advocate the use of Open Source data (Web Map Services or GeoServer) and NDVI data.

5. Conclusions

Successive participatory rounds with key Spanish PA stakeholders, namely PA managers and scientists, have resulted in a highly legitimate and credible, and moderately salient, optimised version of the SIAPA. It is made of the 22 most highly ranked, widely agreed indicators. Greater salience and regular implementation of such an evaluation system or any other system seems challenging without a clear legal mandate in Spain.

The optimised SIAPA could be almost entirely tested in Ordesa NP thanks to collaboration of the NP's staff. In contrast, in Guadarrama NP, lack of collaboration by NP's managers resulted in almost half of the SIAPA indicators not being evaluated. Ordesa NP showed moderate environmental effectiveness, with negative values for 'State of conservation' (mostly due to lack of comparable data for key indicators), and positive values for 'Planning', 'Social perception and valuation', and 'Threats to conservation'. Guadarrama NP scored deficiently in 'Effectiveness', but positively in 'Socioeconomic context' and 'Social perception and valuation', although many indicators could not be evaluated and many partial indices' results are thus estimative. Greater implication of PA managers of the Autonomous Region of Madrid with researchers is a long-lasting [13] and still pending challenge.

Supplementary Materials: The following are available online at www.mdpi.com/2076-3298/4/4/68/s1. Supplementary 1: SIAPA index's and indicator's ratings by protected area managers and scientists, Supplementary 2: Optimised SIAPA's indicator profiles, Supplementary 3: Results by indicator and national park.

Acknowledgments: This paper recognises contributions through the 'sequence-determines-credit' approach. We would like to acknowledge data provision and interpretation support by Ordesa NP's administration staff. We would also like to thank Francisco Fernández Latorre, from the University of Seville, for collaborating in rating the original SIAPA indicator and indices. This study was funded by the Spanish Ministry of Economy, Industry and Competitiveness in the framework of the DISESGLOB project (CSO2013-42421-P).

Author Contributions: David Rodríguez-Rodríguez, Paloma Ibarra, Maite Echeverría & Javier Martínez-Vega collected secondary data for evaluation, rated the original SIAPA's indicators and indices, analysed results and drafted and/or reviewed the article prior to submission. Pilar Echavarría helped to produce some GIS results, such as those on visual impact.

Conflicts of Interest: The authors declare no conflict of interest. The founding sponsors had no role in the design of the study; in the collection, analyses, or interpretation of data; in the writing of the manuscript, and in the decision to publish the results.

References

1. Bhola, N.; Juffe-Bignoli, D.; Burguess, N.; Sandwith, T.; Kingston, N. *Protected Planet Report 2016. How Protected Areas Contribute to Achieving Global Targets for Biodiversity*; UNEP-WCMC: Cambridge, UK, 2016.
2. Butchart, S.H.M.; Walpole, M.; Collen, B.; van Strien, A.; Schalermann, J.P.W.; Almond, R.E.; Baillie, J.E.; Bomhard, B.; Brown, C.; Bruno, J.; et al. Global biodiversity: Indicators of recent declines. *Science* **2010**, *328*, 1164–1168. [CrossRef] [PubMed]
3. Convention on Biological Diversity (CBD). *Programme of Work on Protected Areas (CBD Programmes of Work)*; Secretariat of the Convention on Biological Diversity: Montreal, QC, Canada, 2004.
4. Convention on Biological Diversity (CBD). Aichi Biodiversity Targets. 2010. Available online: https://www.cbd.int/sp/targets/ (accessed on 24 July 2017).
5. Hockings, M.; Stolton, S.; Dudley, N. *Evaluating Effectiveness: A Framework for Assessing the Management of Protected Areas*, 2nd ed.; IUCN: Gland, Switzerland, 2006.
6. Leverington, F.; Lemos, K.; Courrau, J.; Pavese, H.; Nolte, C.; Marr, M.; Coad, L.; Burgess, N.; Bomhard, B.; Hockings, M. *Management Effectiveness Evaluation in Protected Areas—A Global Study*, 2nd ed.; University of Queensland: Brisbane, Australia, 2010.
7. Nolte, C.; Leverington, F.; Kettner, A.; Marr, M.; Nielsen, G.; Bomhard, B.; Stolton, S.; Stoll-Kleemann, S.; Hockings, M. Protected Area Management Effectiveness Assessments in Europe. A Review of Application, Methods and Results. 2010. Available online: http://www.lepidat.de/fileadmin/MDB/documents/service/Skript_271a.pdf (accessed on 3 August 2017).
8. Ervin, J. *Rapid Assessment and Prioritization of Protected Areas Management (RAPPAM) Methodology*; WWF: Gland, Switzerland, 2003.
9. Stolton, S.; Hockings, M.; Dudley, N.; MacKinnon, K.; Whitten, T.; Leverington, F. *Reporting Progress in Protected Areas: A Site Level Management Effectiveness Tracking Tool*, 2nd ed.; World Bank/WWF Forest Alliance and WWF: Gland, Switzerland, 2007.

10. Carranza, T.; Manica, A.; Kapos, V.; Balmford, A. Mismatches between conservation outcomes and management evaluation in protected areas: A case study in the Brazilian Cerrado. *Biol. Conserv.* **2014**, *173*, 10–16. [CrossRef]

11. Cook, C.N.; Carter, R.W.; Hockings, M. Measuring the accuracy of management effectiveness evaluations of protected areas. *J. Environ. Manag.* **2014**, *139*, 164–171. [CrossRef] [PubMed]

12. Rodríguez-Rodríguez, D.; Martínez-Vega, J. Proposal of a system for the integrated and comparative assessment of protected areas. *Ecol. Indic.* **2012**, *23*, 566–572. [CrossRef]

13. Rodríguez-Rodríguez, D.; Martínez-Vega, J. Results of the implementation of the System for the Integrated Assessment of Protected Areas (SIAPA) to the protected areas of the Autonomous Region of Madrid (Spain). *Ecol. Indic.* **2013**, *34*, 210–220. [CrossRef]

14. Rodríguez-Rodríguez, D.; Martínez-Vega, J.; Tempesta, M.; Otero-Villanueva, M.M. Limited uptake of protected areas evaluation systems among managers and decision-makers in Spain and the Mediterranean Sea. *Environ. Conserv.* **2015**, *42*, 237–254. [CrossRef]

15. Rodríguez-Rodríguez, D.; Martínez-Vega, J. What should be evaluated from a manager's perspective? Developing a salient protected area effectiveness evaluation system for managers and scientists in Spain. *Ecol. Indic.* **2016**, *64*, 289–296.

16. Cash, D.W.; Clark, W.C.; Alcock, F.; Dickson, M.N.; Eckley, N.; Guston, D.H.; Jäger, J.; Mitchell, R.B. Knowledge systems for sustainable development. *PNAS* **2003**, *100*, 8086–8091. [CrossRef] [PubMed]

17. Instituto de Economía, Geografía y Demografía (IEGD). DISESGLOB: Diseño de una Metodología de Seguimiento y Evaluación de la Sostenibilidad global de Parques Nacionales y de la Influencia de los Cambios de uso Previstos. 2017. Available online: http://iegd.csic.es/es/research-project/disesglob (accessed on 3 August 2017).

18. Ten Brink, B. A Long-Term Biodiversity, Ecosystem and Awareness Research Network. Indicators as Communication Tools: An Evolution towards Composite Indicators. ALTER-Net, 2006. Available online: http://www.globio.info/downloads/79/Report++ten+Brink+%282006%29+Indicators+as+communication+tools-.pdf (accessed on 3 August 2017).

19. Moldan, B.; Janoušková, S.; Hák, T. How to understand and measure environmental sustainability: Indicators and targets. *Ecol. Indic.* **2012**, *17*, 4–13. [CrossRef]

20. Coney, R.; Dickson, B. *Biodiversity and the Precautionary Principle. Risk and Uncertainty in Conservation and Sustainable Use*; Earthscan: London, UK, 2005.

21. Spanish Government. Declaración del Parque Nacional del Parque Nacional del Valle de Ordesa. *Boletin Oficial Estado Gaceta Madrid Spain* **1918**, *230*, 495.

22. Spanish Government. Ley 52/1982, de 13 de Julio, de reclasificación y ampliación del Parque Nacional de Ordesa y Monte Perdido. *Boletin Oficial Estado* **1982**, *181*, 20627–20629.

23. European Environment Agency (EEA). Biogeographical Regions in Europe. 2017. Available online: https://www.eea.europa.eu/data-and-maps/figures/biogeographical-regions-in-europe-2 (accessed on 26 July 2017).

24. Spanish Government. Ley 30/2014, de 3 de diciembre, de Parques Nacionales. *Boletin Oficial Estado* **2014**, *293*, 99762–99792.

25. Spanish Government. Ley 7/2013, de 25 de junio, de Declaración del Parque Nacional de la Sierra de Guadarrama. *Boletin Oficial Estado* **2013**, *152*, 47795–47852.

26. Ministerio de Agricultura, Pesca, Alimentación y Medio Ambiente (MAPAMA). Memoria de la Red de Parques Nacionales. 2014. Available online: http://www.mapama.gob.es/es/red-parques-nacionales/la-red/gestion/memoria-2014_tcm7-454256.pdf (accessed on 3 August 2017).

27. Instituto Nacional de Estadística (INE). INEbase. Demografía y Población. Padrón. Población por Municipios. Cifras Oficiales de Población de los Municipios Españoles: Revisión del Padrón Municipal. Cifras Oficiales de Población Resultantes de la Revisión del Padrón Municipal a 1 de enero. 2017. Available online: http://www.ine.es/dynt3/inebase/index.htm?padre=525 (accessed on 3 August 2017).

28. Ministerio de Agricultura, Pesca, Alimentación y Medio Ambiente (MAPAMA). Memoria de la Red de Parques Nacionales. 2015. Available online: http://www.mapama.gob.es/es/red-parques-nacionales/divulgacion/memoria-2015_tcm7-454259.pdf (accessed on 3 August 2017).

29. Chape, S.; Spalding, M.; Jenkins, M. *The World's Protected Areas: Status, Values and Prospects in the 21st Century*; University of California Press: Berkeley, CA, USA, 2008.

30. Lü, Y.; Chen, L.; Fu, B.; Liu, S. A framework for evaluating the effectiveness of protected areas: The case of Wolong Biosphere Reserve. *Landsc. Urban Plan.* **2003**, *63*, 213–223. [CrossRef]

31. Struhsaker, T.T.; Struhsaker, P.J.; Siex, K.S. Conserving Africa's rain forests: problems in protected areas and possible solutions. *Biol. Conserv.* **2005**, *123*, 45–54. [CrossRef]

32. Ministerio de Agricultura, Alimentación y Medio Ambiente (MAGRAMA). Primer Informe de Situación de la Red de Parques Nacionales a 1 de Enero de 2007. Tomo I. 2008. Available online: http://www.mapama.gob.es/es/red-parques-nacionales/divulgacion/tomo-1-informe-situacion-red_tcm7-459027.pdf (accessed on 3 August 2017).

33. Institució Catalana d'Història Natural (ICHN). Avaluació del Sistema d'espais Naturals Protegits de Catalunya. Available online: http://ichn.iec.cat/Avaluacio_Espais.htm (accessed on 10 June 2017).

34. Peckett, F.J.; Glegg, G.A.; Rodwell, L.D. Assessing the quality of data required to identify effective marine protected areas. *Mar. Policy* **2014**, *45*, 333–341. [CrossRef]

35. Knowles, J.E.; Doyle, E.; Schill, S.R.; Roth, L.M.; Milam, A.; Raber, G.T. Establishing a marine conservation baseline for the insular Caribbean. *Mar. Policy* **2015**, *60*, 84–97. [CrossRef]

36. Copernicus Land Monitoring Services. Corine Land Cover. Pan-European, 2016. Available online: http://land.copernicus.eu/pan-european/corine-land-cover/ (accessed on 17 July 2017).

37. Lasanta-Martínez, T.; Vicente-Serrano, S.M.; Cuadrats, J.M. Mountain Mediterranean landscape evolution caused by the abandonment of traditional primary activities: A study of the Spanish Central Pyrenees. *Appl. Geogr.* **2005**, *25*, 47–65. [CrossRef]

38. Gartzia, M.; Alados, C.L.; Pérez-Cabello, F. Assessment of the effects of biophysical and anthropogenic factors on woody plant encroachment in dense and sparse mountain grasslands based on remote sensing data. *Prog. Phys. Geogr.* **2014**, *38*, 201–217. [CrossRef]

39. García-Ruiz, J.M.; López-Moreno, J.I.; Lasanta, T.; Vicente-Serrano, S.M.; González-Sampériz, P.; Valero-Garcés, B.L.; Sanjuán, Y.; Beguería, S.; Nadal-Romero, E.; Lana-Renault, N.; et al. Los efectos geoecológicos del cambio global en el Pirineo Central español: Una revisión a distintas escalas espaciales y temporales. *Pirineos* **2015**, *170*, e012. [CrossRef]

40. Gartzia, M.; Pérez-Cabello, F.; Bueno, C.B.; Alados, C.L. Physiognomic and physiologic changes in mountain grasslands in response to environmental and anthropogenic factors. *Appl. Geogr.* **2016**, *66*, 1–11. [CrossRef]

41. Araújo, M.B.; Alagador, D.; Cabeza, M.; Nogués-Bravo, D.; Thuiller, W. Climate change threatens European conservation areas. *Ecol. Lett.* **2011**, *14*, 484–492. [CrossRef] [PubMed]

42. Instituto Nacional de Estadística (INE). INEbase. Censos de Población y Viviendas 2011. Viviendas. Resultados Municipales. Principales Resultados. 2011. Available online: http://www.ine.es/jaxi/Tabla.htm?path=/t20/e244/viviendas/p06/l0/&file=10mun00.px&L=0 (accessed on 3 August 2017).

43. Barrado Timón, D.A. *Actividades de ocio y Recreativas en el Medio Natural de la Comunidad de Madrid*; Comunidad de Madrid: Madrid, Spain, 1999.

44. MolláRuiz-Gómez, M. La Junta Central de Parques Nacionales y la Sierra de Guadarrama. *Ería* **2007**, *73–74*, 161–177.

45. Mollá Ruiz-Gómez, M. "El Grupo de los Alemanes" y el Paisaje de la Sierra de Guadarrama. *Bol. AGE* **2009**, *51*, 51–64.

46. Atauri, J.A.; Bravo, M.A.; Ruiz, A. Visitors' Landscape Preferences as a Tool for Management of Recreational Use in Natural Areas: A case study in Sierra de Guadarrama (Madrid, Spain). *Landsc. Res.* **2000**, *25*, 49–62. [CrossRef]

47. Rodríguez-Rodríguez, D. *Los Espacios Naturales Protegidos de la COMUNIDAD de Madrid. Principales Amenazas Para su Conservación*; Editorial Complutense: Madrid, Spain, 2008.

48. Bianchi, R.V. The Contested Landscapes of World Heritage on a Tourist Island: The case of Garajonay National Park, La Gomera. *Int. J. Herit. Stud.* **2002**, *8*, 81–97. [CrossRef]

49. Ecologistas en Acción. 1.200 Corredores en las Zonas más Sensibles del Parque Nacional de Guadarrama. 2014. Available online: http://www.ecologistasenaccion.org/article28246.html (accessed on 3 August 2017).

50. Pickering, C.M.; Rossi, S. Mountain biking in peri-urban parks: Social factors influencing perceptions of conflicts in three popular National Parks in Australia. *J. Outdoor Recreat. Tour.* **2016**, *15*, 71–81. [CrossRef]

51. EUROPARC-España. *Anuario 2013 del Estado de las Áreas Protegidas en España*; Fundación Fernando González Bernáldez: Madrid, Spain, 2014.

52. Gaston, K.J.; Jackson, S.F.; Cantú-Salazar, L.; Cruz-Piñón, G. The ecological performance of protected areas. *Annu. Rev. Ecol. Evol. Syst.* **2008**, *39*, 93–113. [CrossRef]

53. International Union for the Conservation of Nature (IUCN). IUCN Green List of Protected and Conserved Areas: Standard, Version 1.0. 2016. Available online: https://www.iucn.org/sites/dev/files/iucn_glpca_standard_version_1.0_september_2016_030217.pdf (accessed on 3 August 2017).

54. Negro, M.; La Rocca, C.; Ronzani, S.; Rolando, A.; Palestrini, C. Management tradeoff between endangered species and biodiversity conservation: The case of Carabus olympiae (Coleoptera: Carabidae) and carabid diversity in north-western Italian Alps. *Biol. Conserv.* **2013**, *157*, 255–265. [CrossRef]

55. Tattoni, C.; Ianni, E.; Geneletti, D.; Zatelli, P.; Ciolli, M. Landscape changes, traditional ecological knowledge and future scenarios in the Alps: A holistic ecological approach. *Sci. Total Environ.* **2017**, *579*, 27–36. [CrossRef] [PubMed]

56. Cook, C.N.; Mascia, M.B.; Schwartz, M.W.; Possingham, H.P.; Fuller, R.A. Achieving Conservation Science that Bridges the Knowledge–Action Boundary. *Conserv. Biol.* **2013**, *27*, 669–678. [CrossRef] [PubMed]

57. Centro Nacional de Información geográfica (CNIG). Centro de Descargas. SIOSE Sistema de Información sobre la Ocupación del Suelo de España 1:25,000. 2011. Available online: http://centrodedescargas.cnig.es/CentroDescargas/index.jsp# (accessed on 3 August 2017).

58. Ministerio de Agricultura, Pesca, Alimentación y Medio Ambiente (MAPAMA). Mapa Forestal de España. 2015. Available online: http://www.mapama.gob.es/ide/metadatos/index.html?srv=metadata.show&uuid=ac11b891-6c6c-4458-b89c-2b73f593d019 (accessed on 3 August 2017).

59. Ministerio de Agricultura, Pesca, Alimentación y Medio Ambiente (MAPAMA). Agricultura. Sistema de Información Geográfica de Parcelas Agrícolas (SIGPAC). 2017. Available online: http://www.mapama.gob.es/es/agricultura/temas/sistema-de-informacion-geografica-de-parcelas-agricolas-sigpac-/ (accessed on 3 August 2017).

60. Nitsch, H.; Osterburg, B.; Roggendorf, W.; Laggner, B. Cross compliance and the protection of grassland. Illustrative analyses of land use transitions between permanent grassland and arable land in German regions. *Land Use Policy* **2012**, *29*, 440–448. [CrossRef]

61. Bastin, L.; Buchanan, G.; Beresford, A.; Pekel, J.F.; Dubois, G. Open-source mapping and services for Web-based land-cover validation. *Ecol. Inform.* **2013**, *14*, 9–16. [CrossRef]

MDPI

St. Alban-Anlage 66

4052 Basel

Switzerland

Tel. +41 61 683 77 34

Fax +41 61 302 89 18

www.mdpi.com

Environments Editorial Office

E-mail: environments@mdpi.com

www.mdpi.com/journal/environments

www.ingramcontent.com/pod-product-compliance
Lightning Source LLC
Chambersburg PA
CBHW051841210326
41597CB00033B/5734

* 9 7 8 3 0 3 8 9 7 2 1 2 9 *